Fourier Transform N.M.R. Spectroscopy

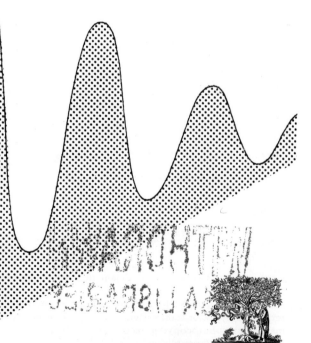

Fourier Transform N.M.R. Spectroscopy

DEREK SHAW

Varian Associates Ltd., Walton - on - Thames

ELSEVIER SCIENTIFIC PUBLISHING COMPANY

AMSTERDAM – OXFORD – NEW YORK 1976

ELSEVIER SCIENTIFIC PUBLISHING COMPANY
335 Jan van Galenstraat
P.O. Box 211, Amsterdam, The Netherlands

Distributors for the U.S.A and Canada:

ELSEVIER NORTH-HOLLAND INC.
52, Vanderbilt Avenue
New York. N.Y. 10017

Library of Congress Cataloging in Publication Data

Shaw, Derek.
    Fourier transform N.M.R. spectroscopy.

    Bibliography: p.
    Includes index.
    1.   Nuclear magnetic resonance spectroscopy.
2.   Fourier transform spectroscopy. I.   Title.
QC762.S45   538.3   76-17121
ISBN 0-444-41466-5

Printed in The Netherlands

CONTENTS

viii

Nuclear magnetic resonance spectroscopy has, over the last ten years or so, developed rapidly into a major spectroscopic technique. In doing so it has assisted chemistry, organic chemistry in particular, in its rapid development. The applications of high resolution n.m.r. in chemistry are enormous and are growing at an almost exponential rate. It is imposible for one book to cover all these applications and this book makes no attempt to do so, especially as there already exist many excellent text books on such topics.

In recent years n.m.r. has undergone a revolutionary change in technique from swept to pulsed excitation. The change has occurred as a result of many contributing factors, firstly the realisation in 1966 by W.A. Anderson and R.R. Ernst that pulsed excitation followed by Fourier transformation could considerably increase the sensitivity of high resolution n.m.r., a technique not noted for its sensitivity! The second major factor was the advances in electronic technology which made computers available at a reasonable price, and developed the techniques of solid state electronics which made commercial pulsed spectrometers feasible. There are many other factors.

Along with the development of Fourier n.m.r. has come the emergence of ^{13}C n.m.r. as an almost equal partner to the well established proton n.m.r. The latent potential of ^{13}C has long been realised, as can be seen in the pioneering work of Lauterbur, Schoolery etc. It required the sensitivity improvement of Fourier techniques to catalyse its growth.

This book is orientated towards technique rather than applications. The basic theory of n.m.r. is dealt with along with Fourier theory in a unified approach which differs from that taken in other works on a high resolution n.m.r. The middle part of the book is concerned with the practical aspects of Fourier n.m.r., both instrumental and experimental. The final chapters deal briefly with the general applications of n.m.r. but concentrate strongly on those areas where Fourier n.m.r. can give information not available by conventional techniques.

The units used throughout the work are S.I. units. For n.m.r. this involves little change from the normal c.g.s. units. The unit of energy in both systems is the Hertz, and chemical shifts are dimensionless. The major difference occurs when expressing magnetic fields, the traditional unit being the Gauss, the S.I. unit the Tesla; one Tesla is 10^4 Gauss, thus protons resonate at 100 MHz/2.3 Tesla. N.m.r. spectroscopists tend to be schizophrenic about units; they report splitting in Hz while sweeping the field etc. and they swap at will between reporting rotational frequencies in radians/s and Hertz. Here, Hertz are used exclusively.

It is hoped that the book will be a comprehensive guide to the theory and practice of Fourier n.m.r. and provide sufficient applications to set the method in the firmament of spectroscopic techniques providing information to the chemist.

Jean Baptiste Joseph Fourier

The author wishes to acknowledge the valuable assistance of many people in the production of this book; Prof. R. Ernst, Dr. J. Feeney, Dr. R. Freeman, Dr. H. Hill and Dr. F. Wehrli for their helpful discussions and comments on the manuscript, Varian Associates and all my colleagues for their support, my wife, Marian, for her artwork and patience, and Mrs. C. Fielder for her typing.

The following Journals and the appropriate authors are thanked for permission to use their material for the figures listed:

Journal of Magnetic Resonance, Figs. 4.9, 4.10, 4.14, 5.2, 5.3, 6.11, 8.11, 9.5, 10.14.

Journal of the American Chemical Society, Fig. 5.10.

Journal of the Canadian Chemical Society, Fig. 9.7.

The Institute of Petroleum, Fig. 3.4.

Journal of Chemical Physics, Figs. 4.13, 7.13, 10.5, 10.8, 10.9, 10.21.

Review of Scientific Instruments, Fig. 5.8.

Molecular Physics, Fig. 5.11.

DEFINITION OF SYMBOLS

a =   (i) Sweep rate
         (ii) Hyperfine (electron-nucleus) coupling constant
         (iii) Line width at $\frac{1}{2}$ height
B_0 = Static magnetic field
$B_{x,y,z}$ = Components of B in the laboratory frame
$B_{x',y',z'}$ = Components of B in the rotating frame
$B_{1,2}$ = R.f. magnetic fields at ϑ_1, ϑ_2 etc.
C = Spin rotation coupling constant
e = Magnitude of the electronic charge
$\mathcal{F}^{\pm}$ = Functional operator for Fourier transformation
$\hbar$ = Planck's constant$/2\pi$
$\hat{H}$ = Hilbert transformation
$\hat{\mathcal{H}}$ = Hamiltonian operator
$\hat{I}_i$ = Nuclear spin operator for spin i
$\hat{I}_{x,y,z}$ = Components of I
$\hat{I}_i^{\pm}$ = 'Raising' and 'lowering' operator
i_i = Magnetic quantum number associated with I_i
i = $(-1)^{1/2}$
n_J = Nuclear spin—spin coupling constant through n bonds
n_K = Reduced nuclear spin—spin coupling constant through n bonds
k = Boltzman's constant
m_i = Eigenvalue of I_{iz} (magnetic quantum number)
$\underline{M}_0$ = Equilibrium macroscopic magnetisation of a spin system in the presence of B_0
M^+ = Magnetisation (in the rotating frame) prior to the initiation of a pulse
M^- = Magnetisation (in the rotating frame) following a pulse
$M_{x,y,z}$ = Components of $\underline{M}$ in the laboratory frame
$M_{x'y'z'}$ = Components of $\underline{M}$ in the rotating frame
$m(t)$ = Magnetisation as a function of time
$M(\vartheta)$ = Magnetisation as a function of frequency
N =   (i) Number of nuclei/unit volume
         (ii) Number of data points
Q = Quality factor of a coil
$r(t)$ = Response of a linear system as a function of time
$R(\vartheta)$ = Response of a linear system as a function of frequency
S = Saturation parameter

t_p	=	Duration of a pulse
T_a	=	Duration of data acquisition
T_d	=	Delay between the end of data aquisition and the next pulse
T_p	=	Total time between two pulses
T_1^x	=	Spin—lattice relaxation time of nucleus x
$T_{1\rho}^x$	=	Spin—lattice relaxation time of nucleus x in the rotating frame
T°	=	Temperature
T_2^x	=	Spin—spin relaxation time of nucleus x
T_2'	=	Inhomogeneity contribution to the dephasing time of $M_{x,y}$
T_2^s	=	Incoherent (stochastic) contributions to the dephasing time of $M_{x,y}$
T_2^*	=	Total dephasing time of $M_{x,y}$
u	=	Dispersion mode signal $(M_{x'})$
v	=	Absorption mode signal $(M_{y'})$
$x(t)$	=	Excitation applied to a linear system as a function of time
$X(\vartheta)$	=	Excitation applied to a linear system as a function of frequency
α	=	(i) Angle of rotation of $\underline{M}_0$ about the x$'$ axis
		(ii) Nuclear spin wave function (eigenfunction of I_z)
β	=	Nuclear spin wave function (eigenfunction of I_z)
γ	=	Nuclear magnetogyric ratio (Hz T^{-1})
δ_x	=	Chemical shift of nucleus x in ppm
δ_{ij}	=	Kronecker delta ($= 1$ if $i = j$; zero otherwise)
δ	=	Small increment of time
Δ	=	Spectral width
ξ	=	Filling factor
η	=	(i) Nuclear Overhauser enhancement
		(ii) Viscosity
μ	=	Magnetic moment of a nucleus
μ_0	=	Permeability of free space
ϑ_c	=	Spectrometer operating frequency (carrier frequency)
ϑ_0	=	Larmor frequency of a nucleus
ϑ_i	=	Frequency of nucleus i from ϑ_0
ϑ_1	=	Frequency of 'observing' r.f. magnetic field
ϑ_2	=	Frequency of 'decoupling' r.f. magnetic field
ϑ_3	=	Frequency of 'locking' r.f. magnetic field
θ	=	Phase angle

τ = Small interval of time between pulses in a sequence
τ_c = Correlation time
τ_e = Exchange life time
χ = Magnetic susceptability

CHAPTER 1

INTRODUCTION

1.1 HISTORICAL

Nuclear magnetic resonance spectroscopy is the study of the magnetic properties of nuclei. The fact that nuclei have magnetic properties was first discovered, surprisingly enough, during the study of optical spectroscopy. With the use of spectrographs of higher and higher resolving power so called hyperfine splitting was observed in these spectra. The study of the hyperfine splitting lead Pauli, in 1924, to suggest that certain nuclei had angular momentum and thus, as a spinning electric charge, a magnetic moment [1]. Nuclear angular momentum, as one would anticipate, is quantised. The quantum number is represented by I which can have any $\frac{1}{2}$ integral value including zero. The angular momentum which corresponds to the nucleus of quantum number I is $I\hbar$. Some nuclei do have a quantum number $I = 0$, hence have no spin angular momentum, and consequently show no n.m.r. spectrum. This negative property of some nuclei e.g. ^{12}C is a blessing in disguise; if $I \neq 0$ for *all* nuclei, n.m.r. in its present form would be impossible! The value for I for a nucleus cannot in general be predicted but some trends do exist. Isotopes having both an even mass and an even atomic number have a spin quantum number $I = 0$. Those with an odd mass number have a $\frac{1}{2}$ integral spin value while those with odd atomic numbers but even mass numbers have I as an integral value. These values are illustrated in Table 1.1.

Earlier in 1921 Stern and Gerlach had demonstrated the existence of quantisation of *atomic* magnetic moments, using molecular beam experiments. In these experiments a beam of molecules was directed through an inhomogeneous magnetic field and their various quantum states distinguished. These molecular beam experiments were later refined to permit the detection of nuclear magnetic moments [2]. A beam of hydrogen molecules was first passed through a large static

TABLE 1.1

Effect of atomic mass and number on magnetic spin quantum number

Atomic number	Atomic mass	Spin quantum number	Example
Even	Even	0	^{12}C, ^{16}O
Even	Odd	$n/2$	^{13}C, ^{17}O
Odd	Even	n	^{14}N, ^{10}B
Odd	Odd	$n/2$	^{1}H, ^{15}N

homogeneous magnetic field prior to being passed through the inhomogeneous magnetic field, and their small nuclear magnetic moments were detected. Further refinement of this experiment led to the addition of an oscillating magnetic field. In this version of the experiment only the molecules which did not change their energy reached the detector; when the energy of the r.f. oscillated field reached the appropriate value there was a sudden drop in the number of nuclei arriving at the detector. Nuclei transition energies could thus be measured [3].

Other techniques were tried unsuccessfully to observe the resonant absorption of radio frequency power. For example in 1936 Gorter tried to detect the absorption of energy by ^{7}Li nuclei in lithium fluoride by calorimetric means [4]. He also later tried along with Broer to observe the same signal using anomolous dispersion techniques [5]. Only the use of unfavourable materials, ironically too pure, hence having too long a relaxation time, was the main cause of failure. However in 1945 two groups of workers simultaneously discovered resonant absorption in bulk matter. Bloch, Hansen and Packard of Stanford University detected a magnetic induction signal from the protons of water at 7.765 MHz using a crossed coil probe [6]. Purcell, Torrey and Pound at Harvard University used a single coil probe and detected the proton magnetic absorption signal of a piece of paraffin wax at 30 MHz [7]. In 1952 Bloch and Purcell were jointly awarded the Nobel prize for the discovery. N.m.r. at this stage was purely an experiment to determine the nuclear magnetic moments of nuclei. Only later with improvements in techniques and instrumentation did it change into a subtle probe of chemical structure.

The proton is the most sensitive naturally occurring nucleus to n.m.r. detection. This property, coupled with its common occurrence in chemistry, particularly organic chemistry, has resulted in the proton being the most studied nucleus. Lower sensitivity of other nuclei, particularly ^{13}C, has until recently made their study less attractive. These barriers have become less important in recent years and the study of nuclei other than the proton is rapidly increasing. The trend will be reflected in later chapters.

1.2 PULSED HIGH RESOLUTION N.M.R.

Very early in the history of n.m.r. Bloch et al. showed that there were many ways of observing the phenomenon [8]. One approach was to sweep either the field or the radio frequency whilst keeping the other fixed. In the sweep experiment two limits can be distinguished. Firstly we have the slow passage limit where the sweep rate is sufficiently slow, on the nuclear spin time scale, that the experiment represents equilibrium conditions. The other limit is a very rapid passage experiment, where adiabatic conditions exist. These terms will be defined in more detail later on. High resolution spectra are obtained using conditions which approximate to the slow passage limit, the adiabatic rapid passage technique now rarely being used, except for rapid location of weak signals. These limits bracket the so-called continuous wave (c.w.) approach to recording an n.m.r. spectrum.

A totally different approach to studying n.m.r. was also proposed by Bloch et al. [8]. This involves subjecting the spin system to a short intense pulse of r.f. power and observing its subsequent behaviour. The pulse or free precession approach was, however, pioneered theoretically by Torrey [9], and practically by Hahn [10].

The pulse and continuous wave approaches to the study of n.m.r. originated at the same time but their usage soon diverged. The pulse technique was applied to the study of time domain phenomena, i.e. relaxation times. Such measurements provide data on time dependent processes such as molecular motion, chemical exchange, and molecular diffusion. However, pulsed experiments were little used in chemistry. The continuous wave approach developed into what is now called high resolution n.m.r. The result of such an experiment is a multi-line spectrum characterised by chemical shifts and coupling constants. These spectra have found great application in chemistry, particularly organic chemistry, where they have revolutionised structural determination techniques.

Pulsed experiments are more efficient than swept experiments with regard to information content, as all the nuclei are simultaneously and not sequentially studied. In 1966 Ernst and Anderson pointed out that by the use of Fourier transformation the more efficient pulsed approach could give the same multi-line spectra as the c.w. method [11]. The two techniques converged and Fourier or pulsed high resolution n.m.r. was born. This technique has the time advantages of pulsed excitation yet produces normal frequency spectra. In addition, it is possible to study the relaxation times of each individual line within the spectrum. However, one is not forced to consider relaxation times as a bulk property, as is done with a conventional pulse spectrometer. This book is concerned with the technique of pulsed high resolution n.m.r. In some ways this is a new type of n.m.r. and yet, at the same time, is really a merger of two much older techniques. Fourier n.m.r. developed surprisingly slowly at first, being mainly applied to proton n.m.r. However, in about 1970, with the realization that despite early pessimism (see later), Fourier n.m.r. was the key to Pandora's box of ^{13}C n.m.r., the development was almost explosive. Initially Fourier transform operation was facilitated by the addition of an expensive accessory to an already expensive research high resolution n.m.r. spectrometer. However as the advantages of operating in the pulsed mode became apparent, pulse-only n.m.r. spectrometers appeared on the market. These spectrometers were considerably less complex than the dual purpose spectrometers previously available, and this simplification, coupled with the constant drop in computer hardware prices, has led to the situation where a Fourier spectrometer is no longer more expensive than its sweep equivalent. This trend can be expected to continue in the future.

1.3 THE FOURIER TRANSFORM

The Fourier transform is a perfectly general mathematical transformation. It was developed by the one-time governor of Lower Egypt, Baron Jean Baptiste Joseph Fourier, strangely enough during work on the conduction of heat (see Appendix 1). The importance of the Fourier transform lies in its ability to relate some very important pairs of physical variables, the most important of which are probably those of time and frequency [12]. Using this technique it is possible to transform results obtained as a function of time and express them as a function of frequency. It was in this form that the Fourier transform has found use in n.m.r. as a link between the pulse and sweep methods.

Other forms of spectroscopy have also benefited from Fourier transform techniques, e.g. X-ray crystallography and far infrared spectroscopy [13]. In the far infrared region radiation where sources are very weak and produce spectra with insufficient sensitivity when conventional scanning techniques are used, much greater sensitivity can be achieved when using a multi-channel spectrometer, such as the interferometer. Here the sample to be investigated is excited by white irradiation and the transmitted light analysed in the interferometer. In the Michleson type, the incoming beam is divided into two separate beams of equal strength which are recombined after having travelled over different pathlengths. It can easily be shown that the output of the detector as a function of pathlength difference (the inter-ferogram) is proportional to the autocorrelation function of the incoming radiation. Its Fourier transforms yield the required spectrum. The interferogram has a similar appearance to the impulse response function found in n.m.r. studies but it is an even function of the pathlength difference [14].

The direct approach, i.e. applying a pulse of energy and following the signal's decay as used in n.m.r., cannot be taken in infrared spectroscopy because of the time scales involved. The spectral width which can be recovered in a time domain experiment is determined by the rate at which data can be sampled. According to the sampling theory the maximum spectral width which can be recovered is half the sampling frequency. Present computer technology allows sampling rates of up to about a MHz, which is enough for most applications in high resolution and wideline n.m.r. Even with the most optimistic prediction, the recording of an optical impulse response signal will remain difficult since it only lasts for a time in the order of the life time of the excited states i.e. $\sim 10^{-13}$ s. The pulse approach, however, may in the future be applicable to electron spin resonance and microwave resonance.

1.4 THE N.M.R. SPECTRUM

Nuclear magnetic resonance spectra can be divided into two main classes, wide line spectra and high resolution spectra. The classification is based mainly on spectral line width and hence experimental techniques. Wide line spectra, as their name suggests, have large line widths (greater than about 10 kHz) resulting from direct dipole—dipole effects, and normally originate from samples in the solid state. High resolution spectra result from liquids or gases where the direct dipole—dipole broadening has been averaged away by rapid molecular motion. High resolution spectra can be obtained from certain solids by the use of multipulse techniques. This book is concerned solely with high resolution spectra.

The n.m.r. spectrum is the behaviour of the nuclear magnetisation as a function of frequency $M(\vartheta)$. The spectrum is complex, in a mathematical sense, i.e.

$$M(\vartheta) = u(\vartheta) + iv(\vartheta) \tag{1.1}$$

where $u(\vartheta)$ is the dispersive, or in phase component, and $v(\vartheta)$ is the absorption, or out of phase component. It is the v mode, which consists of a family of Lorentzian lines, that is normally displayed. The value of a high resolution spectrum lies in the fact that nuclei of the same species absorb at different radio frequencies (with a fixed field) depending on the degree of magnetic shielding resulting from the interaction between their associated electrons and the applied magnetic field. A detailed study of the positions of such a family of Lorentzian lines yields details on the molecular environment of the nuclei being studied.

The fact that nuclei of the same species absorb energy at different frequencies was discovered about four years after the initial experiments as the accuracy in measurement of magnetic moments increased. Such effects were first observed by Knight in metals [15] and later in nonmetallic compounds by Proctor and Yu (^{14}N) [16], Dickinson (^{19}F) [17] and by Lindström [18] and Thomas (^{1}H) [19]. The separations depend on the chemical environment of the nucleus and are called chemical shifts. Fig. 1.1 illustrates the typical chemical shift ranges of the most commonly studied nuclei. The units used to express chemical shift are parts per million (δ) defined thus:

$$\delta_i = \frac{B_i - B_r}{B_0} \times 10^6 = \frac{\vartheta_i - \vartheta_r}{\vartheta_0} \times 10^6 \tag{1.2}$$

where $B_i - B_r$ and $\vartheta_i - \vartheta_r$ are the differences between the reference line in question expressed in field and frequency units respectively. A dimensionless ratio is used, so that when recording chemical

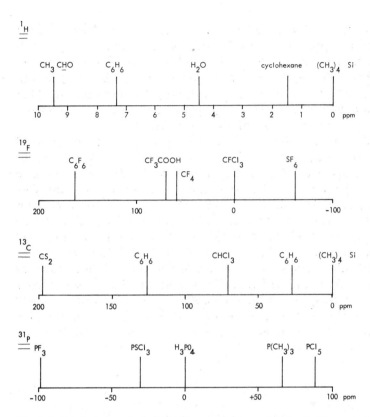

Fig. 1.1 Typical chemical shifts (in ppm) for nuclei studied by n.m.r.

shifts (which are field dependent) the field of the experiment does not affect the value quoted. It is essential to use a reference compound when expressing chemical shifts as they cannot be measured absolutely. The chemical shift is considered in more detail in chapter 8.

A statement on units and scales is necessary at this stage. The original state of chemical shifts was the Tau scales proposed by Tiers [20]. His scale was defined as in eqn. (1.2), except that the reference signals were specified as the proton resonance of tetramethylsilane (TMS). Spectra are normally plotted with the magnetic field increasing from left to right, thus, since TMS is to high field of most resonances, negative shifts result. Tiers overcame this 'difficulty' by defining TMS as 10 on the τ scale. The Tau scale solution is, as can be seen in Fig. 1.1, inapplicable to other nuclei. Chemical shift in this book is reported on the δ scale, that is as given by eqn. (1.2) with the reference taken as zero. The need for a negative sign is 'overcome' by remembering that energy (frequency) in an n.m.r. spectrum increases from right to left, thus, following the example of other spectroscopic techniques, we call a positive chemical shift one to higher energy, lower field, or simply to the left!

The spectrum of compounds containing more than one type of hydrogen atom will, in general, under low resolution, show one line in the n.m.r. spectrum for each type of proton. Moreover, the area of each line is proportional to the number of protons of that type. Consider ethanol: the proton n.m.r., Fig. 1.2a, shows three lines whose relative areas are 1:2:3; they arise from the OH, CH_2 and CH_3 protons respectively. If the resolution of the spectrometer, Fig. 1.2c is increased then further structure is revealed. This fine structure is the result of interactions involving nuclei and the bonding electrons and is called spin coupling [21]. The magnitude of the interaction is expressed in terms of a spin coupling constant J, which is field independent, and hence can be expressed simply in Hertz (i.e. energy units), whereas chemical shifts, being field dependent, are

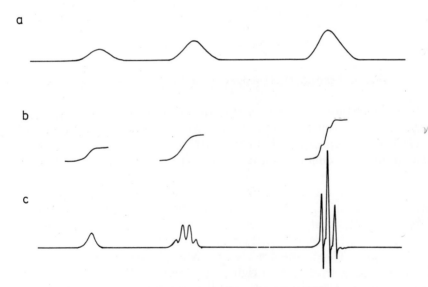

Fig. 1.2. The proton n.m.r. spectrum of ethanol. (a) Medium resolution (linewidth ~ 10 Hz); (b) integral; (c) high resolution (linewidth ~ 0.5 Hz).

expressed as a ratio. The multiplicity of the spin coupling fine structure depends on the number of nuclei involved; thus if the nucleus A
is coupled to say n nuclei of type X (an AX_n system) then A has a
multiplicity of $(2nl + l)$ ($l =$ spin quantum number of nucleus X).
For nuclei like the proton, where $l = \frac{1}{2}$, the intensities are given by
the binomial co-efficients e.g. the signal from the CH_3 protons, coupling to the CH_2 protons (where $n = 2$), is a $1:2:1$ triplet.

Ironically for ^{13}C n.m.r., which due to its low natural sensitivity
developed much later, it is normal to have one line from each carbon
and no spin coupling is seen despite the presence in the system of
two spin $\frac{1}{2}$ nuclei ^1H and ^{13}C. The lack of spin coupling is the result
of two factors, one instrumental and one natural. The instrumental
effect is that for simplicity and sensitivity, it is normal to remove all
proton carbon coupling by means of a double resonance experiment.
The second factor is the low natural abundance of the ^{13}C isotope
($\sim 1\%$) which makes the number of molecules containing two ^{13}C's in
a position to exhibit homonuclear spin coupling very low indeed.

The treatment of spin coupling given above applies only to spectra
arising from molecules where $\delta \gg J$. Such spectra are termed first
order spectra. In the other cases second order spectra result, which
are more complex in nature, and the extraction of J and δ values can
be far from simple. This problem is discussed in more detail in
chapter 8.

Along with its frequency, two relaxation times characterise a line
in an n.m.r. spectrum. These are the spin—lattice relaxation time (T_1)
and the spin—spin relaxation time (T_2). T_1 expresses the rate at
which the nuclei can exchange energy with the surroundings (lattice).
T_2 expresses the rate at which spins can exchange energy with each
other i.e. it is an entropy term; it is the value of T_2 which governs
the width of a line, the line width at half height being equal to
$(\pi T_2)^{-1}$.

Having, in this brief introduction, traced the history of n.m.r., set
pulsed high resolution n.m.r. in the framework, and sketched the
outline of an n.m.r. spectrum, the next chapter of this book will
consider each aspect of the Fourier n.m.r. experiment and the end
product in detail.

REFERENCES

1 W. Pauli, Naturwiss. 12, (1924) 741.
2 O. Stern, Z. Phys., 7, (1921) 249.

3 W. Gerlach and O. Stern, Ann. Phys. Leipzig, 74, (1924) 673.

4 G.J. Gorter, Physica, 3, (1936) 995.

5 G.J. Gorter and L.F.J. Broer, Physica, 9, (1942) 591.

6 F. Bloch, W.W. Hansen and M. Packard, Phys. Res., 69, (1946) 127.

7 E.M. Purcell, H.C. Torrey and R.V. Pound, Phys. Res., 69, (1946) 37.

8 F. Bloch, W.W. Hansen and M. Packard, Phys. Res., 70, (1946) 474.

9 H.C. Torrey, Phys. Res., 76, (1949) 1059.

10 E.L. Hahn, Phys. Res., 80, (1951) 580.

11 R.R. Ernst and W.A. Anderson, Res. Sci. Instrum., 37, (1966) 93.

12 J.B.J. Fourier, Theorie Analytique de la Chaleur, Paris, 1822.

13 E.D. Becker and T.C. Farrar, Science, 178, (1972) 316.

14 M.J.D. Low, Anal. Chem., 41, (1969) 97A.

15 W.D. Knight, Phys. Res., 76, (1949) 1259.

16 W.G. Proctor and F.C. Yu, Phys. Res., 77, (1950) 717.

17 W.C. Dickinson, Phys. Res., 77, (1950) 736.

18 G. Lindström, Phys. Res., 78, (1950) 1817.

19 H.A. Thomas, Phys. Res., 80, (1950) 901.

20 G.V.D. Tiers, J. Phys. Chem., 62, (1958) 1151.

21 W.G. Proctor and F.C. Yu, Phys. Res., 81, (1951) 20.

CHAPTER 2

PRINCIPLES OF MAGNETIC RESONANCE

In this chapter we will consider the basic phenomenon of magnetic resonance. The aim is to provide a knowledge of the basic physical effects which result in an n.m.r. spectrum. This knowledge will form the base from which we will work when studying a system of spins and also when studying the methods of obtaining and using the chemical information such spectra contain. The approach taken is a classical one, both in the strict mathematical sense (i.e. as opposed to a quantum mechanical one) and also from the n.m.r. view point. The quantum mechanical description of resonance will, however, be briefly outlined. The vector description of resonance has been chosen because it more readily lends itself to the explanation of pulsed n.m.r. and relaxation effects, whereas the quantum theory is better suited to spectral analysis problems, which are not the main concern of this book.

2.1 THE NUCLEAR SPIN

As was pointed out in the introduction certain atomic nuclei can be considered as having a 'spin'. The term spin implies that they can be considered as a rotating electrical charge and consequently, along with their electrical properties, they also possess angular momentum. From a purely classical point of view the rotating charge can be considered as an electric current flowing in a loop and will thus behave as a magnetic dipole. The moment of this dipole (μ) can be given in terms of the equivalent current i and the area of the loop around which the current flows (A). Thus

$$\mu = iA \qquad (2.1)$$

From electromagnetism we know that this electric current is equivalent to a charge q divided by t_r, which is the time for one rotation.

$$i = q/t_r \qquad (2.2)$$

Next consider the angular momentum which we have associated with the spinning nucleus. If the mass of the nucleus is m_n, its radius is r and it is rotating at a rate of ω radians per second, then the angular momentum P is given by

$$P = \omega m_n r^2 = 2\pi r^2 m_n / t_r \tag{2.3}$$

As the current is flowing in a circular loop, then the area of this loop is πr^2

$$A = \pi r^2 = \frac{P t_r}{2 m_n} \tag{2.4}$$

and we can thus deduce a value for a magnetic moment from eqns. (2.1), (2.2) and (2.4) which is

$$\mu = -\frac{q}{t_r} \times \frac{P t_r}{2 m_n} = -\frac{qP}{2 m_n} \tag{2.5}$$

This simple model provides a very important basic rule about nulcear magnetism, namely that the vector which represents the magnetic dipole is opposite in sign and co-linear with the nuclear angular momentum vector. The magnitudes of nuclear magnetic moments are very oftex expressed in terms of a constant called the magnetogyric ratio γ which is the ratio of the magnetic moment to

TABLE 2.1

The quantum number and magnetogyric ratio of some common nuclei

Nucleus	Nuclear quantum number	Magnetogyric ratio/MHz T^{-1}	Resonance frequency at 2.35 T/(MHz)
^1H	1/2	42.577	100
^2H	1	6.535	15.35
^{11}B	3/2	13.659	32.08
^{13}C	1/2	10.705	25.14
^{14}N	1	3.074	7.22
^{15}N	1/2	− 4.315	10.13
^{17}O	5/2	5.769	13.55
^{19}F	1/2	40.055	94.07
^{29}Si	1/2	− 8.460	19.86
^{31}P	1/2	17.235	40.48

the associated angular momentum.

$$\gamma = \frac{\mu}{P} \tag{2.6}$$

the value of γ for some commonly studied nuclei are given in Table 2.1.

2.2 THE NUCLEUS IN A MAGNETIC FIELD

Next we must consider the consequences of placing a magnetic nucleus in a magnetic field. A nucleus is a magnetic dipole and when placed in a static field B_0, interaction will take place and it will acquire energy given by the following equation.

$$E = -\mu B_0 \tag{2.7}$$

If the nucleus were simply a bar magnet this would be the end of the story. A nucleus, however, as already seen, is not like a bar magnet in that it also has angular momentum. As a consequence it will

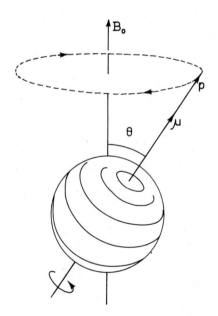

Fig. 2.1. The Larmor precession of the nuclear spin about the applied field.

align itself with respect to B_0, but will also precess at an angle θ about B_0. This precession, the Larmor precession, is illustrated in Fig. 2.1 and is directly analogous to the precession of a gyroscope about a gravitational field. The Larmor frequency of the nucleus is designated by ω_0 radians per second or ϑ_0 Hz ($\omega_0 = 2\pi\vartheta_0$). The precession occurs as a result of the interaction between the torque generated by the rotational motion of the nucleus and the interaction of the magnetic field on the nuclear magnetic moment. Simple magnetic theory tells us that the interaction energy E between the nucleus and the field is given by

$$E = \mu \times B_0 \tag{2.8}$$

Newton's laws of motion say that the rotational torque generated is equal to the rate of change of angular momentum, thus

$$E = \frac{dP}{dt} = \mu \times B_0 \tag{2.9}$$

Now, for angular motion with velocity ω_0, the rate of change of angular momentum is simply given by $P\omega_0$. Therefore

$$\frac{dP}{dt} = P \qquad \omega_0 = \mu \times B_0 \tag{2.10}$$

then using eqn. (2.6) we obtain

$$\omega_0 = \gamma B_0 \tag{2.11}$$

or

$$\vartheta_0 = \gamma B_0 \tag{2.11a}$$

This very fundamental equation is called the Larmor equation, and it describes the basic phenomenon of nuclear magnetic resonance. Field is proportional to frequency and the proportionality constant in the magnetogyric ratio. The equation also shows, through its lack of any term involving the angle of precession, that the energy of the system does not depend on μ, but only on the projection of μ onto the z axis. This condition can be expressed a slightly different way when we consider the quantum mechanical description of resonance where it arises as a consequence of the quantization of the nuclear magnetic moment.

The classical theory used so far says nothing about how we should apply the excitation energy or which transition probabilities to expect. For the latter information we must turn to quantum mechanics. The absorption of energy by nuclei can, however, easily be visualised as a resonance phenomenon as follows. The nuclei are precessing about the magnetic field at the Larmor frequency given by eqn. (2.11). The excitation is applied in the form of a small magnetic field B_1 rotating about the basic field B_0 which is conventionally along the z axis. When the frequency of rotation of B_1 is equal to ϑ_0 and in the same sense, then resonance will occur, energy will be absorbed by the nuclei from the exciting field and the nuclei will precess about B_0 at a larger value of θ. The Larmor condition can be detected by measuring the absorption of energy from the exciting field.

If in practice the exciting field had to be applied as a rotating field around the z axis the experiment would be difficult to implement. The excitation is applied as a linearly polarised field at right angles to the basic field i.e. along the x or z axis. As is illustrated in Fig. 2.2, linearly polarised field can always be resolved into two components, counter rotating about an axis orthogonal to the axis of polarisation. One component of the exciting field will rotate about the z axis in the correct sense, the other has no effect on the nuclei.

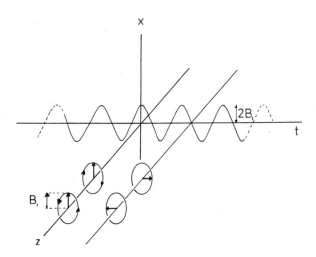

Fig. 2.2. The resolution of a linear oscillating field into its two equivalent rotating fields.

2.3 THE QUANTUM MECHANICAL DESCRIPTION OF MAGNETIC
RESONANCE

Now consider the quantum mechanical approach to nuclear magnetic resonance. The logic is parallel with the classical vector model
already considered and, not surprisingly, the end results are identical.

Quantum theory demands that like any other property, that of
spin angular momentum is quantised. The quantum number is called
I and has half integral values. The angular momentum of a nucleus
can be expressed in units of $h/2\pi$, thus

$$P_1 = [I(I + 1)]^{\frac{1}{2}} \hbar \tag{2.12}$$

The observable magnitudes of this magnetic moment along one axis,
e.g. the z axis, are expressed in terms of magnetic quantum number m,

$$P_z = m_1 \hbar \quad \text{where} \quad m_1 = I, (I-1), (I-2) \ldots -1 \tag{2.13}$$

and changes in P_z are subject to the selection rule that $\Delta m = \pm 1$.
Thus if $I = \frac{1}{2}$ then there are two values of m_1, $\pm \frac{1}{2}$ and two possible
values of the angular momentum of P. If $I = 1$ then there are three
values of m_1; $1, 0, -1$, and so on as I increases. As pointed out in the
introduction, when $I = 0$, m can only be zero and no magnetic interactions are possible, therefore no n.m.r. spectrum results.

When a nucleus is placed in a magnetic field B_0 the Hamiltonian
operator for the interaction is given by

$$\hat{\mathcal{H}} = -\mu B_0 \tag{2.14}$$

If one assumes that B_0 is along the z axis, as is conventional in
n.m.r., and if one writes eqn. (2.6) to take into account the new
quantum mechanical value for the angular momentum, the Hamiltonian then becomes

$$\hat{\mathcal{H}} = -\gamma \hbar \hat{I}_z B_z \tag{2.15}$$

where P_z is a spin angular momentum operator whose expectation
values are m. The eigenvalues of this Hamiltonian are simply multiples
of $(\gamma \hbar B_0)$ and the expectation values of $\hat{I}_z$. The allowed energy values
(i.e. energy levels) for the systems are thus

$$E = \gamma h m B_0 \tag{2.16}$$

For spin $\frac{1}{2}$ nuclei this means that there are two energy levels corresponding to $m = \pm \frac{1}{2}$. The separation between these levels is given by

$$\Delta E = \gamma h B_0 \tag{2.17}$$

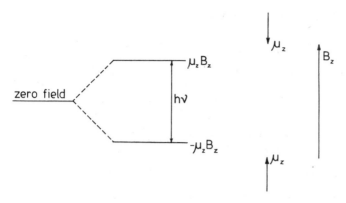

Fig. 2.3. The removal of the degeneracy of the spin stages of a spin $\frac{1}{2}$ nucleus in a magnetic field.

Figure 2.3 shows the two energy levels which can be considered as corresponding to the nuclear moment, being either parallel or anti-parallel to the magnetic field, the former being a lower energy state. If we provide energy which equals the energy difference between these two levels then energy will be absorbed and by using Planck's law we arrive at the same equation for the absorption of energy by the nuclear spins

$$\left.\begin{array}{l} \Delta E \;=\; h\vartheta_0 \;=\; \gamma h B_0 \\[4pt] \vartheta_0 \;=\; \gamma B_0 \end{array}\right\} \tag{2.18}$$

We must now again address ourselves to the problems of transition probabilities and techniques of excitation, and to do so we must consider the angular momentum operators further. A set of spin angular momentum operators can be defined for a nucleus of spin I as follows $\hat{I}_x$, $\hat{I}_y$, $\hat{I}_z$ and $\hat{I}^2$, $\hat{I}_z$ has simple eigenvalues equal to $m\hbar$ whereas $\hat{I}_x$ and $\hat{I}_y$ in general do not have simple eigenvalues. The operator $\hat{I}^2$ is quite often used, as we will see later, because it does have a simple eigenvalue solution in terms of I, $(I + 1)\,\hbar$. If one considers the case of a spin $\frac{1}{2}$ then the eigenvalue of $\hat{I}_z$ are given by $\pm \frac{1}{2}\hbar$ and hence there are two possible energy levels for the system $\pm \frac{1}{2}\hbar B_z$. If the spin eigenfunctions of the appropriate Schroedinger equation are denoted by α and β, then

$$\hat{I}_z \,|\alpha\rangle \;=\; +\tfrac{1}{2}\hbar\,|\alpha\rangle \quad \text{and} \quad \hat{I}_z\,|\beta\rangle \;=\; -\tfrac{1}{2}\hbar\,|\beta\rangle \tag{2.19}$$

The functions α and β must be proper quantum mechanical functions, i.e. orthogonal and normal, i.e.

$$\langle \alpha | \alpha \rangle = \langle \beta | \beta \rangle = 1 \tag{2.20}$$

$$\langle \alpha | \beta \rangle = \langle \beta | \beta \rangle = 1 \tag{2.21}$$

The consequences of the operators $\hat{I}_x$ and $\hat{I}_y$ on these spin functions are given by

$$\hat{I}_x | \alpha \rangle = \tfrac{1}{2}\hbar | \beta \rangle \quad \hat{I}_x | \beta \rangle = \tfrac{1}{2}\hbar | \alpha \rangle \tag{2.22}$$

$$\hat{I}_y | \alpha \rangle = \tfrac{1}{2} i\hbar | \beta \rangle \quad \hat{I}_y | \beta \rangle = \tfrac{1}{2} i\hbar | \alpha \rangle \tag{2.23}$$

We can now consider the probability (W) of a transition between the two levels described by the eigenfunctions α and β being stimulated by an oscillating magnetic field. This probability, according to quantum mechanical theory, is given by

$$W^1 = \langle \phi_m | \hat{I} | \phi_{m+1} \rangle^2 \tag{2.24}$$

If one applies the oscillating magnetic field along the z axis then one cannot stimulate transitions, because from eqns. (2.19) and (2.24)

$$W^1 = \langle \alpha | \hat{I}_z | \beta \rangle = 0^2 \tag{2.25}$$

However if one applies the oscillating magnetic field along either the x or the y axis, then as can be seen by the equation

$$W^1 = \langle \alpha | \hat{I}_x | \beta \rangle^2 = \tfrac{1}{2}\hbar \tag{2.26}$$

a transition can be stimulated.

Both the quantum mechanical and the classical method thus deduce the same resonance condition, and that the same type of excitation, linearly polarised at right angles to the basic field, is necessary to stimulate the nuclei. This result is to be expected; yet comforting!

2.4 THE POPULATION OF SPIN STATES

So far we have considered the behaviour of a single isolated nuclear spin. However, when we measure an n.m.r. spectrum we in fact record the mean behaviour of an ensemble consisting of a large number of similar nuclei. When the system is at thermal equilibrium the spins will be distributed among the various energy levels according to the Boltzman law. Thus for spin $\tfrac{1}{2}$ nuclei, where it has been seen there are two possible energy levels, the ratio of nuclei in the upper level (n_u) to the number in the lower level (n_l) as a function of absolute temperature is given by

$$\frac{n_l}{n_u} = \exp\left(\frac{2\mu B_0}{kT}\right) \approx 1 + \frac{2\mu B_0}{kT} \tag{2.27}$$

where k is the Boltzman constant. The series expansion of the exponential given above is accurate, even when only the one term is used, as at normal temperatures and normal fields the exponent is small ($2\,\mu B$ is in the order of 10^{-3} and $kT\,200\,\mathrm{cm}^{-1}$). For example, for protons at 2.3 Tesla the excess population in the lower state is only about 1 in 10^5. For all other nuclei at lower fields the excess is even smaller. As absorption can only be detected from those, the excess of nuclei in ground state compared with the excited state, n.m.r. is an insensitive technique by spectroscopic standards. For optical spectroscopy, for example, the energy difference involved is much larger, the excess population in the ground state is consequently much larger, hence the sensitivity of the technique is higher.

At ordinary temperatures the populations of the spin states are, as we have seen, almost equal. Simple calculation shows that on average if the total number of nuclei is N then $\frac{1}{2}N(1 + \mu/kT)$ nuclei are in the lower state while $\frac{1}{2}N(1 - \mu/kT)$ are in the upper state. The component of the total magnetic moment resolved along the field direction is either plus or minus μ so the total sample, in thermal equilibrium, acquires a magnetic moment M equal to $N\mu^2 B_0/kT$. The magnetic moment can also be expressed in terms of a bulk magnetic susceptibility (χ).

$$M = \chi B_0 = N\mu^2 B_0/kT \tag{2.28}$$

We therefore see that for nuclei who have a spin $\frac{1}{2}$ the magnetic susceptibility is given by

$$\chi = \frac{N\mu^2}{kt} = \frac{N\gamma^2 h^2}{4kt} \tag{2.29}$$

In a general case the susceptibility of N nuclei with spin 1 is given by the equation

$$\chi = N\frac{\gamma^2 h^2 I(I + 1)}{3kt} \tag{2.30}$$

We have previously compared n.m.r. with optical spectroscopy with respect to the population of the ground of excited states. N.m.r. also differs from optical spectroscopy in one other major characteristic, the mean life time of the excited state. In optical spectroscopy, an excited molecule returns by spontaneous emission to the ground state almost instantaneously (10^{-13} s). The excess energy is dissipated into the lattice as heat. The situation is very different for a nuclear spin which is effectively insulated from its surroundings (lattice). Here the probability of spontaneous emission is negligible [1] (about 1 in 10^{18} years for the proton). The return to the ground state in n.m.r. must be the result of stimulated emission and occurs on a time scale in the order of seconds. The emission is stimulated by locally generated magnetic fields which have components at the Larmor frequency. The origin of these fields is discussed in Chapter 10. The long life time of the excited state has two important consequences; first, by the uncertainty principle, as the excited state exists for a long time, n.m.r. lines are very sharp. The second consequence of this long life time is that 'saturation' can occur. Saturation is the equalisation of the population in the gound and the excited state which occurs because relaxation from the excited state is slow and with a strong exciting field a dynamic equilibrium can be set up. In this equilibrium the number of nuclei in the upper and lower states become equal, and the signal saturates, or disappears. The property of saturation will be returned to in section 5.7 as this is an area where swept and Fourier spectra give differing results.

2.5 RELAXATION EFFECTS

In the absence of a magnetic field the two energy levels available to spin $\frac{1}{2}$ nuclei are of equal energy and are hence equally populated. As we have discussed, in the presence of a magnetic field the energy levels are no longer equal in energy and their populations are no longer equal. The spin populations change to their new equilibrium value by an exponential process, the time constant for which is called the spin—lattice relaxation time (T_1) [2]. This constant reflects the efficiency of the coupling between the nuclear spin and its surroundings (lattice). The shorter this time, i.e. the quicker equilibrium is obtained, the more efficient this coupling is, and vice versa. Spin—lattice relaxation times can range from 10^{-3} to 10^2 s for liquids and the range is even larger for solids [3]. It was a very long spin—lattice relaxation time incident which defeated the first experiments to detect n.m.r. in the solid state. In the samples chosen the signal saturated before it could be detected.

We have previously assumed that absorption and emission are equally probable. Once we have introduced the concept of spin—lattice relaxation this is no longer true. If W_1 and W_2 are the probabilities of an absorption or emission transition respectively, and, as previously, we defined the total number of spins as N (equal to the sum of the number in the upper and lower states), then the approach to equilibrium is described by the following differential equations,

$$\frac{dn_1}{dt} = n_u W_2 - n_1 W_1 \qquad (2.31)$$

$$\frac{dn_u}{dt} = n_1 W_1 - n_u W_2 \qquad (2.32)$$

at equilibrium

$$\frac{dn_1}{dt} = \frac{dn_u}{dt} = 0 \qquad (2.33)$$

If we now introduce n as the population difference between the two nuclear spin states then we can derive the following equation for the two above

$$\frac{dn}{dt} = 2W_2 n_u - 2W_1 n_1 = N(W_2 - W_1) - n(W_1 + W_2) \qquad (2.34)$$

This equation can itself be re-written into the following form

$$\frac{dn}{dt} = \frac{n_0 - n}{T_1} \qquad (2.35)$$

if n_0 is defined as

$$n_0 = N \frac{W_2 - W_1}{W_1 + W_2} \qquad (2.36)$$

and $1/T_1$ given the following value

$$\frac{1}{T_1} = W_1 + W_2 \qquad (2.37)$$

Equation (2.35) is of great importance to us as it describes in simple terms the return to equilibirum of the z component of the magnetization after it has been disturbed, e.g. by the application of a pulse. In order to describe accurately the behaviour in a swept n.m.r. experiment an extra term should be included in eqn. (2.35) to allow for radiation-induced transitions. We will not consider this additional complication.

In addition to interaction with their lattice, nuclear spins can interact with each other. Each nuclear spin, since it is a dipole, generates its own local field B_{loc}, which can be calculated from classical electromagnetic theory and is given by the following equation.

$$B_{\text{dipolar}} = \frac{\mu_z}{4r^3}(3\cos^2\theta - 1) \tag{2.38}$$

where r is the distance of the first dipole from the second and θ is the angle between the axis of the first dipole and the line joining their centres. This direct dipole—dipole effect, due to the term in r^3, is a very local one and the average value of the B_{dipolar} field is zero; as the average of $\cos^2\theta$ is a third, the term in brackets reduces to zero. However, at any particular point in time, the dipolar field is not necessarily zero.

Spin—lattice relaxation occurs via transitions which are stimulated by components of the local magnetic field seen by a particular nucleus which fluctuates at its Larmor frequency. Fluctuations in the local magnetic field are generated by variations in B_{dipolar} caused by changes in r and θ resulting from Brownian motion. There are other possible sources of fluctuating magnetic fields which are discussed in the chapter on relaxation.

Spin—spin relaxation also occurs through local magnetic fields, but in a different way. If one nucleus undergoes a transition from one spin state to another then B_{loc} changes at the correct frequency to induce a transition in a second nucleus. If at this point in time a second nucleus of the same type and the opposite spin state is close by, then the two nuclei can in effect exchange energy. Such a process is not of the spin—lattice type as the total system energy is unchanged. A process of this type, however, does affect the life time of the excited state, hence the resonance line width; these effects are called spin—spin effects. Spin—spin relaxation is an entropy effect, as opposed to spin—lattice relaxation which is an energy effect. The spin—spin relaxation time can be approximately defined as the life time or phase memory of the excited state.

The effect of local magnetic fields on line widths can be visualised in another way. If we have at any instant a $B_{dipolar}$ which is not zero, the local magnetic field seen by one nucleus (the sum of B_z and $B_{dipolar}$ at that point), is not the same as that seen by other nuclei in a different part of the sample. Instead of one resonance condition there are many; they are close in frequency and consequently the spectral line is broadened. As defined above this is the spin—spin process. There are two other effects which lead to not all the nuclei being in the same magnetic field and consequently contribute behaviour which we would consider to be of a spin—spin type. The first of these is due to static dipolar fields originating from other nuclei within the sample. In the case of liquids and gases this effect does not arise because the molecular motion leads to the averaging out of such fields to zero. However, this is not the case in the solid state where fields of this type become very important. The second effect arises from inhomogeneities in the basic magnetic field due to

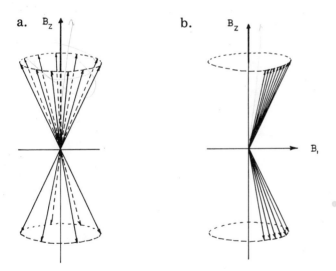

Fig. 2.4. (a) Precession of an ensemble of vectors with random phase in the x', y' plane produces a net magnetisation only in the z direction. (b) Application of B_1 field, which rotates at the resonant frequency, causes the vectors to attain phase coherence and a net x', y' magnetisation results.

instrumental imperfections. Broadening of n.m.r. lines due to magnetic inhomogeneities is not of course truly a T_2 process: however, it has consequences which are very hard to distinguish from those originating from genuine T_2 processes. We will return to this problem in chapter 10.

The relationship between T_1 and T_2 can be visualised by a vector model shown in Fig. 2.4. In this model the nuclei are considered as vectors. In the presence of only the static magnetic field B_z, the spin vectors precessed about B_z with a random phase (Fig. 2.4a), there being slightly more vectors parallel to the field than opposed to it, as was previously discussed. The average value of the magnetization in the x and y directions is zero, i.e. $\bar{M}_x$ and $\bar{M}_y = 0$. If now an exciting field B_1 is applied precisely at the Larmor frequencies of the nuclei, then resonance occurs, i.e. they obtain phase coherence as is shown in Fig. 2.4b. M_x and M_y become finite and M_z decreases appropriately. An alternative description, and perhaps a simpler one, can be given when using the idea of a rotating frame of co-ordinates, which will be discussed later. On the removal of B_1, M_z returns to its original value by spin—lattice relaxation, however M_x and M_y relax exponentially at a different rate, responding to any process which changes the vector's phase coherence but not the total energy of the system. The constant T_2 can thus be defined from the following equation.

$$\frac{dM_y}{dt} = -\frac{M_y}{T_2} \quad \text{and} \quad \frac{dM_x}{dt} = -\frac{M_x}{T_2} \tag{2.39}$$

Inspection of the above equation, along with eqn. (2.35), illustrates the origin of the alternative names for T_1 and T_2 which are, respectively, the longitudinal and transverse relaxation times [2].

More detail will be given later as to the exact mechanism involved in T_1 and T_2 processes, but certain important conclusions can be drawn at this stage. Local magnetic fields fluctuating at the Larmor frequency can contribute to both T_1 and T_2 relaxation processes. Static magnetic fields, however, can only contribute to transverse relaxation. It is thus not surprising that, as will be shown more rigorously later, T_2 must be equal to, or shorter than, T_1. In normal liquids T_1 and T_2 are often equal, though this depends in detail on the molecular motions involved. In solids, however, particularly at low temperatures, T_1 may be very long due to the lack of any suitable molecular motion, while T_2 may be very short due to static dipolar fields. In between liquids and solids there is a gradual change in the absolute and relative values of T_1 and T_2.

2.6 THE N.M.R. LINE SHAPE

The line shape of an n.m.r. resonance signal depends on a number of different factors. If there were no instrumental effects, or relaxation effects, the excited state would have an infinite life, and the line shape would be a delta function. As discussed in the previous section, nuclear spin relaxation does occur, hence the excited state has a finite life time, and the resonance line a finite width, which is in the order of the reciprocal of the life time of the excited state. A resonance line shape is normally defined in terms of the shape function $g(\vartheta)$. There are two common line shape functions found in spectroscopy, illustrated in Fig. 2.5. They are the Lorentzian line shape defined thus

$$g(\vartheta)_{\text{Lorentz}} = \frac{1}{a}\left[\frac{1}{1 + (2\pi\Delta\vartheta/a)^2}\right] \tag{2.40}$$

and the Gaussian line shape defined thus

$$g(\vartheta)_{\text{Gauss}} = \exp\left[-(2\pi\Delta\vartheta)^2/a^2\right] \tag{2.41}$$

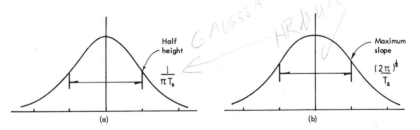

Fig. 2.5. The two line shapes commonly found in spectroscopy. (a) Lorentzian, (b) Gaussian.

where $\Delta\vartheta$ is the offset from the resonance frequency and a is the half line width at half height. High resolution n.m.r. line shapes, as the next sections will show, are almost always of the Lorentzian type which is characteristic of damped oscillatory motion. The Gaussian line shape, which drops off much more rapidly towards the wings, is often found in the n.m.r. of crystallised solids. As has been seen, in n.m.r. the line width at half height can be related to the spin—spin relaxation time $T_2 (a = 1/\pi T_2 \, \text{Hz})$. The Fourier transform of the Lorentzian line is, as will be seen, a simple exponential (whereas that of a Gaussian line is also Gaussian). Considerable use will be made of the Lorentzian line shape and its Fourier transform pair when discussing the relationship between pulsed and swept n.m.r. spectra.

2.7 THE BLOCH EQUATIONS IN THE LABORATORY FRAME

The earliest, and still one of the most convenient, treatments of the magnetic resonance phenomenon is that of Professor Bloch [2, 4]. His approach was to use a vector model and to treat the assembly of nuclear spins in macroscopic terms. When the assembly is placed in a magnetic field, a total magnetic moment per unit volume M_0 is produced. This arises because, as we have seen, more nuclei align themselves parallel to the field than anti parallel, or, in other words, population of the lower state n_1 is larger than the upper state (n_u)

$$\underline{M}_0 = \gamma h (n_u - n_1) = \gamma h n \tag{2.42}$$

As we have seen in section 2.4 this magnetization can be related to the applied magnetic field via the nuclear magnetic susceptibility χ_0.

$$\chi_0 = \underline{M}_0 / B_0 \tag{2.43}$$

If we consider the static magnetic field as being along the z axis, M_0 can be resolved into three components as shown in Fig. 2.6. At equilibirum, in the absence of any exciting field, the z component

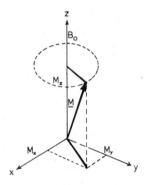

Fig. 2.6. The resolution of total macroscopic magnetisation $\underline{M}$ into its components in the laboratory frame. $\underline{M}$ precesses about z at the Larmor frequency.

of $\underline{M}_0$ will remain constant.

$$\frac{dM_z}{dt} = 0 \tag{2.44}$$

The magnitudes of the x and y components will vary as $\underline{M}_0$ precesses about B_0 at the Larmor frequency, but will have a zero mean. Also, as can be seen from Fig. 2.6, M_x has a maximum value when M_y has a zero value and vice versa. They are, in other words, 90° out of phase with each other. This situation can be represented in the following two equations.

$$\frac{dM_x}{dt} = \gamma M_y B_0 \tag{2.45}$$

$$\frac{dM_y}{dt} = -\gamma M_x B_0 \tag{2.46}$$

We must next consider the addition of an exciting field. As seen previously, in order to stimulate transitions this magnetic field must rotate about the z axis in the correct sense. As is illustrated in Fig. 2.2, excitation is achieved by a linearly polarised field of amplitude $2B_1$ which can be resolved into two components rotating around the z axis, one of which will have the correct sense to stimulate transitions. The x and y components of B_1 as it precesses about the z axis are given by

$$(B_1)_x = B_1 \cos 2\pi \vartheta t \qquad (B_1)_y = -B_1 \sin 2\pi \vartheta t \tag{2.47}$$

Now apply the same considerations to the effects on the nuclear magnetization of the additional field B_1 that were used to describe the effect of B_0; for example, the magnetization along the x axis will now not only contain a component generated by M_y rotating about the static magnetic field B_0, but it will also contain a component due to M_z rotating around B_1. Following these arguments we can deduce the following equations.

$$\frac{dM_x}{dt} = \gamma [M_y B_0 - M_z (B_1)_y] \tag{2.48a}$$

$$\frac{dM_y}{dt} = -\gamma [M_x B_0 + M_z (B_1)_x] \tag{2.48b}$$

$$\frac{dM_z}{dt} = -\gamma[M_x\,(B_1)_y - M_y\,(B_1)_x]\tag{2.48c}$$

To complete the description, one must take into account the effects of relaxation discussed in the previous section. If any of the components of $M_0\,(M_x, M_y, M_z)$ are disturbed from their equilibrium value $(0, 0, M_z)$, they will return to these equilibrium values exponentially with the appropriate time constant (T_2, T_2, T_1). This situation can be described by the following set of equations.

$$\frac{dM_x}{dt} = -\frac{M_x}{T_2}\tag{2.49a}$$

$$\frac{dM_y}{dt} = -\frac{M_z}{T_2}\tag{2.49b}$$

$$\frac{dM_z}{dt} = -\frac{M_z - \underline{M}_0}{T_1}\tag{2.49c}$$

If one now combines eqns. (2.48) with eqns. (2.49), the full Bloch equations are obtained:

$$\frac{dM_x}{dt} = \gamma[M_y\,B_z - M_z\,B_1 \sin 2\pi\vartheta t] - \frac{M_x}{T_2}\tag{2.50a}$$

$$\frac{dM_y}{dt} = -\gamma[M_x\,B_z - M_z\,B_1 \cos 2\pi\vartheta t] - \frac{M_y}{T_2}\tag{2.50b}$$

$$\frac{dM_z}{dt} = -\gamma[M_x\,B_1 \sin 2\pi\vartheta t + M_y\,B_1 \cos 2\pi\vartheta t] - \frac{M_z - M_0}{T_1}\tag{2.50c}$$

Whereas it is not possible to give an analytical solution to this set of differential equations, they can be solved under various limited conditions, the most useful of which are the adiabatic slow passage or steady state conditions. In this limit the absorption of r.f. energy is balanced by the transfer of energy to the lattice, so that the system is at equilibrium at all times, i.e. $dM_z/dt = 0$. However, before considering the solution of these equations in detail, we will consider an alternative set of co-ordinate axes in which to describe the spin system. This new set of co-ordinates is called the rotating frame of reference and permits a much easier solution to the differential eqns. (2.50).

2.8 THE ROTATING FRAME OF REFERENCE

To date all our descriptions of the n.m.r. experiment have used a set of Cartesian co-ordinates fixed with respect to the laboratory. Within this frame we have seen that the nuclei precess about the z axis of the Larmor frequency. Now consider a new set of Cartesian axes (x′, y′ and z′) rotating about the static magnetic field after the Larmor frequency.

$$x' = x \cos 2\pi\vartheta t + y \sin 2\pi\vartheta t \qquad (2.51a)$$

$$y' = -x \sin 2\pi\vartheta t + y \cos 2\pi\vartheta t \qquad (2.51b)$$

$$z' = z \qquad (2.51c)$$

As inhabitants of the planet Earth, we are familiar with the concept of a rotating co-ordinate system; we live in one. Our rotating frame with respect to a fixed point in space, is rotating at 1.157×10^{-5} Hz ($2\pi/24$ radians hour^{-1}) and revolving at 3.171×10^{-8} Hz. What to us looks like a very simple motion, e.g. walking in a straight line, would appear to a truly stationary observer a very complex circular motion. By a parallel process, the complex behaviour of the nuclear magnetization can be simplified by rotating with it and watching it.

If one is to use the rotating frame approach quantitatively, one must consider it carefully. In doing so some consequences will emerge which are not immediately obvious. To illustrate this we will consider in detail the behaviour of $\underline{M}_0$. This is a vector and thus can be expressed in terms of its components.

$$\underline{M}_0 = M_x i + M_y j + M_z k \qquad (2.52)$$

where i, j and k are the unit vectors in the x, y and z directions respectively. In the same nomenclature $\underline{B}_1 = iB_1$. Now let us differentiate $\underline{M}$ with respect to time.

$$\frac{d\underline{M}}{dt} = \frac{\delta M_x}{\delta t} i + M_x \frac{\delta i}{\delta t} + \frac{\delta M_y}{\delta t} j + M_y \frac{\delta j}{\delta t} + \frac{\delta M_z}{\delta t} k + M_z \frac{\delta k}{\delta t}$$

$$= \left[\frac{\delta M_x}{\delta t} i + \frac{\delta M_y}{\delta t} j + \frac{\delta M_z}{\delta t} k \right] + \left[M_x \frac{\delta i}{\delta t} + M_y \frac{\delta i}{\delta t} + M_z \frac{\delta k}{\delta t} \right] \qquad (2.53)$$

The meaning of the terms in the first bracket is fairly easy to visualise. They represent the rate of change of M_x etc. within the rotating frame. It is not easy to see what the terms in the second bracket of the type $\delta i/\delta t$ mean as i, j and k are unit vectors hence cannot change their magnitude with time. They can, however, rotate. Rotation is represented by the vector cross product, therefore we can understand what is meant by the rate of change of i with respect to time. It is the product of a rotation frequency and the unit vector. Expressed mathematically this is

$$\frac{\delta i}{\delta t} = 2\pi\vartheta \times i \qquad \frac{\delta j}{\delta t} = 2\pi\vartheta \times j \qquad \frac{\delta k}{\delta t} = 2\pi\vartheta \times k \qquad (2.54)$$

Using this concept we can now reformulate the eqn. (2.53), in terms of the rate of change of the magnetization with respect to time in the rotating frame itself. Thus

$$\frac{\mathrm{d}\underline{M}}{\mathrm{d}t_{\mathrm{fixed}}} = \frac{\mathrm{d}\underline{M}}{\mathrm{d}t_{\mathrm{rot}}} + 2\pi\vartheta \times (M_x i + M_y j + M_z k) \qquad (2.55)$$

The terms in the bracket can be recognised as being $\underline{M}$. We thus arrive at equation

$$\frac{\mathrm{d}\underline{M}}{\mathrm{d}t_{\mathrm{fixed}}} = \frac{\mathrm{d}\underline{M}}{\mathrm{d}t_{\mathrm{rot}}} + 2\pi\vartheta \times \underline{M} \qquad (2.56)$$

which describes the behaviour of the magnetization in the rotating frame in terms of that in the fixed or laboratory frame. Throughout the book we will use the prime notation to indicate a rotating frame value, thus $M_{x'}$ and $M_{y'}$ are the components of M along the x and y axes of the rotating frame. M_z is of course the same in both coordinate systems. Now using the expression for $\mathrm{d}\underline{M}/\mathrm{d}t$ in the laboratory frame we derived previously we obtain

$$\frac{\mathrm{d}\underline{M}'}{\mathrm{d}t} = (\gamma\underline{M} \times B_0) - (2\pi\vartheta \times \underline{M}) \qquad (2.57)$$

This equation can be rearranged to a more convenient form as given in the following equation

$$\frac{\mathrm{d}\underline{M}'}{\mathrm{d}t} = \gamma\underline{M} \times B_{\mathrm{fic}} \quad \text{where} \quad B_{\mathrm{fic}} = B_0 - \frac{\vartheta}{\gamma} \qquad (2.58)$$

This situation is depicted in Fig. 2.7. The term ϑ/γ has the dimensions of a magnetic field and can be considered as the fictitious field arising from the rotation. These equations tell us that the ordinary equations of motion used in the laboratory frame are valid in the

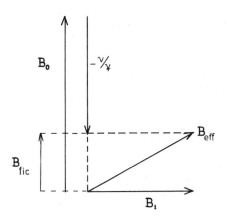

Fig. 2.7. The various fields present within the rotating frame. The effective field is derived, from B_1, and the fictitious field as shown.

rotating frame provided that B_0 is replaced by B_{fic} defined by eqn. (2.58). Thus in the rotating frame nuclei precess about the effective just as they precess about B_0 in the laboratory frame. The advantage of the rotating frame description is that when we are precisely at resonance, i.e. the rotation frequency of the reference frame is made ϑ_0, then as ϑ_0 divided by γ is equal to B_0, the fictitious field disappears, and with it all precession. Thus in the rotating frame of reference, $\underline{M}$, instead of precessing about B_0, is static, pointing along the z axis.

As an example, consider the normal field sweep experiment described from a rotating point of view as opposed to a fixed laboratory frame of reference, B_{eff} is now the vector sum of the exciting field B_1 and the residual field difference between B_0 and the fictitious field described above. Thus

$$B_{eff} = [(B_0 - \vartheta/\gamma)^2 + B_1^2]^{\frac{1}{2}} \tag{2.59}$$

This relationship is illustrated in Fig. 2.7. When B_0 is far above the resonance condition, the z component of B_{eff} is large, and under these conditions, M lies along B_0. As B_0 sweeps towards resonance then $B_0 - \vartheta/\gamma$ decreases and B_{eff} tips away from the z axis towards the x' axis. Provided that the rate of change from B_{eff} is slow enough, the magnetization follows it and at resonance is aligned along x'; as B_0 decreases from resonance, the magnetization follows B_{eff} back along to the z axis. The path traced out by the magnetization as B_0

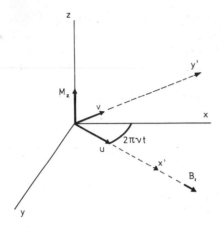

Fig. 2.8. The relationship between the two components of the transverse magnetisation (the u and v mode) in both the laboratory (x, y, z) and rotating (x', y', z') co-ordinate systems.

is swept through resonance is that of a Lorentzian absorption line shape.

2.9 THE BLOCH EQUATIONS IN THE ROTATING FRAME

Having introduced the concept of a rotating frame of reference we will now use it to help solve the Bloch equations (2.50). If we define two new symbols, as illustrated in Fig. 2.8, for the x and y components of M_0 in the rotating frame thus

$$M_{x'} = u \quad \text{and} \quad M_{y'} = v$$

using eqn. (2.56) the Bloch equation can be re-written thus:

$$\frac{du}{dt} = (\vartheta_0 - \vartheta)v - \frac{u}{T_2} \tag{2.60a}$$

$$\frac{dv}{dt} = -(\vartheta_0 - \vartheta)u - \frac{v}{T_2} + \gamma B_1 M_z \tag{2.60b}$$

$$\frac{dM_z}{dt} = -\gamma B_1 v - \frac{(M_z - M_0)}{T_1} \tag{2.60c}$$

where ϑ_0 is the frequency of the rotating frame. The excitation is assumed to be applied along the x axis and of frequency ϑ_0 thus in the rotating frame it appears simply as a static field along the x′ axis, see Fig. 2.8. The effect of the exciting field is thus simple to describe; it causes rotation of M_z about the x′ axis, therefore only contributing to the magnetisation along the y′ axis. It is common to refer to the x′ and y′ components of the magnetisation as the u and v mode signals.

The steady state solution of the above equations, i.e. where the sweep is slow enough that equilibrium is always maintained, can be shown to be

$$u = M_0 \frac{\gamma B_1 T_2^2 (\vartheta_0 - \vartheta)}{1 + 4\pi^2 T_2^2 (\vartheta_0 - \vartheta)^2 + \gamma^2 B_1^2 T_1 T_2} \tag{2.61a}$$

$$v = M_0 \frac{\gamma B_1 T_2}{1 + 4\pi^2 T_2^2 (\vartheta_0 - \vartheta) + \gamma^2 B_1^2 T_1 T_2} \tag{2.61b}$$

$$M_z = M_0 \frac{1 + 4\pi^2 T_2^2 (\vartheta_0 - \vartheta)^2}{1 + 4\pi^2 T_2^2 (\vartheta_0 - \vartheta) + \gamma^2 B_1^2 T_1 T_2} \tag{2.61c}$$

Under conditions where $\gamma B_1^2 T_1 T_2 \ll 1$, i.e. the r.f. power applied is sufficiently low that saturation does not occur, then further simplification takes place and the v mode becomes

$$v = M_0 \frac{\gamma B_1 T_2}{1 + 4\pi^2 (\vartheta_0 - \vartheta)^2 T_2^2} \tag{2.62}$$

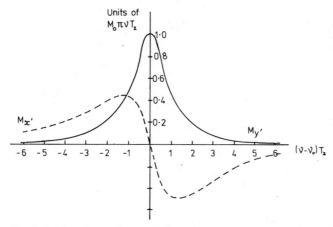

Fig. 2.9. The absorption and dispersive components of the nuclear magnetisation as a function of frequency.

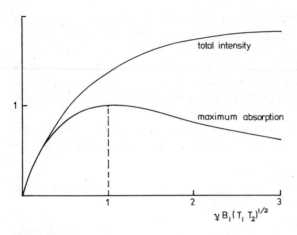

Fig. 2.10. The effect of increasing r.f. power in the sweep experiment on both the peak height and total intensity of a line.

which by comparison with eqn. (2.40) can be seen to be a Lorentzian line centered at ϑ_0 with a line width at half height of $1/\pi T_2$. The u mode is the associated dispersive component. Figure 2.9 illustrates the shape of the u and v mode signals. The amplitude of the v mode signal below saturation is proportional to $\gamma B_1 T_2$ and it has a maximum when the $\gamma^2 B_1^2 T_1 T_2$ (called the saturation parameter, S) equals 1. Once S exceeds 1 the amplitude decreases and the line broadens, as shown in Fig. 2.10. The maximum signal to noise ratio thus occurs when $S = 1$.

The Bloch theory therefore predicts that the n.m.r. signal has two components which are $90°$ out of phase. The u mode component is in phase with the exciting field, i.e. along the same axis; the v mode is out of phase with the excitation, i.e. along the orthogonal axis to both the excitation and the static field. Bloch introduces the concept of a complex susceptibility $(\chi(\vartheta))$ to describe these modes [2], the real part being associated with the u mode, the imaginary part with the v mode. It is the latter component which is usually detected and plotted, as it is an aborption line whose area below saturation is proportional to the number of nuclei generating the signal; the u mode has a zero integral.

It may at first sight seem rather strange to concentrate on the rotating frame components of the nuclear magnetization rather than the laboratory frame signals, which incidentally have the same basic behaviour. The reason is that, as seen later, n.m.r. spectrometers detect the signal coming from the sample with respect to the excitation, using phase sensitive detectors (see later). They therefore produce signals which correspond to the u and v modes rather than the behaviour of M_x and M_y. The output from the sample contains both modes; the v mode signal is selected for the reasons given above.

REFERENCES

1 E.M. Purcell, Phys. Rev., 69 (1946) 681.
2 F. Bloch, Phys. Rev., 70, (1946) 460.
3 C.J. Gorter and L.F.J. Bruer, Physica, 9, (1942) 591.
4 R.K. Wangsness and F. Bloch, Phys. Rev., 81, (1958) 728.

CHAPTER 3

THE MATHEMATICS OF FOURIER N.M.R.

In this chapter the mathematical concepts behind the application of Fourier transforms to high resolution n.m.r. will be considered. The aim is not to give rigorous proofs of various theorems, for which the reader is referred to standard text books on the subject (see refs. 1, 2), but to discuss certain key concepts associated with Fourier transforms which make these particularly useful in the study of systems, such as an assembly of nuclear spins. These ideas will recur in various guises throughout the book; in this chapter they are grouped together within a unified mathematical format.

3.1 FOURIER SERIES AND FOURIER TRANSFORMS

It is well established that if any property varies periodically with time, it can be analysed 'into its harmonic components'. Provided the property repeats itself with basic frequency, then no matter how complicated it is within that period, it can in principle be represented by a set of harmonically varying functions, each having repetition frequencies equal to multiples of the base frequency. A classical example of this is to be found in the properties of sound waves. The complex wave form emitted by a specific instrument can be imitated by mixing together the output of many oscillators, all harmonically related to the note in question but varying in amplitude and phase. This is an example of a Fourier synthesis. The opposite process is possible; on hearing the sound of an orchestra the human ear can identify individual instruments and even specific notes. The ear in this case is performing a Fourier analysis. On the other hand, the eye, for example, is incapable of such an analysis; white light is white, and the eye cannot resolve it into it's frequency components, i.e. colours. The simulation of a musical note is really a use of Fourier series. The required note is built up by oscillators representing the terms of the series. An example of a Fourier series representing a

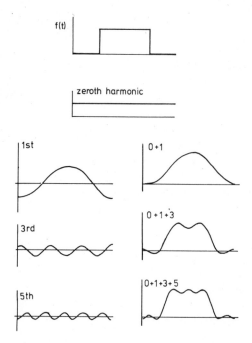

Fig. 3.1. The Fourier synthesis of a rectangular waveform. (Note the progressive improvement as higher order terms are included).

square wave is given in Fig. 3.1; the larger the number of terms the better the representation. Fourier series allows a periodic function to be represented as an infinite series of harmonic oscillations at discrete frequencies equal to multiples of the base frequency. The Fourier transform extends these ideas; the series is in effect a special case of the transform, to allow an aperiodic function to be expressed as an integral sum over a continuous range of frequencies.

Indeed, the original idea of representing a periodic function by a series of related trigonometrical functions originated in 1807 during work on the conduction of heat, and when published in 1822 by Fourier it was controversial to the point of almost being heretical [3]. In deference to their importance, most of the rest of this chapter will use time and frequency as the related variables but it must be stressed that these are, in fact, only a spectral case and all the results quoted are quite general.

There is no universally accepted definition to the term 'Fourier transform' or it's inverse. The definition we will adopt is as follows:

$$f(t) = \int_{-\infty}^{\infty} F(\vartheta) \exp\left(-i2\pi\vartheta t\right) dt \equiv f(t) = \mathcal{F}^- F(\vartheta) \tag{3.1}$$

and inversely

$$F(\vartheta) = \int_{-\infty}^{\infty} f(t) \exp\left(+i2\pi\vartheta t\right) dt \equiv F(\vartheta) = \mathcal{F}^+ f(t) \tag{3.2}$$

The units here are second and Hertz, if, as is sometimes the case, radians/s are used in place of Hertz, the transform is usually defined thus:

$$F(\omega) = \int_{-\infty}^{\infty} f(t) \exp\left(+i\omega t\right) dt \tag{3.3}$$

and inversely

$$f(t) = (2\pi)^{-1} \int_{-\infty}^{\infty} F(\omega) \exp\left(-i\omega t\right) dt \tag{3.4}$$

which lacks symmetry. In mathematical texts symmetry is achieved by placing a factor of $(2\pi)^{-1/2}$ in both the transform and its inverse; the units can no longer be simply s and radian/s; hence this definition is little used in physics. It is convention to use large case letters for functions in the frequency domain and small letters for functions in the time domain. We will abide by this convention and that of defining the transform as $\mathcal{F}^-$ and the inverse as $\mathcal{F}^+$ whenever possible.

Having defined the Fourier transform it is profitable to consider some simplifications which can be deduced if certain limits are placed on the properties of $f(t)$.

(1) If $f(t)$ is real then

$F(\vartheta)$ is complex and Hermitian

$$F(\vartheta) = F^*(-\vartheta); \quad F^*(\vartheta) = F(-\vartheta) \tag{3.5}$$

(2) If $f(t)$ is even (i.e. $f(t) = f(-t)$) then

$$F(\vartheta) = F(-\vartheta) \tag{3.6}$$

(3) If $f(t)$ is odd (i.e. $f(t) = -f(-t)$) then

$$F(\vartheta) = -F(-\vartheta) \tag{3.7}$$

and further if $f(t)$ only exists for $t > 0$, as is the case in pulse n.m.r. (this limitation is called causality) then further simplifications exist.

$$f(t) = \int_0^\infty F_c(\vartheta) \cos 2\pi\vartheta t \, dt \tag{3.8}$$

$$f(t) = \int_0^\infty F_s(\vartheta) \sin 2\pi\vartheta t \, dt \tag{3.9}$$

where

$$F_c(\vartheta) = \int_0^\infty f(t) \cos 2\pi\vartheta t \, dt \tag{3.10}$$

$$F_s(\vartheta) = \int_0^\infty f(t) \sin 2\pi\vartheta t \, dt \tag{3.11}$$

$F_c(\vartheta)$ and $F_s(\vartheta)$ are known as the cosine and sine transforms of $f(t)$. As we will see in pulsed n.m.r., the cosine transform of the time domain signal which is real and only exists for positive values of t can be equated with the v mode signal and the sine transforms with the u mode. These assignments can be checked using properties 2 and 3 above; the v mode signal is even (see Fig. 2.9). We therefore know that the time domain signal is also even, which means a cosine function and vice versa. Since from property one we know that an n.m.r. spectrum is Hermitian, we can also deduce that its amplitude (magnitude) spectrum is even.

3.2 FOURIER PAIRS

Throughout the rest of the book we will frequently use functions which are Fourier transform pairs interchangeably, depending on which 'domain' we are working in. One is said to be working in the 'time domain' when time is the variable, and in the 'frequency domain' when frequency is the variable, etc. In this section we state, without proof, some relationships between functions that are Fourier pairs. These relationships will be found to be useful when working with such functions. Table 3.1 contains a list of functions commonly encountered in high resolution n.m.r. along with their Fourier 'twins'. The variables used are time and frequency but it must be repeated that the relationships are perfectly general; in fact, the two columns can be interchanged if desired.

The following rules govern the manipulation of Fourier pairs.

(i) Addition and subtraction;

$$\mathcal{F}^+[f(t) \pm g(t)] \equiv \mathcal{F}^+f(t) \pm \mathcal{F}^+g(t) \tag{3.12}$$

(ii) Scaling:

$$f(at) \xrightarrow{\mathcal{F}^-} \frac{1}{|a|} F(2\pi\vartheta/a) \tag{3.13}$$

TABLE 3.1

Fourier pairs frequently encountered in n.m.r. [1].

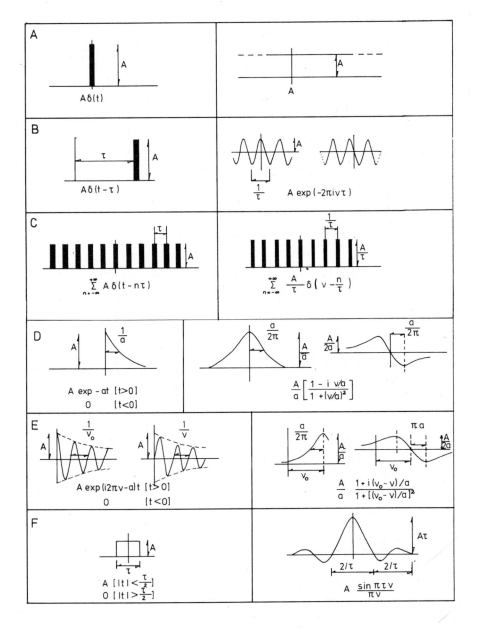

Table 3.1 (continued)

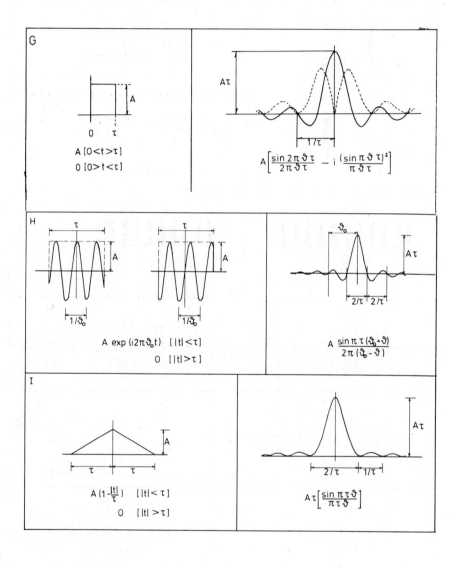

G

$A \; [0 < t > \tau]$

$0 \; [0 > t < \tau]$

$$A\left[\frac{\sin 2\pi\vartheta\tau}{2\pi\vartheta\tau} - i\,\frac{(\sin\pi\vartheta\tau)^2}{\pi\vartheta\tau}\right]$$

H

$A \exp(i2\pi\vartheta_0 t) \quad [|t| < \tau]$

$0 \quad [|t| > \tau]$

$$A\,\frac{\sin\pi\tau(\vartheta_0 - \vartheta)}{2\pi(\vartheta_0 - \vartheta)}$$

I

$A\left(1 - \frac{|t|}{\tau}\right) \quad [|t| < \tau]$

$0 \quad [|t| > \tau]$

$$A\tau\left[\frac{\sin\pi\tau\vartheta}{\pi\tau\vartheta}\right]$$

Table 3.1 (continued)

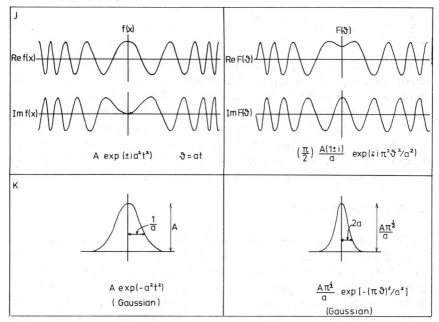

J	
f(x)	F(ϑ)
Re f(x)	Re F(ϑ)
Im f(x)	Im F(ϑ)
$A \exp(\pm i a^2 t^2)$ $\vartheta = at$	$\left(\frac{\pi}{2}\right) \frac{A(1\pm i)}{a} \exp(\mp i \pi^2 \vartheta^2/a^2)$

K	
$\frac{1}{a}$ A	$2a$ $\frac{A\pi^{\frac{1}{2}}}{a}$
$A \exp(-a^2 t^2)$ (Gaussian)	$\frac{A\pi^{\frac{1}{2}}}{a} \exp[-(\pi\vartheta)^2/a^2]$ (Gaussian)

(iii) Constant multiplication:

$$\mathcal{F}^\pm af(t) = a\mathcal{F}^\pm f(t) \tag{3.14}$$

(iv) Change of origin;
If the zero of a function is changed by an amount δt (or $\delta\vartheta$) then the twin function is multiplied by $\exp \pm (i(\delta t)2\pi\vartheta)$ (see Table 3.1, line B)

$$f(t + \delta t) - \mathcal{F}^\pm \to \exp + (i2\pi\vartheta\delta t) F(\vartheta) \tag{3.15}$$

(v) Area;
The area under a function is equal to the value of its transform at the origin. N.B. areas are not simply related.

$$\int_{-\infty}^{\infty} F(\vartheta) \, d\vartheta = f(0) \tag{3.16}$$

(vi) Symmetry;
If $f(t)$ is even (or odd) then $F(\vartheta)$ is even (or odd).

(vii) Digitisation;
If a function $f(t)$ is sampled and can thus be considered as a series of δ functions τ s apart then its transform is also a series of δ functions

$1/\tau$ Hz apart.

$$\sum_{n=-\infty}^{\infty} \delta(t - n\tau)f(t) = \frac{1}{\tau} \sum_{n=-\infty}^{\infty} F(\vartheta - n/\tau) \qquad (3.17)$$

Note that if the digitisation is done sufficiently finely, $\mathcal{F}^-$ is unaffected except for the repetitive adding of replicas which may be made so far removed as to be negligible. Thus if we digitise the time domain signal and transform it the resultant is simply a digitised frequency domain function. This is of course a very important result; if it were not so the digital techniques would be of no use to us. The Fourier pair of the pulse is of the form $\sin x/x$; this latter function will occur many times and is therefore abbreviated to sinc x.

3.3 THE TWO DOMAINS

The rules defined in the previous sections show the consequence in one domain of doing an operation in the co-domain. These rules are useful when we consider any arithmetic operations performed on a free induction decay prior to transformation; for example, addition is the same in both domains. This apparently trivial result is important as it means that one can time average in either domain with the same results. Similar simple results occur on scaling and constant multiplication e.g. if one doubles the size of a f.i.d. the size of the frequency signal is doubled.

The choice of origin in both domains is important. The point we define as zero time has a complex effect on frequency spectra. Also important to note is that the value of a function at the origin in one domain is the area under the Fourier pair function in the co-domain. Consequently the integral of the n.m.r. spectrum is determined solely by the value of the time domain spectrum at the origin. Later events have no effect on the area, neither do factors like weighting functions, resolution and acquisition time. These factors do of course have a significant effect on peak heights.

One final important concept, whose behaviour has been indicated above in property, iv, is that of 'phase'. The concept of phase is easy to understand in the basic domain for wave forms with which we are dealing in n.m.r. A general sine wave can be written as (see Fig. 3.2)

$$x(t) = a \sin (2\pi\vartheta t + \theta) \qquad (3.18)$$

where a is its amplitude, ϑ its frequency and θ is its phase. Phase

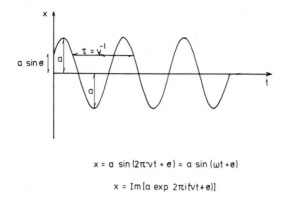

$$x = a \, \sin(2\pi\nu t + \theta) = a \, \sin(\omega t + \theta)$$

$$x = \text{Im}[a \, \exp \, 2\pi i(\nu t + \theta)]$$

Fig. 3.2. A generalised sine wave.

is a measure of the time difference between the origin and the first positive zero slope value on the wave form. For a sine wave with zero at the origin, θ is zero; despite being a time, θ is normally referred to as an angle, this being the time interval expressed as a fraction of periodicity of the signal τ, i.e. $\tau s = 360°$. Thus a sine wave with a $90°$ phase shift is a cosine wave. If in the course of our experiments we shift the origin of our time domain spectrum either for the whole spectrum or differentially for one frequency, e.g. by frequency dependent delays in the electronics, we introduce a 'phase shift'. On Fourier transformation into the frequency domain phase shifts can be seen from property iv to result in a mixing of what, without these shifts, would have been the real and imaginary parts to form 'new' real and imaginary functions. We will return to this aspect of Fourier theory many times.

3.4 THE REPRESENTATION OF PHYSICAL QUANTITIES

To date the question of functions and their Fourier transforms have been treated quite generally, and complex numbers have been used without comment. It is in order at this point to ask why we use these general, in some cases abstract, ideas to describe physically measured functions by the real part of a complex function and not simply as a uniquely real function (see ref. 1, page 61), e.g. a simple harmonic oscillation is the real part of $\exp(i\vartheta t)$. However, the transform of a real function, as we have seen, is not just a real function; it is a complex and Hermitian function.

The result of each transform must be considered as an aid in describing a physical property and its parts assigned accordingly. From this point of view n.m.r. is a nice example where the real and imaginary parts of transform of the free induction decay (which is itself a real function of time) can be equated with the u and v mode spectra in the frequency domain. Both are 'real' in the sense that they can be measured, but they result from different phases, with respect to the excitation, and hence are distinguishable. An interesting comparison here is found in infrared Fourier transform spectroscopy. Here, there is no available phase-coherent source, hence only the magnitude of the radiation at a particular frequency is of interest. The output from the Fourier transform which is meaningful in this case is simply its magnitude (the square root of the sum of the squares of the real and imaginary part), which as we discussed previously is an even function.

There are two further properties of real functions of time which need to be considered, namely their energy and/or power spectra. The energy spectrum of a function describes the energy available within that function as a function of frequency. In some systems it is only practical to consider the mean energy available, as the function only exists for a limited time and in these cases only a so-called power spectrum exists. The power and energy spectra are defined for a function $f(t)$ thus:

Energy spectrum $S_f(\vartheta) = F(\vartheta) F^*(\vartheta)$ (3.19)

Power spectrum $P_f(\vartheta) = \lim_{T \to \infty} \frac{1}{2T} F(\vartheta) F^*(\vartheta)$

$$f(t) = 0 \text{ if } |t| > T \qquad (3.20)$$

Energy and power spectra will be considered again in Chapters 4 and 10.

The latter is often also referred to as the spectral density function and there is an important theory, the Wiener-Khintchine theorem [4], which shows that the Fourier transform of a power spectrum is the 'autocorrelation function' (see later).

3.5 CONVOLUTION

Convolution and correlation (see later) are two mathematical processes of considerable power which can be simplified by the use

of Fourier transform techniques. Convolution is a specific transformation which is used to describe many physical processes. Probably
the most important of these is the filtering process in which one
preferentially handles the required signal with respect to the unwanted noise. Other important physical processes which can be
represented by a convolution are the diffraction of a beam of light
by a grating and the modification (or spreading out) of an impulse
function by a system.

The convolution $f_0(t)$ of two functions $f_i(t)$ and $h(t)$ is a broadening of one function by the other; when the process being described
is filtering, one function $(h(t))$ is called the weighting function, the
other functions being considered as input and output functions. The
transform is defined thus

$$f_0(t) = \int_{-\infty}^{\infty} h(\tau) \cdot f_i(t - \tau) d\tau$$

$$f_0(t) = h(t) \otimes f_i(t) = f_i(t) \otimes h(t) \qquad (3.21)$$

The convolution process can be thought of as a running average, (see
Fig. 3.3, which illustrates the broadening of a Lorentzian line by a
triangle). Each point on the function $f_0(t)$ is the weighted mean of
the function $f_i(t)$ around the point τ. The corresponding weights
form the weighting function $h(t - \tau)$. For the calculation of each
point on $f_0(t)$, the integral of the weighted input function is evaluated; the weighting function is then moved and the calculation
repeated for the next point, and so on over all $f_0(t)$.

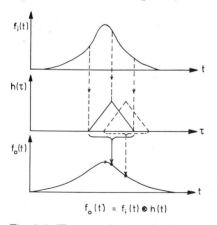

$$f_0(t) = f_i(t) \otimes h(t)$$

Fig. 3.3. The convolution of a Lorentzian line with a triangle.

The usefulness of Fourier techniques when studying convolution processes lies in the convolution theorem which states that 'the Fourier transform of a convolution of two functions is proportional to the products of the individual Fourier transforms, and conversely the Fourier transform of a product of two functions is proportional to the convolution of their individual Fourier transform'. Put into symbolic form:

if $\qquad h(t) \xrightarrow{\mathscr{F}^-} H(\vartheta)$ $\hspace{3cm}$ (3.22)

and $\qquad f(t) \xrightarrow{\mathscr{F}^-} F(\vartheta)$ $\hspace{3cm}$ (3.23)

then $\qquad f(t) \cdot h(t) \xrightarrow{\mathscr{F}^-} H(\vartheta) \otimes F(\vartheta)$ $\hspace{2cm}$ (3.24)

and $\qquad f(t) \otimes h(t) \xrightarrow{\mathscr{F}^-} H(\vartheta) \cdot F(\vartheta)$ $\hspace{2cm}$ (3.25)

Thus if one has a convolution to perform in one domain then by transforming the functions into the co-domain the process becomes the much simpler process of multiplication.

To illustrate the value of these ideas we will consider the example of filtering in more detail. We can filter a signal in order, among many things, to improve either it's signal to noise ratio or it's resolution. The weighting function used depends on the line shape of the signal and the characteristics of the noise. The optimum filter is called the 'matched filter'. It can be shown that the matched filter is one which has the same band shape as the signal to be recovered (see ref. 5); thus for an n.m.r. signal the matched filter should have a Lorentzian band shape. If the filter has a band shape $H(\vartheta)$, the filtering of an input signal $F_i(\vartheta)$ is described thus

$$F_0(\vartheta) = F_i(\vartheta) \otimes H(\vartheta) \hspace{3cm} (3.26)$$

In C.W. spectrometers filters are made from simple resistor capacitor networks and have far from ideal properties i.e. a Lorentzian band pass. In principle, if the raw signal were digitised, a computer could be used to carry out the filtering process by calculating the convolution of the signal with a function representing the filter as is shown in Fig. 3.4. However, convolution as such is not a simple calculation.

If we apply the convolution theorem to the problem, we can replace the convolution process by two Fourier transforms, one

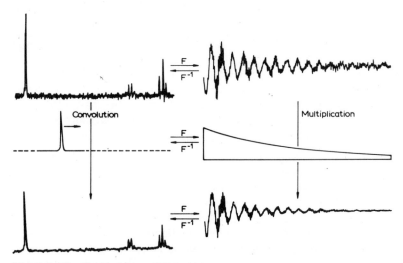

Fig. 3.4. The interrelationship between convolution in one domain and multiplication in the co-domain illustrated with the proton spectrum of ethyl benzene. Taken from R.R. Ernst, ref. 9.

inverse transform and a multiplication (as is again shown in Fig. 3.4). The process may seem rather circuitous, and indeed it is. The advantage of this approach is that Fourier transforms can be easily and efficiently performed in modern computers. If a matched filter is being used, one of the transforms is unnecessary, the weighting function simply being $\exp(-t/Tc)$, Tc is a time constant representing the bandwidth of the filter. Furthermore in pulsed n.m.r. the raw data is already in the time domain so filtering with a matched filter is achieved simply by multiplying by $\exp(-t/Tc)$ before transformation. It can easily be shown that if the filtering process is applied in order to enhance resolution, a process impossible by analogue means, then the weight function is $\exp(t/Tc)$.

3.6 CROSS-CORRELATION

Correlation has a mathematical similarity to convolution. The cross-correlation between the two functions $f(t)$ and $h(t)$ is defined by

$$\rho_{fh}(t) = \int_{-\infty}^{\infty} f^*(\tau)h(t+\tau)\,d\tau \equiv f(t)*h(t) \tag{3.27}$$

in this case, unlike convolution, the order of the functions does matter in that

$$\rho_{fh}(t) = \rho_{hf}^*(-t) \qquad (3.28)$$

As it's name implies, the cross-correlation function is a measure of whether any correlation exists between two functions. The concept is especially useful and easy to grasp for periodic functions, and some noise functions, e.g. pseudo-random noise. Two entirely different and unrelated functions will have zero cross-correlation for all values of their common variable, which is, in the above definition, time. On the other hand, two functions associated with physical quantities which have a direct causal connection will often have a finite cross-correlation for some or all values of τ.

As an example of a cross-correlation process we will consider the phase sensitive detector, an electronic device widely used in high resolution n.m.r. spectrometers. The phase sensitive detector can be used for many purposes, such as measuring the magnitude of a harmonic signal, normally buried in noise, when a synchronous wave form is available. The detector functions is described by eqn. (3.27) and illustrated in Fig. 3.5. The input and the reference are first multiplied together and their product is then averaged (integrated). If the input and the reference have the same frequency, a d.c. output will result whose magnitude depends on the phase of the input with respect to the reference. If their phase difference is zero then the output has a maximum value; on the other hand, if they are 90° out of phase then, as can easily be visualised, the output is zero. In the general case the output is a d.c. signal whose magnitude is proportional to the cosine of the phase angle between the input and reference signals. If the frequencies of the input and reference are not equal, the output of the multiplier is two frequencies, one equal to the sum and one the difference between

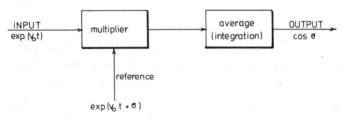

Fig. 3.5. Diagrammatical representation of a phase sensitive detector (cross correlation).

the input and reference frequencies. The output from the integrator depends on its time constant and the magnitude of the frequencies involved.

In n.m.r. the reference frequency is the r.f. used to excite the spin system, and the output from the phase sensitive detector is the difference in frequency between the response and the excitation. The sum of their frequencies, being a radio frequency, is filtered out. The output of the phase detector is thus the magnetisation in the rotating frame. By introducing a phase shift into the reference frequency either the u or v mode signals can be observed.

Where the reference wave form is known, but is not harmonic, e.g. in stochastic excitation, cross-correlation techniques can still be used to detect the response in the presence of noise by utilising the coherence between the excitation and the response. However, instead of a simple phase detector, a computer is necessary to perform the cross-correlation. It is in this case that Fourier transformation techniques become advantageous. In the previous section we saw that convolution in one domain is simple multiplication in the co-domain; in the case of cross-correlation the corresponding process in the co-domain is complex conjugate multiplication i.e.

if $\qquad \rho_{f.h.} = f(t) * h(t)$ $\hfill (3.29)$

then $\qquad \mathcal{R}_{F.H.} = F^*(\vartheta) \cdot H(\vartheta)$ $\hfill (3.30)$

where $\mathcal{R}_{F.H.} \xrightarrow{\;\mathscr{F}^-\;} \rho_{f.h.}$ $\hfill (3.31)$

If we wish to cross-correlate the response of the nuclear spin system with the excitation we simply multiply the response by the complex conjugate of the Fourier transform of the excitation and then transform this product into the frequency domain. Although such a process seems complex and circuitous, it is easy to program in a digital computer.

The two functions which are cross-correlated will in practice both be real functions in the time domain. On transformation to the frequency domain, however, the resultant function is complex, i.e. we have information about *both* the u and v mode signals simultaneously. When using a simple phase detector we only obtained information about either the u *or* v mode. We will return to this problem with some examples in the next chapter.

3.7 AUTO-CORRELATION

Auto-correlation is 'cross-correlation' of a function with itself and is defined thus

$$\rho_{f.f.}(t) = \int_{-\infty}^{\infty} f^*(\tau) \cdot f(t+\tau) \, d\tau = f^*(t) * f(t) \tag{3.32}$$

The auto-correlation function is a measure of the correlation between the value of a function at times differing by τ. As with cross-correlation, if there is a high degree of coherence, the averaging (integration) will yield a finite signal; however, if there is no coherence, the average tends to zero. The situation is illustrated in Fig. 3.6 for the case of a function, $f(t)$ which is a mixture of noise and a sine wave. For τ zero there is coherence for both the noise and the sine wave, both therefore have positive auto-correlation functions; however as τ is increased their behaviour differs. The sine component is periodic, i.e. it has the property that

$$f(t-n\tau) = f(t)$$

for certain values of τ and as τ increases a cosine wave is the result of the averaging process. The random noise on the other hand has no

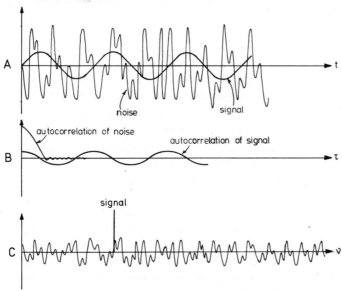

Fig. 3.6. (a) A signal consisting of a sine wave and random noise. (b) The auto-correlation of (a). (c) The Fourier transform of (b).

coherence, so as τ is increased, the auto-correlation function decays with a time and shape which is characteristic of the noise. Figure 3.6c shows the Fourier transform of the auto-correlation function which, as we have seen previously, is called the power spectrum of the original function. The auto-correlation function is especially useful in characterising the properties of irregular functions for which no reference function is known (or even exists).

The obvious advantage of auto-correlation is that one does not need to know the input function. For this reason it is much used in other spectroscopic techniques and radio astronomy where a detailed knowledge of the excitation is impractical. The main use in pulsed n.m.r. is with true (as opposed to pseudo random [6]) stochastic excitation [7] where the output of the system has to be auto-correlated rather than cross-correlated with the undefined exciting noise. The disadvantage of auto- with respect to cross-correlation is that on subsequent transformation no phase information is available. N.m.r. is unique when compared with other spectroscopies such as i.r., u.v., etc. in that a phase coherent source is available. These spectroscopies yield spectra which represent only the magnitude of the property under investigation, whereas n.m.r. can produce information on the phase properties, i.e. u and v modes can be distinguished as well as simply the magnitude of the magnetism as a function of frequency $(u^2 + v^2)^{1/2}$.

The lack of phase information with auto-correlation can be visualised by drawing a parallel with rational numbers. From the previous section it follows that the Fourier transform of any auto-correlation function is the product of the transform of that function, and its complex conjugate

if $\qquad \rho_{ff} = f(t) * f(t)$ $\qquad\qquad\qquad\qquad$ (3.33)

then $\qquad \mathcal{R}_{FF} = F^*(\vartheta) \cdot F(\vartheta)$ $\qquad\qquad\qquad$ (3.34)

which (from eqn. (3.19)) is it's energy (or power, if a mean is taken) spectrum. To recover the function of interest it is necessary to take the square root of the energy spectrum which yields the so called 'magnitude spectrum'. In taking the square root one loses one degree of freedom, i.e. the sign in the case of a simple number, and the phase information in the case of spectra.

Auto-correlation techniques are of greatest value when studying random process. One important usage is in characterising the motion

54

of molecules in solution. Such motion is basically random; it is possible, however, to define some meaningful properties (which will depend on molecular size), such as the mean time between collisions, solution viscosity etc. Correlation theory can be used to generate an energy spectrum based on the properties which will in turn express the energy present at a particular frequency. Knowledge of this type forms the basis for relaxation theory, in which we need to know the energy generated at the Larmor frequency by a specific molecular motion.

3.8 LINEAR SYSTEMS

A final concept must be introduced at this stage, that of the linear system. The concept is again a very general one. By a system we mean a 'black box' to which we apply an input $x(t)$ and from which comes a response $r(t)$ (Fig. 3.7). In this case the black box is an assembly of nuclear spins, but it could just as well be an electronic circuit, or a diffraction grating.

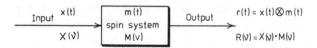

Fig. 3.7. An n.m.r. sample can be considered as a linear system when the input and response are as shown.

One method of finding the response of a system is to solve the detailed differential equations (Bloch equations) for that system. Such an approach is in theory possible for any spin system but in practice only possible for a few simple cases, and of course is impossible for an unknown case. The real experiment is to measure the response from a system and deduce from it the properties of that system in the form of it's n.m.r. spectrum. A detailed knowledge of the response of the system to an impulse $h(t)$ provides all the necessary information to calculate the spectrum, as long as the system is linear and shift invariant. A system is linear when, by increasing the input by a certain factor, one increases the output by the same factor, e.g. doubling the input doubles the output. Below saturation, n.m.r. samples fulfil this condition. A system is shift invariant providing that delaying the input has no effect other than to delay the output by the same amount. In general n.m.r. systems also fulfil this condition.

The key concept when using Fourier transform techniques to study a linear system is in the intuitively reasonable idea that the output is the result of the input being 'spread out' by the system. This 'spreading' process is, of course, convolution. Thus the response $r(t)$ of a linear system to an input function $x(t)$ is given by

$$r(t) = h(t) \otimes x(t) \qquad (3.35)$$

where $h(t)$ is a property characteristic of the system. The function $h(t)$ is in fact the response of the system to a delta function as can easily be seen if $x(t)$ in eqn. (3.35) is set equal to a delta function thus

$$r(t) = \int_{-\infty}^{\infty} h(\tau) \cdot \delta(t - \tau) \, d\tau \qquad (3.36)$$

which reduces to

$$r(t) = h(t) \qquad (3.37)$$

We have therefore shown that the behaviour of any linear system can, in the time domain, be characterised by the impulse response function. This function is sometimes called Green's function. If we Fourier transform eqn. (3.35) we obtain (see eqn. (3.20))

$$R(\vartheta) = H(\vartheta) \cdot X(\vartheta) \qquad (3.38)$$

where $H(\vartheta)$ is called the 'Transfer function' of the system. We could have used this approach to deduce eqn. (3.37). The transform of a delta function is unity, therefore back transforming eqn. (3.38) with $X(\vartheta)$ set of unity would also give us eqn. (3.37). Within the frequency domain $H(\vartheta)$ will completely describe the properties of the system just as $h(t)$ did in the time domain. The transfer function imparts the change in amplitude and phase of the output, with respect to the input, produced by the system we are studying.

The transfer function defined, as above, is obviously the same as 'the n.m.r. spectrum', we discussed in chapters 1 and 2, the 'system' is an assembly of nuclear spins. Fourier theory of linear systems therefore relates that if one determines the impulse response of an assembly of nuclear spins, one has all the necessary information to obtain the complete n.m.r. spectrum.

Before leaving this topic it should be noted that even if both the input and output are real quantities, the transfer function is, in the mathematical sense, a complex quantity. The two parts of $H(\vartheta)$, as was discussed in section 3.4, are in the case of n.m.r. assigned to the two components of the nuclear magnetisation. A real input can, however, be represented by a real (or imaginary) part of a complex function $x(t)$; the output is then by the linearity of the spectrum the real (or imaginary) part of $r(t)$ which will also be a complex number. Thus complex numbers can arise, and be used, in two different ways, but confusion can arise if this is not realised.

3.9 DISCRETE OR DIGITAL FOURIER TRANSFORMS

So far it has been assumed that one is dealing with continuous analytical functions. When a computer is used the function must be converted into a regular series of discrete values. This is equivalent to redefining the function as a series of N equally spaced delta functions thus

$$m(t) = \sum_{n=0}^{N-1} m_n \delta(t - n\tau) \tag{3.39}$$

As will be seen later, if $m(t)$ is the f.i.d. from a spectrum Δ Hz wide then $\tau = (2\Delta)^{-1}$. As shown in eqn. (3.17) there is also a series of delta functions on the Fourier pair. A satisfactory transform of discrete, digitised values is thus possible.

Using digital techniques also imposes further contrasts. The integration must be replaced by a sum over a finite series of values. The transformation is thus reformulated as

$$M(\vartheta) = 1/N \sum_{k=0}^{N-1} m(t) \exp(-i2\pi k\vartheta/N) \tag{3.40}$$

The methods of performing this calculation are discussed in section 6.12g. The finite nature of the sampling time results in finite resolution in the transformed spectrum.

Suppose there are N data points available to define a function of time, e.g. the decay of nuclear magnetisation after an excitation pulse, and these are taken at regular intervals over a period T_a, i.e. a data rate of N/T_a sample/s. Transformation of the data will result in a frequency spectrum defined by $N/2$ real and $N/2$ imaginary points with a resolution of T_a^{-1} Hz. If all the original data points are considered as real, since the imaginary part of the time domain signal is undefined, in the frequency domain the resultant real and imaginary parts are not independent. If, as is the case in n.m.r., only one component of the frequency spectrum is utilised, this interdependence results in a loss of information, which in turn results in a lowering of the signal to noise ratio. A further consequence of having no imaginary part of the time domain signal is the inability within one mode to record signals of both plus and minus a specific frequency.

The real and imaginary parts of the frequency spectrum are interdependent because the principle of causality, which applies to n.m.r. spectra, has not been fully utilised. The function is defined from 0 to T_a but not from $-T_a$ to 0, which is necessary in order to produce independent components after transformation. Causality says that the function must be zero between $-T_a$ and 0 [8]. If, therefore, an array of N zero's is added to the data and the total i.e. $2N$ data points are transformed, N real and N independent imaginary points are obtained. The spectrum obtained using the real data now contains all the available information and has a signal to noise ratio that is higher by a factor of approximately $2^{1/2}$. For a more detailed discussion of causality in n.m.r. (see section 4.6) the reader is referred to the work of Bartholdi and Ernst. However, since no data is yet available concerning the imaginary part of the time domain function, ambiguity concerning ± frequencies still exists.

In n.m.r., as will be seen, it is possible, using two phase detectors, to obtain two components from each sample of the time domain signal. These two components can be assigned as the real and imaginary parts of the time domain signal. Sampling would now take place at a rate of $N/2T_a$, each sample requiring two data locations. Transformation again yields $N/2$ real and imaginary points and since we have input information about both components of the time domain signal, the ambiguity concerning plus and minus frequencies has been removed and they can both be displayed in one mode.

The practical value and realisation of these abstract concepts will be discussed later.

REFERENCES

1 D.C. Champeney, Fourier Transforms and Their Physical Applications, Academic Press, New York, 1973.
2 R. Bracewell, The Fourier Transform and Its Physical Applications, McGraw-Hill, New York, 1955.
3 J.B.J. Fourier, Theorie Analytique de la Chaleur, Paris, 1822.
4 See appendix J of ref. 1.
5 R.R. Ernst, Advances in Magnetic Resonance Vol. 2, No. 1, Academic Press, New York, 1967.

6 R. Kaiser, J. Magn. Res., 3, (1970) 28.
7 R.R. Ernst, J. Magn. Res., 3, (1970) 10.
8 E. Bartholdi and R.R. Ernst, J. Magn. Res., 11, (1973) 9.
9 R.R. Ernst, The Applications of Computer Techniques in Chemical Research, Institute of Petroleum, London, No. 61, 1972.

CHAPTER 4

EXCITATION TECHNIQUES IN N.M.R.

4.1 INTRODUCTION

Now return to consider the n.m.r. experiment. Chapter 2 discussed the behaviour of an assembly of nuclear spins in a magnetic field and described their response as a function of the frequency of the exciting radiation. When the correct component in the rotating frame was chosen, that which is out of phase with the exciting frequency (v mode), a Lorentzian absorption signal was obtained. As we will see later, the behaviour of a typical chemical compound, as opposed to the single nuclear spin considered so far, consists of a family of Lorentzian lines. This family is called the n.m.r. spectrum of the compound, and from this spectrum is derived information of chemical interest.

As has been seen, the n.m.r. spectrum is a plot of the absorption of energy by the nuclear spin system as a function of exciting frequency, i..e the behaviour of the nuclear magnetism $M(\vartheta)$. We must now be more precise. When we solved the Bloch equations there were two components, the u mode and the v mode, to the magnetisation in the rotating frame. These modes are in and out of phase with respect to the rotating frame and the exciting frequency, the latter defining the former. We therefore expand our definition of $M(\vartheta)$ to include phase thus

$$M_{(\vartheta)} = A_{(\vartheta)}[\cos \varphi_{(\vartheta)} + \sin \varphi_{(\vartheta)}] \tag{4.1}$$

A is the amplitude of the magnetisation at frequency ϑ and φ is a phase angle defined with respect to the exciting field, which may be a function of the frequency of the exciting field. If one chooses φ to be $0°$, the result is a spectrum which is the u mode; $\varphi = 90°$ gives the v mode. Another way of expressing these ideas is to consider the n.m.r. spectrum $M(\vartheta)$ as a mathematically complex function, where

60

the u mode is the real part and the v mode is the imaginary part, i.e.

$$M_{(\vartheta)} = A_{(\vartheta)} \exp (i\varphi\vartheta) \tag{4.2}$$

Using the above ideas, an n.m.r. spectrum can be defined in terms of the input and the response function

$$R(\vartheta) = M_{(\vartheta)} \cdot X_{(\vartheta)} \tag{4.3}$$

By analogy with the ideas outlined in the previous chapter (see eqn. 3.38) the n.m.r. spectrum of a nuclear spin system is its transfer function. The possible methods of obtaining the transfer function of a system will be discussed in the next chapter, together with the possible types of excitation that can be, and are, used in n.m.r.

4.2 THE N.M.R. SPECTRUM IN TWO DOMAINS

The previous chapter introduced the concept of the two Fourier related domains; the time domain and the frequency domain. The previous section showed that the n.m.r. spectrum, which is what we are interested in, can be identified with the frequency domain transfer function of the nuclear spin system. In this section we assume that the assembly of nuclear spins can be described as a linear system. We will return to this topic in more detail in section 4.11, where we consider the deviation of a nuclear spin system from linearity. In chapter 3 it was demonstrated that for a linear system the transfer

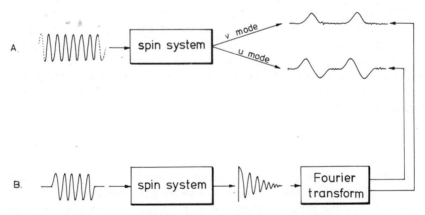

Fig. 4.1. The two basic methods of obtaining an n.m.r. spectrum.
(a) By applying a continuous excitation and varying its energy;
(b) by applying a pulse of energy and Fourier transforming the result.

function of a system in one domain is equivalent to the impulse response (Green's function) in the co-domain. Thus one can obtain an n.m.r. spectrum by determining the spin system's response to an impulse and Fourier transforming it [1]. This is the basis of pulsed n.m.r. and, as can be seen, is a more efficient approach than frequency sweep, yielding a higher signal to noise ratio. Pulsed n.m.r. provides the closest approximation generated to an impulse, and one can Fourier transform the resultant free induction decay. The two experiments of either obtaining the spin system's transfer function directly (adiabatic slow passage) or measuring the impulse reponse, represents two extremes, shown diagrammatically in Fig. 4.1.

Now consider a generalised experiment, as is shown in Fig. 4.2. All n.m.r. experiments are covered by this scheme. The spin system (sample) is subjected to an excitation, the phase of the response is unscrambled, and finally the resultant is Fourier transformed. The phase unscrambling process for swept and pulsed excitation is

Fig. 4.2. The general n.m.r. experiment.

trivial, as opposed to the case when using more complex excitation techniques where the phase of each individual frequency component is unknown and different.

At first sight the adiabatic slow sweep experiment would not appear to be covered by Fig. 4.2; in fact it is. The unscrambling of the phases and the Fourier transform are performed by a simple phase detector. The detector identifies the phase and amplitude of each frequency component as the excitation slowly sweeps the spectrum; this is a sequential Fourier transform, and is the principle on which analogue Fourier transform devices e.g. wave analysers, are based.

There is often considerable confusion in the meaning of the variable ϑ as it can occur in two different places with two different meanings [2].

(1) In frequency sweep methods ϑ is used to indicate the variable in the time domain. The independent variable is still time, being related to the frequency via the sweep rate (a) thus

$$\vartheta = a \cdot t \tag{4.4}$$

(2) The Fourier transform, $F(\vartheta)$, of a time function $f(t)$ is the frequency spectrum of $f(t)$ in the frequency domain. ϑ is now the true independent variable.

The 'spectrum' leaving a spectrometer is always a function of time; it is in the time domain. The term spectrum is confusing in this sense. The 'spectrum' from a spectrometer has a corresponding spectrum in the frequency domain. It is the latter which is the true n.m.r. spectrum. In the slow sweep experiment the sweep rate is made slow enough that the system is at equilibrium at all times; the practical variable now does become ϑ and the 'spectrum' plotted onto the recorder bed is effectively the true n.m.r. spectrum. If the sweep is 'too fast' then the plotted spectrum contains distortions, evidence of transient effects, ringing etc.

To return to our generalised experiment, the excitation used must have certain general properties. Firstly, it must have a flat power spectrum, that is, it must excite all the nuclei within the spectrum equally. There are three simple time domain excitation functions which produce a flat power spectrum. One is an impulse, another is white noise. The condition of a flat power spectrum is not rigorously met in pulsed n.m.r., as in practice a *pulse* (as opposed to an impulse) is used as the exciting function (see section 6.5) and only an impulse can give a flat power spectrum (see Table 3.1). The application noise modulation in n.m.r. has been used for some time in the proton noise decoupling technique used to facilitate the measurement of ^{13}C spectra. Noise modulation is the basis of stochastic resonance which will be discussed in section 4.6. The third excitation function which has a uniform power spectrum is the linear frequency sweep. As we have seen, if the sweep is slow enough, the frequency spectrum can be obtained directly. If the sweep is speeded up, transient effects are observed, which can be corrected for by cross-correlation, giving rise to the method of recording n.m.r. spectra called correlation spectroscopy. Finally, one can abandon the idea of a simple power spectrum altogether and use Fourier techniques to calculate a series of pulses which correspond to any desired frequency spectrum. This is tailored excitation n.m.r. However, even in tailored excitation each region has a flat power spectrum; it is just the power within the region which is changed.

4.3 TWO BASIC F.T. EXPERIMENTS

In the rest of this chapter we will compare and contrast these various possible excitation functions from a rather general point of

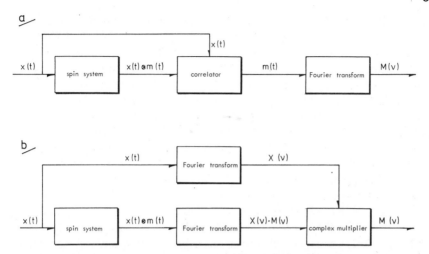

Fig. 4.3. The general n.m.r. experiment can be achieved in two ways:
(a) using a correlator to unscramble the phases or equivality;
(b) using a second Fourier transformation and multiplication.

view, in later chapters we will consider the case of pulsed excitation
in much more detail, but we must next decide how to extract the
spectrum from response in our generalised experiment. Since the
output is caused by the input and the system is linear, one can use
cross-correlation techniques to unscramble the phases. Cross-correlating the input with the output yields impulse response function
which on transform yields the spectrum (see Fig. 4.3a). The process
of cross-correlation can be considered as 'sorting out the phases'. If
the input is a simple harmonic oscillation, a simple phase sensitive
detector is adequate as a correlator; with more complex excitations
computation is necessary. The pulsed excitation is the only other
case where cross-correlation is straight forward. Here the responses
from all the nuclei start out in phase, hence there is no need to sort
out the phases; they are coherent at zero time and cross-correlation
is unnecessary. We will discuss this point more rigorously later.

The form of the n.m.r. experiment discussed so far is illustrated
in Fig. 4.3a. There is, however, an alternative, mathematically
equivalent method, shown in Fig. 4.3b. Here use is made of the
theorem that correlation in one domain is complex multiplication
in the co-domain, i.e.

$$\mathcal{F}[x(t)*r(t)] = X(\vartheta)^* \cdot R(\vartheta) \qquad (4.5)$$

This is because with digital computers Fourier transformation is straight forward, whereas correlation in general is not. The former method however has the advantage in that it produces $m(t)$, as an intermediate function, which can be filtered by simple multiplication; this thus used whenever possible. Conceptually, however, it is simpler to consider all the experiments as proceeding by the second route. In this chapter homodyne r.f. detection is assumed, i.e. the response of the system is detected (correlated) with respect to the excitation. The output from the receiver is therefore proportional to the magnetisation in the rotating frame. If only one detector is used then only one component of $m(t)$ is recovered; this is identified with the real part of $m(t)$, the imaginary part being considered as zero. After transformation both components of $M(\vartheta)$ are produced. If two detectors 90° out of phase to each other (in quadrature) are used then $m(t)$ can be measured as a complex quantity; the two outputs are then identified with the real and imaginary parts of $m(t)$. Whether $m(t)$ is considered as real or complex does not affect the logic of the experiment, although it will be shown that in the latter case the signal to noise ratio is increased by $2^{1/2}$.

4.4 ADVANTAGES OF THE TIME DOMAIN

Before considering the merits of the various time domain techniques used in n.m.r., let us consider why such techniques are used at all. Looking at the question from the other view point, what is wrong with the 'conventional' field or frequency sweep experiment? The simple answer is that it is inefficient. Firstly, in order to produce a distortion free spectrum, a simple sweep experiment must be slow on the nuclear spin time scale, as the nuclear spin system must be at equilibrium with its surroundings at all times. The processes which allow the system to maintain equilibrium are characterised by the appropriate spin lattice relaxation times since these are typically seconds in high resolution n.m.r. Sweep rates must be in the order of Hertz or tenths of a Hertz per second.

The efficiency per unit time of the sweep approach can be improved by departing from the equilibrium condition, using rapid sweeps and cross-correlation techniques. Speeding up the sweep improves the efficiency of the experiment but does not attack the basic limitation of a sweep approach, i.e. that the resonances are excited sequentially. It is possible to imagine a way of improving

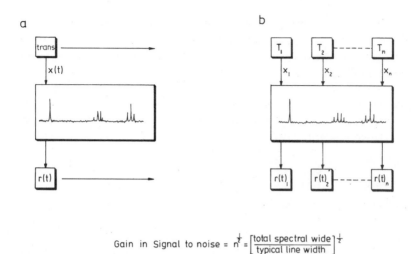

Fig. 4.4. Fellgett's principle; sensitivity of an experiment can be increased by changing from a single channel experiment. (a) To a multiple channel experiment; (b) the maximum number of channels in n.m.r. is one per line width.

the conventional swept spectrometer by using more than one transmitter. A multi-channel spectrometer is visualised in Fig. 4.4. In such a spectrometer each channel will contribute coherent information i.e. signal and also incoherent information i.e. noise. The coherent signal from the spectrometer will increase as the number of channels n, the coherent noise will increase as the square root of n. The sensitivity thus increases as the square root of the number of channels used. If quadrature detection is used, then both components of the f.i.d. are detected, and by the above argument the gain in sensitivity is $(2n)^{1/2}$. The increase is signal to noise ratio achievable by using multi-channel observation is a general phenomenon first enunciated by P. Fellgett in 1951 [3]. If n channels are used simultaneously in an experiment, then providing the dominant source of noise is not the excitation used, a gain in sensitivty of $(n)^{1/2}$ is available. Infrared interferometry takes advantage of the Fellgett principle to improve the sensitivity, though this gain is not available in visible interferometry where photo-noise dominates. In n.m.r. the excitation contributes effectively no noise to the final experiment and Fellgett principle applies.

The gain in sensitivity outlined above obviously cannot go on indefinitely; there must be a limit to the value of n. This limit is reached when there is one channel to each spectral element throughout the total spectrum. In n.m.r. the spectral element can be equated to a resolvable line. The optimum number of channels is thus the spectral width being studied, divided by a typical line width. The possible sensitivity enhancement available by a multi-channel approach increases with the resolution of the spectrum being studied and can be expressed;

$$\frac{S/N \text{ multi-channel}}{S/N \text{ single}} \propto \left[\frac{\text{Spectral width}}{\text{Typical line width}} \right]^{\frac{1}{2}} \tag{4.6}$$

In the time domain experiments under consideration we excite all the spins simultaneously using one of the type of excitation functions discussed later, a flat power spectrum being the equivalent of a multiple transmitter. The spectrometer obtains information from all the spins simultaneously; the Fourier transformation becomes the equivalent of the multi-channel receiver.

While the gain in sensitivity outlined above is the main attraction of working in the time domain it is by no means the only one. A second one is the ease by which filtering can be achieved mathematically. As was discussed in the previous chapter, filtering is a convolution process in the frequency domain. In pulsed n.m.r. the data is collected in the time domain. In order to filter this data it is only necessary to multiply by a suitable function; convolution in one domain is multiplication in the co-domain. The function that should be used for this multiplication is the Fourier transform of the frequency domain line. Another advantage of working in the time domain is gained when using pulsed excitation, where there is a clear starting point from which to measure dynamic effects. This is particularly useful when studying relaxation effects.

4.5 PULSED EXCITATION AND THE IMPULSE RESPONSE FUNCTION

The use of a sequence of pulses of high power radio frequency energy as an excitation source in n.m.r. is the central theme of this book. It will be considered next from a general view point so it may form a basis when considering other available methods. In all the comparisons the one shot experiment will first be discussed, then the more general case of a series of pulses.

The purpose of pulsed excitation is to simulate an impulse and record the impulse response of the spin system. A pulse differs from an impulse, in that it has a finite duration and a frequency domain spectrum of sinc ϑ shape as opposed to the latter's flat power spectrum. Nevertheless, if the pulse is made short enough it can be considered as an impulse, and no significant relaxation can occur during it. As pulses are micro-seconds and relaxation times are seconds this constraint is trivial. The theoretical and practical aspects of this approximation are considered in sections 5.3 and 6.6 respectively. If the excitation is indeed an impulse, then as was proved in eqns. (3.35) and (3.37) the response measured is the impulse response function of the system. This, by definition, on Fourier transformation will give us the transfer function of the spin system i.e. its n.m.r. spectrum. If we follow the generalised experiment described previously in Fig. 4.3a, we are cross-correlating the response with a δ function. Such correlation is trivial as the transform of a δ function is unity. The use of pulse excitation is hence conceptionally and practically a simple experiment to handle with respect to extracting the n.m.r. spectrum from the spectrometer's output. One simply Fourier transforms it. The reason for such a simple cross-correlation in pulsed n.m.r. is that the response is phase coherent, i.e. there is a definite zero point in time for all the components of the response: the point in time when the impulse occurred. The response is measured from this reference point in time, when all the nuclei are in phase. This condition is not met in stochastic excitation and from one view point this is the fundamental distinction between the two excitation techniques. Detailed study of the response of a spin system and its Fourier transform is profitable at this point because the impulse response is easy to visualise and the deductions made also apply to the more complicated cases like stochastic excitation.

4.6 THE FREE INDUCTION DECAY

Immediately after the impulse, the detected response of each group of magnetically equivalent nuclei will be $M_{y'}^{+}$, which is the projection of their magnetisation onto the y axis of the rotating frame (rotating at the pulse frequency ϑ_0). The superscript $+$ indicates the magnetisation just after an impulse, a superscript $-$ indicates the magnetisation just prior to an impulse. If the nuclei of type i have a chemical shift of ϑ_i from ϑ_0 then their magnetisation will process at ϑ_i, as viewed from the rotating frame. The detected response will thus be given by

$$r_i(t) = M_y^+ \cos 2\pi\vartheta_i t \cdot \exp(-t/T_2) \tag{4.7}$$

where the exponential function allows for the decay of the function by spin—spin relaxation.

The function $r(t)$ is real and even; it can however be generalised as the real part of a complex function.

$$r_i(t) = M_y^+ \exp - [(T_2)^{-1} - i2\pi\vartheta] t \tag{4.8}$$

As can be seen from Table 3.1 on Fourier transformation

$$R_i(\vartheta) = M_y^+ T_2 \left[\frac{1 - 2\pi i T_2(\vartheta_i - \vartheta)}{1 + 4\pi^2 T_2^2(\vartheta_i - \vartheta)^2} \right] \tag{4.9}$$

$R_i(\vartheta)$ has a real part which is a Lorentzian line of amplitude $M_y^+ T_2$, offset ϑ_i from the pulse frequency, with a line width at half height of $(\pi T_2)^{-1}$ (the v mode signal) and an imaginary part which is the corresponding dispersion signal (u mode signal). It is worth noting that the integral and also the peak height of the absorption signal (providing that any experimental factors, specifically truncation, do not effectively reduce T_2) depend only on the value of the f.i.d. at zero time. This property which follows from the properties of Fourier pairs must be borne in mind when studying time-dependent phenomena such as gated decoupling experiments.

The above discussion has led to the identification of the real part of $R(\vartheta)$ with the v mode spectrum whereas in section 2.9 we associated the imaginary part of $M(\vartheta)$ with the v mode. This discrepancy can be explained if we consider the generalised experiment again. $R(\vartheta)$ can only be equated to $M(\vartheta)$ after cross-correlation with the exciting function

$$M(\vartheta) = R(\vartheta) * X(\vartheta) \tag{4.10}$$

in this case since $x(t)$ is a delta function all that is involved is to multiply $R(\vartheta)$ by i which interchanges the real and imaginary parts in $R(\vartheta)$ when forming $M(\vartheta)$. The real and imaginary part of $M(\vartheta)$ are consequently interchanging. The two methods of defining the components of $M(\vartheta)$ now agree. This step of cross-correlation is not normally considered and in spectra obtained by Fourier transform the v mode signal is associated with the real part of $M(\vartheta)$.

The impulse response of a real linear system has an interesting and obvious property, that of causality. The principle of causality simply states that the response is caused by the impulse and does not exist before the impulse is applied i.e.

$$m(t) = 0 \quad \text{for} \quad t < 0$$

One of the consequences of the property of causality is that the real and imaginary parts of the spin system's transfer function are not independent; they are related by Hilbert transform. A Hilbert trans-formation within the frequency domain is convolution with $1/\vartheta$. Thus for n.m.r. within the frequency domain the u and v mode can be related [4]. By definition

$$M(\vartheta) = u + v \tag{4.11}$$

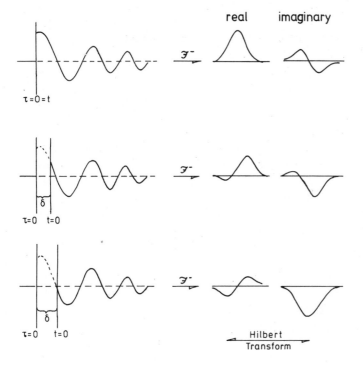

Phase shift induced = 360.ν.δ°

Fig. 4.5. The effect of delaying the start of data acquisition is to mix both absorption and dispersive elements into both the real and imaginary parts of $M(\vartheta)$. The two components are related by the Hilbert transform.

from the principle of causality

$$u \xrightarrow{\hat{H}} v \qquad (4.12)$$

which is equivalent to

$$u = v \otimes \vartheta^{-1} \qquad (4.13)$$

The value of eqn. 4.13 is that it provides a method of phase correcting spectra when only one part of $M(\vartheta)$ have been determined e.g. in a c.w. spectrum. The working of the method can be better understood by moving into the time domain where convolution is simply multiplication. The transform of $M(\vartheta)$ is $m(t)$; the transform of $(\vartheta)^{-1}$ is i sgn(t), where sgn(t) is 1 for $t > 0$ and -1 for $t < 0$. Thus to convert a v mode spectrum into a u mode spectrum one simply transforms the v spectrum into the time domain, multiplies by i and transforms back into the frequency domain. Such a procedure was used in the preceding paragraph to rationalise the two 'versions' of $M(\vartheta)$ in common use. These ideas, summarised in Fig. 4.5, are in fact general and a consequence of the n.m.r. system obeying the so called Kramers-Kronig relationships [5]. For a more detailed discussion of this aspect of n.m.r. theory the reader is referred to the work of R.R. Ernst [4].

The above analysis represents an ideal situation; specifically it assumes that there are no phase shifts induced in $r(t)$ by the spectrometer's electronics and that the f.i.d. can be studied from $t = 0$. The causes of phase shifts in n.m.r. spectrometers are many and unavoidable. The second condition is impractical as the receiver must be protected from the exciting pulse and cannot start detecting the f.i.d. for some finite time after the termination of the pulse. Both these effects can be considered in the same mathematical way, that is by changing the origin of the time axis from 0 to δt. The consequence is that $r(t)$ is no longer a pure cosine function but a mixture of sine and cosine functions (see section 5.2).

One further factor needs consideration at this stage, that of the consequences of application of more than one impulse. In practice, as the experiment involves the time averaging of many f.i.d.'s in order to enhance the signal to noise ratio, a regular series of impulses are applied. If the period between the impulses is T_p then the excitation function becomes

$$x(t) = \sum_{k=-\infty}^{\infty} \delta(t - kT_p) \qquad (4.14)$$

which on Fourier transformation becomes

$$X(\vartheta) = \frac{1}{T} \sum_{j=-\infty}^{\infty} \delta\left(\vartheta - \frac{j}{T}\right) \tag{4.15}$$

This frequency spectrum is a set of equidistant frequencies $1/T_p$ Hertz apart. The effect of using pulses instead of impulses is to multiply the excitation frequency by a pulse shape which is equivalent to convoluting $X(\vartheta)$, given above, with the Fourier transform of a pulse. The frequency spectrum is now a series of frequencies $1/T_p$ Hz apart whose amplitudes are given by a $\mathrm{sinc}(\pi\vartheta t_p)$ function where t_p is the pulse width as shown in Fig. 4.6a. Solution of the Bloch equations for pulsed excitation will be given in section.5.1.

4.7 CONTINUOUS WAVE EXCITATION

In a 'conventional' spectrometer the excitation is applied to the spin system in the form of continuous linearly oscillating r.f. field

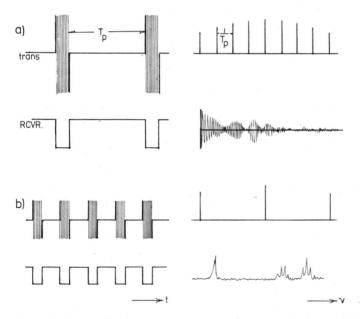

Fig. 4.6. Pulsing the exciting frequency generates side bands. If the pulse frequency is slow (a) multichannel excitation results; if fast (b) only a single channel excites the sample.

ϑ_0. This excitation can be represented by the general expression

$$x(t) = B_1 \exp (i2\pi\vartheta_0 t) \tag{4.16}$$

the real part corresponding to the component of the excitation rotating in the correct sense (about the z axis) to induce transitions. The output for the sample is

$$r(t) = M\vartheta_0 \exp (i2\pi\vartheta_0 t) \tag{4.17}$$

The spectrometer detects $r(t)$ with respect to $x(t)$ by the use of a phase sensitive detector, which outputs the value of either the u or v mode signal at frequency ϑ_0, depending on the phase of the reference signal used in the detector compared with the exciting frequency.

In order to obtain the n.m.r. spectrum, i.e. the magnetisation as a function of frequency, $M(\vartheta)$ as opposed to the value at ϑ_0, $M\vartheta_0$, the frequency is slowly swept. The input is now

$$x(\vartheta(t)) = B_1 \exp (i2\pi\vartheta(t)t) \tag{4.18}$$

If, as is usually the case, the sweep is linear with a sweep rate a, ($\vartheta = at$) then

$$x(t) = B_1 \exp (iat^2) \tag{4.19}$$

If the sweep rate is large, the output from the spectrometer will contain evidence of transient effects. We will return to this case later as it forms the basis of correlation spectroscopy. If, however, a is small, such that the spectrum represents an equilibrium situation, then the excitation and the response become functions of frequency and the output is $M(\vartheta)$.

Continuous wave excitation can be thought of as a special case of pulse excitation. Equations (4.14) and (4.15) showed that a train of equal impulses Ts apart is equivalent to a series of discrete frequencies separated by $1/T$ Hz. In pulsed n.m.r. the time between successive pulses is in the order of seconds; the discrete frequencies, or Fourier side bands as they are often called, are therefore tenths of a Hertz apart. If the repetition frequency is much smaller, the side bands become further apart until only one of them is within the spectrum being studied; this is then c.w. irradiation (see Fig. 4.6b). The central frequency is slowly swept and only this invites any response. As the mean energy fed into the system has to be comparable in the two approaches, the pulse amplitude is decreased in the c.w. limit.

In all pulse experiments it is arranged that the receiver and transmitter are never on at the same time. The time is shared between them, hence the alternative name for the above experiment, 'time share'. The advantages of a time sharing are that as the receiver and transmitter are never on at the same time, direct leakage of energy into the receiver, along with the associated problems of base line drift, etc. are eliminated (ideally anyway). The switching on and off of the receiver is a form of modulation. In a conventional spectrometer, modulation techniques are employed by applying an audio frequency to the magnetic field and using a phase sensitive detector referenced to this audio frequency as a correlator. In a time share spectrometer audio phase sensitive detection using the switching frequency is often used, as opposed to the pure r.f. phase detection implied in this chapter.

The response of an n.m.r. system to a continuous wave excitation was described in section 2.9 using the Bloch equations.

4.8 STOCHASTIC EXCITATION [6, 7]

In the previous section we considered the experiment where the nuclear spin system was excited by a chain of identical pulses, the output being received during the interval between these pulses. In stochastic n.m.r. the same basic experimental technique is used, except that the pulses are subjected to random changes in either phase, intensity or duration. The power spectrum of such an excitation is basically flat, as required. The output of the system will now contain the consequences of the random changes in the excitation function; these can be corrected for by cross-correlation and $m(t)$ obtained, which on Fourier transform yields the n.m.r. spectrum. Since we are working in the time domain, exciting all the nuclei at the same time we gain the same multiplex advantage associated with pulse excitation.

The noise function applied to the excitation should have a frequency-independent (flat) power spectrum and in order to avoid any d.c. bias have a zero average; such noise is called 'white noise'. If white noise were to be applied to the system, cross-correlation would have to be carried out continuously, the output from the correlator being $m(t)$. Such an experiment, although possible, is not very practical. A more practical noise source is an appropriately coded binary sequence. This form of excitation is well defined and periodic, and as such, cross-correlation need not be done continuously;

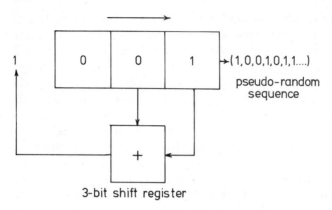

Fig. 4.7. A 3 bit pseudo-random shift register.

it need only be performed once, on the time average spectrum, resulting from many passes through the excitation sequence.

Shift register codes are periodic sequences of binary numbers (0 or 1) generated by means of shift registers made up of a series of n flip-flops connected with a suitable feed back loop. Figure 4.7 shows such a shift register containing 3 flip-flops. On receipt of a timing pulse the state of each flip-flop is changed to that of the flip-flop on the left. If the first flip-flop is set to state '1' as the timing pulses are received, the '1' state is progressively shifted to the right along the register. In order to use the shift register to generate a sequence some feed back is necessary; in this case, the state of the third flip-flop is 'added' to that of the second and the result used as an input. The output of the register is an equally spaced series of binary numbers given by the state of the last flip-flop. The sequence generated by the register shown in Fig. 4.7 repeats after 7 shifts and shows pseudo random sequence $2^n - 1$ bits long. Shift register codes generated in this way are a suitable excitation function for stochastic n.m.r. If the time required to generate a full sequence is T s and each 1 state is translated into a pulse t_p s long then the resultant power spectrum is identical in form to that produced by a sequence of pulses t_p long, regularly spaced every T s i.e. one obtains the same power spectrum as one would using simple pulsed excitation. The difference between the two forms of excitation is their phase distribution. For the excitation to appear random to the nuclei the sequence must extend for longer than T_1. In practice the duration of the sequence is made equal to the data acquisition time required to characterise the spectrum with sufficient resolution; in other words, duration of the sequence is the same as the acquisition time in simple pulsed excitation.

The pseudo random noise can in principle be applied to the excitation in many ways e.g. as r.f. phase, amplitude or magnetic field modulation. The first two are commonly used, the 0 and 1 states being made to correspond to either a $\pm 90°$ phase shift or the presence or absence of a pulse in the chain. The data from the spectrometer is sampled by the receiver in between the pulses in a time sharing manner. There is one experimental complication which must be borne in mind. The excitation sequence is 2^{n-1} pulses long and cannot be changed by even one pulse without drastically affecting the spectrum, whereas for computation efficiency (see section 6.12g) 2^n data points per sequence are required. The relevant amount of zero filling must be carried out before transformation [6].

The difference between pulse and stochastic excitation is illustrated in Fig. 4.8. Each form of excitation corresponds to a series of side bands T^{-1}s apart with the same sinc ϑ amplitude envelope. In the case

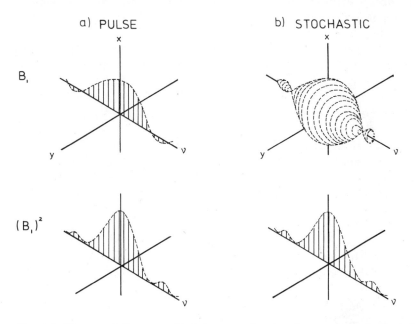

Fig. 4.8. The excitation provided by a pulse is phase coherent; that from stochastic excitation is random. They however have the same power spectrum.

of pulsed excitation (Fig. 4.8a) all these side bands have the same phase; for stochastic excitation they have random phases (Fig. 4.8b) and phase unscrambling is a necessary process, not a trivial one as in the former case. The power spectra in both cases however are the same provided that the same mean power per sequence is applied. The pulses used in pulsed excitation have therefore to be much more powerful than those used in stochastic excitation, which process therefore puts less demand on the spectrometer. The difference in phase characteristics of the two forms of excitation leads to different saturation behaviour. In stochastic n.m.r. no point in time within the sequence is more significant than any other; resolution and sensitivity can therefore be optimised independently, resolution being governed by the length of the sequence, sensitivity by the mean power applied. In order to obtain the n.m.r. spectrum from the response produced by stochastic excitation, one of the two schemes shown in Fig. 4.3 must be used. If true white noise were used, one would have to calculate the auto-correlation function. However, when using the output from a shift register the 'noise' is known and cross-correlation is possible. As discussed previously, although the upper scheme in Fig. 4.3 is preferable as $m(t)$ is produced, the lower scheme is normally used because of the problems of calculating a cross-correlation. If a binary pseudo-random sequence is used as the excitation, cross-correlation within the time domain can be achieved by means of the Hadamard transform. The Hadamard transform is a very efficient algorithm which unscrambles the phases and produces $m(t)$ which can then be transformed into the required n.m.r. spectrum [8].

A detailed solution to the problem of the response of an n.m.r. system to stochastic excitation is outside the scope of this book. A limited solution in terms of the Bloch equation is informative provided the excitation power is considered as being low. If sufficient power is used to cause significant saturation to occur, then, unlike pulsed excitation, since not all the nuclei are in the same rotating frame, double resonance effects can occur and density matrix methods must be used in order to describe the experiment fully.

The Bloch equations as given in eqn. (2.50) can be modified for stochastic excitation thus [6]

$$\frac{dM_{x'}}{dt} = -\frac{M_{x'}}{T_2} \qquad (4.20)$$

$$\frac{dM_{y'}}{dt} = -\frac{M'_y}{T_2} + \sigma s(t)M_z \tag{4.21}$$

$$\frac{dM_z}{dt} = -\frac{M_z - M_0}{T_i} - \sigma s(t)M_{y'} \tag{4.22}$$

where the excitation is a frequency ϑ_0, amplitude modulated by $\sigma s(t)$, a Gaussian time-independent process with unit power spectral density. In this case the assumption of a rotating frame at a definite frequency ϑ_0 is not necessary, all frequencies being equally probable; $s(t)$ has a zero ensemble (time) average.

These Bloch equations may be solved for $M_{y'}$ and $M_{z'}$ noting that $M_{x'}$ is independent of $M_{y'}$, M_z and $s(t)$. The solution is obtained by expanding the stochastic parts in powers of σ thus

$$M_{y'}(t) = \bar{M}_{y'}(t) + \sigma M_{y'}(t) + \sigma^2 M_{y'2}(t) + \text{etc.} \tag{4.23}$$

$$M_z(t) = \bar{M}_z(t) + \sigma M_z(t) + \sigma^2 M_{z2}(t) + \text{etc.} \tag{4.24}$$

The ensemble average of magnetisation which is independent of time is made. Ignoring higher powers M_z is then given by

$$\bar{M}_z = M_0(1 + T_1\sigma^2/2)^{-1} \tag{4.25}$$

It is interesting to note that saturation, i.e. the behaviour of M_z, depends only on T_1 and not on $T_1 \cdot T_2$, as is the case in c.w. excitation (see eqn. 2.61). With stochastic excitation there is no average magnetisation in the x'y' plane hence T_2 processes are ineffectual. The cross-correlation of $s(t)$ and $r(t)$ can be shown to be to first order

$$R_{sr}(\tau) = M_z \exp(-\tau/T_2) U(\tau) \tag{4.26}$$

where $U(\tau)$ is a unit step function. Fourier transform of $R_{sr}(\tau)$ yields the transfer function thus

$$M(\vartheta) = \frac{M_z T_2}{1 + i2\pi T_2\vartheta} \tag{4.27}$$

which is a complex Lorentzian line with a line width at half height of $1/\pi T_2$ Hz. In contrast to continuous wave excitation where line broadening occurs at high power, for stochastic excitation the line width is independent of the applied power. From eqn. (4.27) it follows that the maximum signal height is obtained for $\sigma = (2/T_1)^{1/2}$ where $M_z = \frac{1}{2}M_0$.

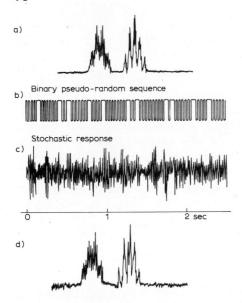

a)

b) Binary pseudo-random sequence

Stochastic response

c)

0   1   2 sec

d)

Fig. 4.9. Stochastic excitation for 2, 4-difluorobenzene [6]. (a) The ^{19}F spectrum obtained using a 250 s sweep experiment. (b) The pseudo random sequence applied for 2.5 s. (c) The response to (b). (d) The spectrum obtained from (c) by cross correlation and transformation.

Analysis of the maximum achievable signal to noise ratio by stochastic excitation shows that it is, not surprisingly, the same as for pulsed excitation. However, fulfilling the conditions for maximum sensitivity, does not result in a compromise, as these conditions are now independent of the length of the excitation sequence used. The ^{19}F spectrum of 2, 4-difluorotoluene obtained by stochastic resonance along with the excitation used is shown in Fig. 4.9.

4.9 CORRELATION SPECTROSCOPY [9, 10]

As will become apparent later on, considerable experimental problems can occur when spectra are digitised. These problems arise because of the limited number of data points available and their limited word length (accuracy). The former limits the resolution obtainable, the latter the dynamic range of samples which can be studied: a particular problem in proton n.m.r. where large solvent signals are often encountered. It would be nice, therefore, to be able to excite uniformly a limited region of the spectrum. A way of achieving such an excitation has been considered by Dadok and involves simply a rapid linear frequency sweep on the time scale of T_1 of the area of interest. The resultant transient effects are removed by correlation techniques.

The idea is based on the concept that a linear sweep has a constant power spectrum over the range it sweeps; this can easily be proved using the nomenclature of section 4.5 thus

$$X(t) = B_i \exp \left(-\frac{iat^2}{2} \right) \qquad (4.28)$$

on Fourier transform this gives (see Table 3.1j)

$$X(\vartheta) = B_1 \left(\frac{2\pi}{a} \right)^{\frac{1}{2}} \exp \left(\frac{\pi i \vartheta^2}{a} \right) \qquad (4.29)$$

$X(\vartheta)$ has a constant power spectrum $\left(|X(\vartheta)|^2 = \frac{2\pi}{a} B_1^2 \right)$. The above is strictly true only for a sweep from $+$ to $-$ infinity; if the sweep is limited to between ϑ_p and ϑ_q, the spectrum is approximately constant between these frequencies and zero outside this frequency range. There will be small transition regions near ϑ_p and ϑ_q.

The sweep is repeated many times and the experiment's time averaged in the normal way. The output of the spin system is, as usual, the convolution of the input excitation and $m(t)$.

$$r(t) = \int_{-\infty}^{\infty} m(\tau) \exp \tfrac{1}{2} (\tau^2 - 2\tau t) \, a \, d\tau \qquad (4.30)$$

Cross-correlation and transformation thus yields $M(\vartheta)$ which can be manipulated as outlined previously to give the required spectrum.

In correlation spectroscopy a slightly different process from that shown in Fig. 4.3a is used in processing the data, although in this case it is mathematically equivalent (as $\exp(\pm x^2) \xrightarrow{\mathscr{F}} \exp(\mp y^2)$). The response of the spectrometer is Fourier transformed and then multiplied point by point with the excitation function, i.e. cross-correlation in the co-domain. The result of this multiplication is analogous to the f.i.d. pulsed n.m.r. and can be manipulated with weighting function to enhance either the signal to noise or the resolution. The data is then Fourier transformed for a second time

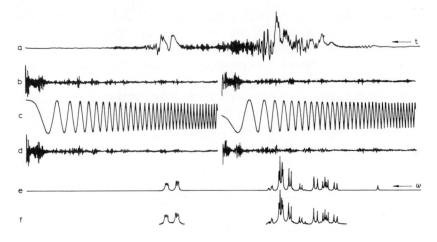

Fig. 4.10. Correlation spectroscopy on *ortho*-chloronitrobenzene
(a) rapid scan response
(b) the Fourier transform of a, cosine terms first
(c) the Fourier transform of the excitation
(d) b × c*
(e) Fourier transform of d
(f) conventional spectrum. From J. Dadok and R.F. Sprecher, J. Magn. Res., 13 (1974) 243.

back into the frequency domain and the transfer function obtained. The main advantage of the approach lies in having a function analogous to the f.i.d. which does not occur in the 'conventional scheme' (Fig. 4.3b). Signals from all parts of the experiment are shown in Fig. 4.10.

Rather than cross-correlating with the exciting function it is possible to use the response from a reference line recorded under identical conditions. Both techniques have pros and cons. The advantages in using a reference line are that the spectra are always correctly phased; induced phase shifts are automatically corrected for, and referenced to the reference line. The draw-backs to using a reference line are that its line width is added to all the lines within the spectrum along with any noise contained in the reference line. TMS is not a good reference as it contained visible ^{29}Si satellites; a sample of degassed benzene or cyclohexane is preferable [10]. Simplicity is the advantage of using the excitation function as it can be calculated and does not have to be measured. The disadvantage is that the spectrum has to be phase corrected in the normal way to compensate for instrumental effects such as the sweep being non-linear or its rate not constant. The efficiency of correlation spectroscopy compared with either pulse or stochastic has been evaluated by Gupta et al. [10]. If the signal to noise ratio attainable in a unit time is compared they show that

$$\frac{(S/N)_p}{(S/N)_{cs}} = \left[\frac{(\Delta/a + 3T_2^*)(1 - E_1^p)(1 + E_1^{cs})}{3T_2^*(1 + E_1^p)(1 - E_1^{cs})}\right]^{\frac{1}{2}}$$

where Δ is the spectral width, $E_1^p = \exp(-3T_2^*/T_1)$ and $E_1^{cs} = \exp(-(\Delta/a + 3T_2^*)/T_1)$. The assumption is made that the acquisition used in the pulse experiment is $3T_2$ and that a time required for a single scan is $(\Delta/a + 3T_2^*)$. For short T_1 (negligible E_1) the pulse method is always more efficient. However, for $T_1 = T_2^* = 1$, $\Delta = 1000\,\text{Hz}$ and $a = 250\,\text{Hz/s}$ correlation spectroscopy is only 35% less efficient than pulse excitation and the difference decreases as T_1 increases.

4.10 TAILORED EXCITATION [11]

All the previous excitation functions that have been discussed have aimed to produce a uniform power spectrum, be it over a limited range in the case of the sweep techniques. In general such an excitation may not be the most useful, two such examples being firstly the case of a solute in a solvent containing the same nuclear species where the strong solvent line can become a problem, and secondly when double resonance effects are being studied. The former case requires a uniform power spectrum with a notch in it at the solvent frequency. In the latter case a spectrum with a spike in it would be necessary at the frequency of the multiplet to be decoupled.

The two cases cited above are only two of many; in general it would be advantageous to be able to reproduce any desired power spectrum. As has been pointed out by Tomlinson and Hill, this end can be achieved by the use of Fourier transformation. The desired power spectrum is first defined as a function of frequency $(F(\vartheta))$ and then Fourier transformed. (An example is given in Fig. 4.11). As the transform is normally carried out in a digital computer the result is series of N discrete values in the time domain. The number of values is dictated by the size of the computer used. This time domain function can be superimposed (modulated) onto a regular sequence

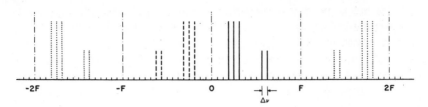

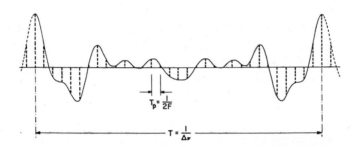

Fig. 4.11. The Fourier synthesis of a time domain excitation function to match the required power spectrum. The power spectrum is only specified between 0 and F but it must be considered as continuous periodic function [13].

of r.f. pulses (T_p s apart) in such a way that the series of N points forms a sequence of pulses T s long, where T is the time required to achieve the desired resolution. The excitation function is thus a modulated series of delta functions (ideally anyway).

$$x(t) = \sum_{n=1}^{N} B_1 f(t) \cdot \delta(t - n\tau) \qquad (4.31)$$

where

$$f(t) \xrightarrow{\mathcal{F}^-} F(\vartheta)$$

and

$$\tau = \frac{T}{N} \qquad (4.32)$$

The modulation can be applied to the pulse chain by changing either the phase, amplitude or length of the pulses. The magnitude of $f(t)$

is normally represented by the pulse length and a ± 90° phase shift represents the sign. The sequence can be repeated and the output time averaged in the normal way.

Stochastic and pulsed excitation are obviously special cases of tailored excitation. For stochastic excitation $f(t)$ would be the output from a pseudo-random generator. For pulsed n.m.r. $f(t)$ is unit for the first impulse and zero for all the others. As was pointed out earlier the energy per sequence must be the same in both limits, hence the peak pulse power used in the stochastic limit is lower than the 'pulse' limit. If one can arrange to stay in the stochastic limit the experimental requirements will be less demanding. Fortunately by a very simple procedure this is possible.

In section 4.6 it was shown that the principal difference between pulsed and stochastic excitation lies in their phase relationships. If one specifies $X(\vartheta)$ as being phase coherent, as in practice one must, then the pulse sequence calculated will tend to the pulse limit. If, however, one adds a random phase angle to each point in $X(\vartheta)$ before transformation then the excitation function synthesized will tend to the stochastic limit. This is illustrated in Fig. 4.12 for the case of a power spectrum containing a notch. Each point m, of the transfer function calculated from such an experiment will contain

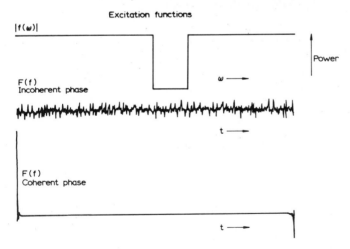

Fig. 4.12. Power spectrum for peak suppression synthesised with and without phase scrambling. Both the excit ition functions are plotted on the same scale demonstrating the larger dynamic range required without phase scrambling [11].

the same random phase angle as was added to the mth point of the excitation function before transformation. Simple subtraction of this angle results in the output of a normal spectrum. The stages of a tailored excitation are shown in Fig. 4.13.

The techniques of tailored excitation are still in their infancy and appear to have a very bright future. There are many other possible excitation spectra possible. For example, one could produce a frequency power spectrum which was a rectangle, by modulating the pulse chain a sinc ϑ function. With such a function specific areas of a spectrum could be examined without fold-over problems. A spectrometer functioning along this line can also achieve double and triple

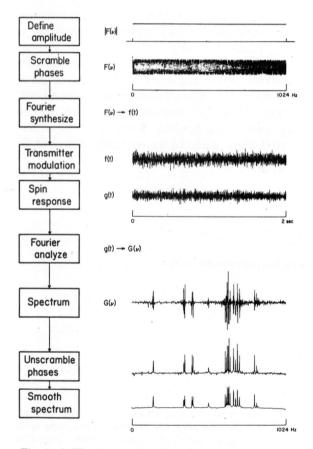

Fig. 4.13. The stages of a tailored excitation experiment [13].

resonance using one r.f. transmitter (see Fig. 9.12) moreover most of the present day experiments are merely special cases of the most general n.m.r. experiments.

4.11 THE EQUIVALENCE OF SWEPT AND FOURIER N.M.R. [12]

The previous sections implicity assumed that the spectrum obtained by the adiabatic slow passage technique was the same as that obtained by pulse and stochastic excitation. Fortunately this is normally so; however, as in the former case the spins are excited sequentially and in the second case simultaneously, differences may be expected when non-equilibrium states are involved.

Before considering the general question of equivalence further one must exclude effects introduced by the measurement technique itself. Specifically:

(i) Any sweep experiment is only an approximation to the slow passage experiment; some asymmetry effects will be present.

(ii) Deviations of the specific experiment from linearity: the most important example of this type is saturation, which is considered in detail in the next chapter.

(iii) Fourier spectra involved digitisation: due to the necessarily limited number of data points a totally faithful representation of the spectrum may not be produced.

(iv) Fourier spectra are usually the time average of a series of experiments. If a sufficient time is not left between experiments then interference effects (echoes) will distort the spectrum.

(v) The excitation function used may not have a uniform power spectrum.

These essentially experimental artifacts are dealt with in other parts of the book.

For the spectrum obtained by Fourier transformation of the impulse response to be equivalent to that obtained by a true slow passage experiment certain conditions must be fulfilled. A detailed deviation of these conditions requires a full density matrix treatment and will not be given here. The main conditions for equivalence are [12]:

(1) The system must be time independent. In other words it must always give the same output, no matter when the pulse is applied. This will in general be true for small pulse angles but can be violated if 'steady state' conditions are stabilised. Further violations of this

condition can occur in connection with modulation effects whose behaviour is not coherent with the excitation timing. A good example of the latter type of process is spinning side bands. For a single pulse experiment the phases of spinning side bands will differ from those recorded in a c.w. experiment. Time averaging removes these anomolies and the f.t. and c.w. spectra become equivalent.

(2) The system must be linear. In this case the term linearity must be qualified. We have already excluded the inherent non-linearity of the spins due to saturation effects. In this context the *form* of the response does not depend on the pulse power used. This condition is in general met.

(3) All nuclear spins under consideration must be excited equally. The need for this condition is self evident.

(4i) Each strongly coupled subsystem of the spin system must be in internal thermodynamic equilibrium prior to the excitation. Additional spins which are not affected by the excitation and are only weakly coupled to the subsystem being studied may be in an arbitrary non-equilibrium state.

(4ii) The high temperature approximation must be valid i.e. $(RT^0)^{-1} \ll 1$ (of eqn. 2.27) where T^0 is the spin temperature of the system.

(4iii) The high field approximation must be valid i.e. the Zeeman terms are the dominant interactions.

The conditions 4i — iii are three practical limitations derived for somewhat abstract conditions imposed by the symmetry of the density matrix required to study the problem fully.

For normal samples in single resonance experiments all the above conditions are met and complete equivalence exists between the spectra obtained. The conditions which are the most easily violated are conditions 1 and 4i. Condition one is questionable for time dependent experiments e.g. chemical exchange and double resonance experiments where the nuclear Overhauser effect is time dependent. A detailed analysis shows that in the case of chemical exchange, spectra obtained by swept and Fourier techniques are equivalent. This is not, however, the case in some double resonance experiments and in experiments in which condition 4i is not met. In slow passage experiments on homonuclear coupled spin systems the signal intensities are directly related to the population differences of the connected transitions. In pulsed experiments with larger flip angles, the relations are considerably more complex and in general intensities depend on all the spin populations. The most important case is the CIDNP experiment.

The CIDNP experiment (see section 8.7) obviously does not meet condition 4i as the molecules giving rise to the spectrum result from the spin selective combination of radicals with strongly polarised unpaired electrons. The non-equilibrium state, in most cases, only affects the spin populations and not the energy levels themselves, and is manifested by selectively enhanced absorption lines or by the occurrence of emission lines. In CIDNP experiments on weakly coupled systems two effects can be distinguished, the 'net effect' which determines the total intensity of a multiplet, and the 'multiplet effect' which affects the relative intensities within the multiplet. The 'net effect' is, of course, not affected by the excitation technique but the 'multiplet effect' in the case of homonuclear coupling is; with pulse excitation it shows a dependence on pulse angle, as can be seen in Fig. 4.14 and indeed, it disappears altogether for $\alpha = 90°$. For small angles the spectra are equivalent, as α increases and intensity error (ϵ) between the intensity of line observed by c.w. (I_i^0) and pulsed excitation (I_i^P) appears, with ϵ defined as

$$\epsilon = \frac{\sin L_i^0 - L_i^P}{\sin L_i^0} \qquad (4.33)$$

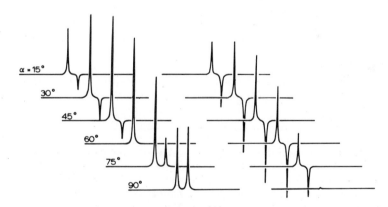

Fig. 4.14. Computer simulated CIDNP spectra for an AX system with a triplet precursar for different angles α. The parameters used are $B_0 = 1$ Tesla, difference in the g factors is 10^{-3}, hyperfine couplings $A_1 = 3.5 \times 10^{-3}$ and $A_2 = 4.5 \times 10^{-3}$ Tesla; chemical shift difference $= 100/\pi$ Hz and $J_{AX} = \pi - 1$ Hz. From S. Schäublin, A. Hohener and R.R. Ernst, J. Magn. Res., 13 (1974) 196.

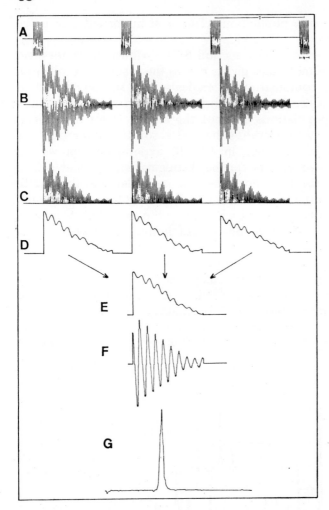

Fig. 4.15. Schematic representation of the principles of Fourier difference spectroscopy. (A) Exciting pulse sequence; (B) response of sample and reference; (C) signal after diode detector; (D) low frequency components of C; (E) ensemble average of D; (F) signal with eliminated reference response; (G) Fourier transform of F. From R.R. Ernst, J. Magn. Res., 4 (1971) 280.

it can be shown to depend on the pulse angle thus

$$\epsilon = 1 - \cos \alpha \tag{4.34}$$

The errors stay below 10% up to 26° at which point the signal intensity (proportional to sin α) has only 44% of its maximum value. A reasonable compromise between sensitivity and intensity error suggests the use of pulse angles of $10° - 20°$.

The above discussions effectively only concern spectra where homonuclear coupling is involved e.g. ^{1}H. As stated in condition 4i, for cases like ^{13}C where the r.f. pulse does not affect the spin states of the nuclei causing the coupling, i.e. the protons, this equilibrium state is of no consequence; pulse and swept spectra are always equivalent.

Similar effects and conclusions result from the study of selective saturation effects and Overhauser studies. Whenever a frequency-selective perturbation is applied to a coupled spin system, the Boltzmann distribution is destroyed and relative intensities show a dependence on pulse angle.

4.12 DIFFERENCE SPECTROSCOPY

An interesting variant on pulse excitation has been proposed by Ernst [14], known as difference spectroscopy. Here one strong single line in the spectrum is used as a 'carrier' and reference line; the spectrum is reference to this line by a subtraction process. Difference spectroscopy is illustrated in Fig. 4.15. The sample is subjected to a repetition sequence of pulses as in normal pulse excitation and the resulting free induction is detected with a linear (diode) detector instead of the normal phase sensitive detector; use of a linear detector produces a positive envelope of the response (Fig. 4.15c). This signal is next routed through a low pass filter, the output of which contains the audio difference frequencies between the sample and the strong reference line (Fig. 4.15d). These are digitised in the normal way. The signal originating from the reference line itself is removed by subtracting a suitable function, typically generated from a Legendre polynomial of the 4th and 6th order (Fig. 4.15f). The remaining signal is transformed into the frequency domain to produce the required spectrum.

The reference compound used in difference spectroscopy must give a strong single line to one side of the spectral range of interest, as in its simple form difference spectroscopy cannot distinguish positive and negative frequencies. Water has been used and trifluoro-acetic acid is also a possibility for proton work. As the reference line can easily be 100 or 1000 times the intensity of the sample line, complications can occur due to spinning sidebands etc. The reference compound is therefore best kept in a capillary and degassed etc. to ensure the longest possible T_2^*. Since a reference signal must obviously be present at the detector all the time data is being accumulated, a long T_2^* is essential in order to have a signal present at the detector as long as possible.

As all the data used in the final spectrum is obtained via a reference line, the spectrum is automatically referenced to it, and no other form of field/frequency stabilisation is required, provided, of course, that the field/frequency ratio does not drift so far that the pulse no longer excites the reference signal sufficiently. Indeed the spectrum is insensitive to any homogeneous modulation effect; it is, however, sensitive to any modulation which is inhomogeneous over the sample. An example of the latter type of this effect is the field inhomogeneities which give rise to spinning sidebands; these therefore appear in difference spectra. Side bands due to homogeneous effects, like ripple on the magnetic field, on the other hand, are not seen. An interesting example of the use of difference spectroscopy in the study of fast reactions is given in section 8.7f.

REFERENCES

1 I.J. Lowe and R.E. Norberg, Phys. Rev., 107, (1957) 46.
2 R.R. Ernst, Advan. Magn. Res., 2, (1967) 1.
3 P. Fellgett, Ph.D. Thesis, University of Cambridge, 1951, and J. Phys. Radium, 19, (1958) 187.
4 R.R. Ernst, J. Magn. Res., 1, (1969) 7.
5 A. Abragam, The Principle of Magnetic Resonance, Oxford University Press, London, 1961, p. 36.
6 R.R.Ernst, J. Magn. Res., 3, (1970) 10.
7 R. Kaiser, J. Magn. Res., 3, (1970) 28.
8 D. Ziessow, On-line Reckner in der Chemie, de Gruyter, Berlin, 1973.
9 J. Dadok and R.F. Sprecher, J. Magn. Res., 13, (1974) 243.
10 R.K. Gupta, J.A. Farretti and E.D. Becker, J. Magn. Res., 13, (1975) 275.
11 B.L. Thomlinson and H.D.W. Hill, J. Chem. Phys., 59, (1973) 1775.
12 S. Schäublin, A. Hohener and R.R. Ernst, J. Magn. Res., 13 (1974) 196.
13 H.D.W. Hill, personal communication.
14 R.R. Ernst, J. Magn. Res., 4 (1971) 280.

CHAPTER 5

PULSED N.M.R.

Until 1966 high resolution n.m.r. spectra had been obtained exclusively by sweep techniques, initially by sweeping the magnetic field, latterly the radio frequency. In that year Ernst and Anderson demonstrated that it was more efficient to excite all the spins simultaneously and use Fourier transform techniques to unscramble the resultant signal [1]. The next two chapters are given over to the theoretical and practical aspects of repetitive pulsed excitation which, since its introduction, has grown rapidly at the expense of the more traditional methods of recording spectra. In this chapter we will first consider the solution of the phenomenalistic Bloch equations for pulsed excitation using a sequence of identical pulses, and the discussion will later be extended to include the use of multipulse sequences.

5.1 THE BLOCH EQUATIONS FOR PULSED EXCITATION

Consider a system of isolated nuclear spins in magnetic field B_0 which are subjected to a regular sequence of rectangular pulses. These pulses are t_p seconds long, of amplitude $2B_1$ and occur every T_p seconds. The static magnetic field B_0 defines the z axis of the co-ordinate system which is a rotating frame (see section 2.8) precessing at the pulse radio frequency ϑ_c.

During the presence of the second radio frequency the nuclei will precess about the effective magnetic field B_{eff}. This field can be calculated as illustrated in Fig. 2.7 using eqn. (2.59). One has to consider the effective field rather than simply B_1 because in general there will be nuclei present in the sample whose Larmor frequency is ϑ_0, not ϑ_c. Such nuclei are not synchronised with the rotating frame and therefore the consequence must be allowed for. If, however, one uses a 'strong' pulse, which is defined as a pulse with γB_1 much larger than the total width of the spectrum under investigation, i.e.

$$\vartheta_0 - \vartheta_c \ll \gamma B_1 \quad \text{for all } \vartheta_0 \text{ being considered}, \tag{5.1}$$

then B_{eff} becomes equal to B_1 for all nuclei. For the rest of this section we will assume all the pulses fulfil this inequality, and return in section 5.3 to the situation where, for experimental reasons, strong enough pulses cannot be used.

When the pulse is over, the magnetisation vector, which has been tipped towards the $x'y'$ plane by the rotation about the x' axis induced by B_1, will relax back to its equilibrium position along the z axis, precessing about that axis at ϑ_0. The main advantage of the pulsed excitation is to give sensitivity, and so the second pulse should ideally be initiated as soon after the first pulse as is consistent with the desired resolution (see section 6.10). The consequence of this, particularly in ^{13}C n.m.r. where T_1 and T_2 may be long, is that all the magnetisation perturbed by the first pulse may not have returned to the z axis or even have decayed within the $x'y'$ plane by the time the second pulse occurs. If the $x'y'$ magnetisation does not decay to zero between pulses, the situation becomes quite complex and so called spin echoes are formed. However, the system eventually reaches an equilibrium condition which is independent of the initial conditions.

If the pulse length t_p is short compared to T_1 or T_2, which it always is in high resolution work, then relaxation during the pulse is negligible. Each pulse rotates all the nuclei about the x' axis by an angle α which is defined thus

$$\alpha = \gamma B_1 t_p \tag{5.2}$$

and is illustrated in Fig. 5.1b.

The magnetisation immediately after the pulse $\underline{M}^+$ is then related to the magnetisation before the pulse $\underline{M}_0$ by rotation operator $\hat{R}_x(\alpha)$ thus

$$\underline{M}^+ = \hat{R}_x(\alpha)\underline{M}_0 \tag{5.3}$$

In between the pulses the magnetisation will precess through an angle Θ about the effective field $(\vartheta_0 - \vartheta_c)/\gamma$ (cf. eqn. 5.2)

$$\Theta = 2\pi(\vartheta_0 - \vartheta_c) \cdot t = 2\pi\vartheta_i \cdot t \tag{5.4}$$

Θ will usually consist of more than one revolution; however, the observable result is only that of a rotation where

$$\Theta = 2n\pi + \theta \tag{5.5}$$

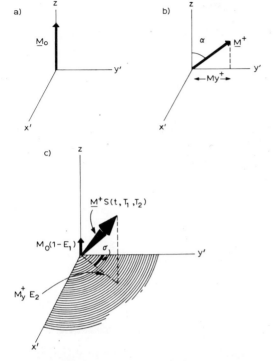

Fig. 5.1. The components of the macroscopic nuclear magnetisation. (a) At equilibrium; (b) immediately after a pulse; (c) t s after the pulse.

During the time T_p the nuclear spins relax. The $x'y'$ magnetisation decays exponentially with the time constant T_2 (ignoring magnet inhomogeneity effects) and the magnetisation along the z axis recovers with time constant T_1. This relaxation can be represented by an operator $S(t \cdot T_1 \cdot T_2)$ thus

$$S(t \cdot T_1 \cdot T_2) = \begin{vmatrix} E_2(t) & 0 & 0 \\ 0 & E_2(t) & 0 \\ 0 & 0 & E_1(t) \end{vmatrix} \quad \text{where } \begin{aligned} E_1(t) &= \exp - (t/T_1) \\ E_2(t) &= \exp - (t/T_2) \end{aligned}$$

$$(5.6)$$

If $\underline{M}_0$ is the equilibrium magnetisation along the z axis, the magnetisation at time t after the pulse, is given by the sum of the magnetisation that has recovered onto the z axis and that remaining in the $x'y'$ plane, see Fig. 5.1c.

$$\underline{M}(t) = \underline{M}_0(1 - E_1(t)) + \underline{M}^+ R_z(\theta) S(t \cdot T_1 \cdot T_2) \tag{5.7}$$

For a steady state condition, the magnetisation just before the next pulse $(\underline{M}^-)$ is equal to the value of $\underline{M}(t)$ at $t = T_p$. Solving these equations for all the components of $\underline{M}$ it can be shown that [2]

$$M_{x'}^- = \underline{M}_0(1 - E_1)E_2 \sin \alpha \sin \theta / D \tag{5.8}$$

$$M_{y'}^- = \underline{M}_0(1 - E_1)E_2 \sin \alpha (\cos \theta - E_2)/D \tag{5.9}$$

$$M_z^- = \underline{M}_0(1 - E_1)[1 - E_2 \cos \theta - E_2 \cos \alpha (\cos \theta - E_2)]/D \tag{5.10}$$

$$M_{x'}^+ = M_{x'}^- \tag{5.11}$$

$$M_{y'}^+ = \underline{M}_0(1 - E_1)(1 - E_2 \cos \theta) \sin \alpha / D \tag{5.12}$$

$$M_z^+ = \underline{M}_0(1 - E_1)[E_2(E_2 - \cos \theta) + (1 - E_2 \cos \theta) \cos \alpha]/D \tag{5.13}$$

where

$$D = (1 - E_1 \cos \alpha)(1 - E_2 \cos \theta) - (E_1 - \cos \alpha)(E_2 - \cos \theta)E_2 \tag{5.14}$$

and

$$E_1 = \exp - (T_p/T_1) \quad \text{and} \quad E_2 = \exp - (T_p/T_2) \tag{5.15}$$

T_p = time between the pulses during a repetitive pulse sequence.

These rather formidable equations can be greatly simplified if the time between the pulses is long compared with T_2 (not T_2^*), in which case E_2 tends to zero. The constant D now simply becomes $(1 - E_1 \cos \alpha)$ and M_z^+ and M_y^+ can be simplified to

$$M_z^+ = \frac{1 - E_1}{1 - E_1 \cos \alpha} M_0 \tag{5.16}$$

$$M_y^+ = \frac{(1 - E_1) \sin \alpha}{1 - E_1 \cos \alpha} M_0 \tag{5.17}$$

Under this simplification the signal immediately after the pulse has a maximum value when α is chosen such that $\cos \alpha = E_1$. As will be shown, this value of α has a more general significance and will be used when considering the question of an optimum pulse angle. With this choice of α, M_y^+ is given by

$$M_y^+ = \frac{\sin \alpha}{1 + E_1} M_0 \tag{5.18}$$

The behaviour of nucleus i during a repetitive pulse sequence is thus completely described in terms of the equilibrium magnetisation, the two angles α and θ, the two relaxation times T_1 and T_2 and the offset frequency of the i spin from the pulse frequency. The detectable signal in such an experiment, i.e. the y' component of $\underline{M}(t)$, can now be calculated thus (see Fig. 5.1).

$$M_{y'}(t) = M_{y'}^+ \cos \theta \, E_2(t) + M_{x'}^+ \sin \theta \, E_2(t) \tag{5.19}$$

which using eqn. (5.4) becomes

$$M_{y'}(t) = M_{y'}^+ \cos (2\pi\vartheta_i t)E_2(t) + M_{x'}^+ \sin (2\pi\vartheta_i t)E_2(t) \tag{5.20}$$

where $M_{y'}^+$ and $M_{x'}^+$ are given by eqn. (5.11) and (5.12) respectively.

The above equation is of the form expected from consideration of the generalised experiment in Chapter 4. It shows that the detected time domain signal has two components 90° out of phase with respect to each other which decay exponentially with a time constant T_2. If one considers the case of an isolated pulse, as in section 4.4, then as there is no transverse magnetisation before the pulse, i.e. $M_x^+ = M_x^- = 0$, eqn. (5.20) reduces to eqn. (4.7). The value of generalising eqn. (4.7) to eqn. (4.8) now becomes apparent; by making time a complex variable we can describe the behaviour of both components of $M_{y'}(t)$. It should be noted that in practice with older spectrometers only one component of $M_{y'}(t)$ is recorded and phase correction techniques are used to obtain the desired cosine term. All the information in the other component is lost; it can be recovered by the use of a second phase detector working in quadrature (i.e. 90° out of phase with respect to the first).

Equation (5.20) shows that in the steady state situation the phase of the signal after each pulse depends on the offset ϑ_i of that signal from the carrier frequency. If, as was shown above, the delay between the pulse is longer than $3T_1$, each pulse is independent of the preceding pulses, and a constant phase spectrum results where the initial magnetisation is given by

$$M_{y'}^+ = \underline{M}_0 \sin \alpha \quad \text{and} \quad M_{x'}^+ = 0 \tag{5.21}$$

which has a maximum value, independent of T_1, when $\alpha = 90°$. The main aim of using pulsed excitation is to increase the sensitivity of the experiment. It is therefore important to consider in detail the conditions which give rise to the maximum detectable signal. In this optimisation there are two free parameters, the time between the pulses and the pulse angle. The former is normally fixed by consideration of the resolution desired (achievable resolution is equal to 1/time taken to acquire data, see section 6.10); we will therefore consider optimisation of the experiment with α as the independent variable. It can be shown that the optimum angle has a complex dependence on the T_1, T_2 and offset (via θ) which is given by [1]

$$\cos \alpha_{opt} = \frac{E_1 + E_2 (\cos \theta - E_2)/(1 - E_2 \cos \theta)}{1 + E_1 E_2 (\cos \theta - E_2)/(1 - E_2 \cos \theta)} \qquad (5.22)$$

Such an equation is too complex to be of practical value, especially as for an unknown sample most of the variables are unknown. However, three special cases are informative.

The first special case is the obvious one, when the time between the pulses is long compared with T_1 and T_2. Equation (5.22) then yields 90° as the optimum pulse angle as one would expect. The second case is the inverse of the first, i.e. the duration between the pulses is very much less than T_1 and T_2; this is shown in Fig. 4.6 and is equivalent to the experiment with continuous wave excitation. If one now defines an average r.f. amplitude thus

$$(\gamma B_1)_{opt} (t_p/T_p) = \alpha_{opt}/T_p \qquad (5.23)$$

then eqn. (5.22) produces the optimum signal when [1]

$$[(\gamma B_1 t_p)_{opt}/T_p]^2 T_1 T_2 = 1 \qquad (5.24)$$

provided one of the side bands coincides with the line under consideration. The above equation is an exact parallel with the condition obtained in section 2.9 for maximum signal intensity. Such a result is not surprising.

The third case is obtained by averaging over excess precision angle θ which varies from line to line within the spectrum. An average optimum pulse angle is then given by [1]

$$\overline{\cos \alpha_{opt}} = E_1 \qquad (5.25)$$

This angle is commonly known as the Ernst angle and will be given the symbol α_E. Using a pulse which satisfies eqn. (5.25) results in a spectrum where the intensity of the line is independent of its offset (ignoring the effect of finite pulse width). The value of $M_{y'}^+$ is now simply given by

$$M_{y'}^{+} = \frac{M_0 \sin \alpha_E}{1 + E_1} \tag{5.26}$$

To obtain the frequency spectrum which results for pulsed excitation we must Fourier transform eqn. (5.20). Such a transformation is of the type illustrated in Table 3.1E. The procedure can be simplified to a cosine transform (eqn. 3.10) as we know that the free induction decay is even and only exists for positive time. The result of cosine transforming eqn. (5.20) is

$$M(\vartheta) = \frac{KM_0}{(2T_a)^{1/2}} \left\{ \frac{T_2}{1 + (2\pi\vartheta_i T_2)^2} \left[1 - E_2^* \cos\theta - E_2 \cos(2\pi\vartheta_i T_a) + \right. \right.$$

$$E_2 E_2^* \cos(\theta + 2\pi\vartheta_i T_a) \right] + \frac{2\pi\vartheta_i T_2^2}{1 + (2\pi\vartheta_i T_2)^2}$$

$$\left. [E_2^* \sin\theta + E_2 \sin(2\pi\vartheta_i T_a) - E_2 E_2^* \sin(\theta + 2\pi\vartheta_i T_a)] \right\} \tag{5.27}$$

where

$$K = \frac{(1 - E_1)\sin\alpha}{(1 - E_1\cos\alpha)(1 - E_2\cos\theta) - (E_1 - \cos\ \alpha)(E_2 - \cos\theta)E_2} \tag{5.28}$$

$$E_2^* = \exp(-T_a/T_2) \qquad E_2 = \exp(-T_p/T_2)$$

The first term in the above equation represents the absorptive part and the second part the dispersive part of the spectrum. As we have previously mentioned, if the acquisition time is set long compared to T_2, the dispersive part can be neglected. This occurs as now both E_2 and E_2^* tend to zero and eqn. (5.27) simplifies to

$$M_{(\vartheta)} = \frac{M_0 \sin\alpha}{(2T_a)^{1/2}} \left[\frac{T_1}{1 + (2\pi\vartheta_i T_2)^2} \right] \tag{5.29}$$

which is simply, as one would expect, a Lorentzian line with a line width of $(\pi T_2)^{-1}$ Hz and whose amplitude is proportional to $M_0 \sin\alpha$.

So far all the functions studied have been considered as continuous analytical functions. In practice they are finite signals which have been digitised for a time T_a s. There are certain consequences of this digitisation that it is profitable to study at this point. Suppose we use N samples to define the time domain spectrum; these will be taken regularly every T_a/N s thus defining $M(\vartheta)$ at regular frequency intervals of $\vartheta_s = N/2T_a$. Now if $T = T_p$, $E_2^* = E_2$ then from eqns. (5.4) and (5.5) it follows that [2]

$$\theta + 2\pi\vartheta_i T_p \;=\; 2\pi\vartheta_c T_p - 2\pi(\vartheta_s - \vartheta_c)T_a \;=\; 2\pi\vartheta_s T_a \;=\; N\pi \qquad (5.30)$$

thus

$$\sin\theta \;=\; -\sin 2\pi\vartheta_i T_a \qquad (5.31)$$

and

$$\sin(\theta + 2\pi\vartheta_i T_a) \;=\; 0 \qquad (5.32)$$

Under these conditions the second term in eqn. (5.27) vanishes and only the absorption part of the spectrum is recovered independent of offset. The intensities of the signals are, however, still a function of offset via the dependence of K on the excess precession angle θ. In any real spectrometer T_a and T_p are not exactly equal as some time must be left after the end of a pulse before digitisation can commence to allow the receiver to equilibrate. Even slight differences between T_a and T_p can cause $\sin\theta$ and $\sin 2\pi\vartheta_i T$ to get out of phase; the second term in eqn. (5.27) is now no longer negligible and a pure absorption spectrum is no longer obtained. We have discussed this problem previously in section 4.6 from a more abstract view-point, and obtained an equivalent result. By saying that the acquisition time does not exactly equal the time between pulses, one is effectively displacing the zero time point by the difference. As illustrated in Fig. 4.5, this change in zero induces a phase shift in the resulting spectrum which will depend on the frequency. The time discrepancies need only be small to cause quite large phase shifts; for example, a difference of $50\,\mu s$ induces a $90°$ phase shift across a 5 kHz wide spectrum.

5.2 THE STEADY STATE, PHASE AND INTENSITY ANOMOLIES [2]

We have so far derived an analytical equation for the behaviour of a nuclear spin system subjected to a regular series of pulses, but we have only considered the expression for cases where the time between the pulses has been long compared with the relaxation times of the nuclei in question. Now consider what happens when, as is quite common in practice, this is not the case. Under these conditions, a steady-state is built up where so-called spin echoes appear. These echoes are the consequence of the net transverse magnetisation

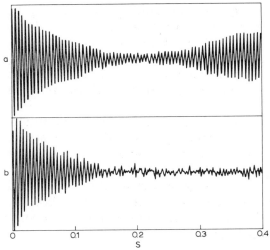

Fig. 5.2. Steady state echoes in the f.i.d. from ^{13}C enriched methyl iodide, coherently decoupled from protons. (a) Under stable pulse timings; (b) the time averaged echo suppressed by introducing a random delay. From R. Freeman and H.D.W. Hill, J. Magn. Res., 4 (1971) 366.

not being zero before the initiation of the next pulse. If instrumental factors — such as instability in the field/frequency ratio, or the incoherent effects arising from the use of a heteronuclear noise decoupler — irreversibly reduce the transverse magnetisation between one pulse and the next, then the effects of such echoes will be masked. However, with instruments of high stability, particularly in proton work, echoes are often seen as a free induction decay which starts to increase towards the end (see Fig. 5.2). Fourier transform of a free induction decay leads to a spectrum which exhibits phase and intensity errors. A detailed study of these spin echoes has been given by Freeman and Hill.

The effect of magnetic field inhomogeneity must be taken into account when considering the steady-state regime. This can be achieved by imposing a Gaussian distribution onto M_x^+ and M_y^+ in eqns. (5.11) and (5.12). An interesting effect emerges; even though magnetic field inhomogeneity causes dephasing of the magnetisation in the $x'y'$ plane, a repetitive chain of pulses has the property of refocussing the magnetisation at the time of the next pulse. This situation was first observed by Hahn in 1950 using a sequence of two $90°$ pulses [3]. In such a sequence negative going echos appear at time $2T_p$. In a continuing sequence the third pulse occurs at the centre of this echo. Two further echoes are generated at $3T_p$ and $4T_p$ etc, until a steady state is built up where transverse magnetisation dephased by magnet inhomogeneity is effectively refocussed prior to each pulse. The presence of spin echoes leads to an increase in the total integrated n.m.r. signal and hence improves the sensitivity of repetitively pulsed n.m.r.

It is a matter of some concern to realise that a nuclear spin system subjected to a regular series of identical pulses settles into a steady state, where, unless the pulse interval is long compared with the relaxation times involved, phase and intensity anomalies appear, dependent on the offset. The phase errors can usually be corrected for (see section 6.12h); the intensity errors cannot. They are minimised by the choice of the Ernst angle as the pulses power when the range of intensities can be shown to be

$$\frac{M(\vartheta)\ \text{max}}{M(\vartheta)\ \text{min}} = \frac{1 + E_2}{1 - E_2} \tag{5.33}$$

and, as discussed above, effects of this type can be neglected once T_p exceeds $3T_2$, at which point they cause errors of about $\pm 3\%$. On the other hand one is always trying to obtain the maximum sensivity: consequently it is normal to use the minimum time between the pulses acceptable on resolution grounds. Under these conditions, steady states frequently arise from $M_{y'}^+$ not being zero. Moreover, the existence of a steady state in itself increases the sensitivity of the experiment. Fortunately there is a satisfactory solution to this dilemma.

There are two basic remedies to this problem, either to destroy the $x'y'$ magnetisation just prior to the next pulses, or to average steady state signals in such a way that the anomolies too are averaged out. The first remedy can itself be approached in two ways. Firstly, a homogeneity spoiling pulse (h.s.p.) can be applied, i.e. the magnet homogeneity is degraded for a fraction of a second. The use of homogeneity spoiling pulses is not very satisfactory as it throws away valuable signal information. The second way of destroying this transverse magnetisation is by the use of a $90°$ pulse to return it to the z axis. This is the basis of the d.e.f.t. multiphase sequence which is discussed in section 5.7.

The second basic remedy may be achieved by averaging the signals over all values of the excess precision angle θ. Such averaging can be achieved by introducing a small random delay in the order of a few milliseconds into the interval between each pulse. The result of this process is that any line with an offset from the carrier of greater than a few hundred Hertz has essentially a random value of θ (see eqn. 5.4). The disadvantage of the approach is that a steady state is never achieved. A more satisfactory technique is to allow a steady state to build up for a certain period of time, which should be at least $4T_1$, then introduce a small random change in the pulse interval and allow another steady state to build up. If during the course of an experiment many steady states exist, and are averaged, the resultant spectra will have been obtained from data which has effectively been averaged over all values of θ. Phase and intensity anomalies will have been minimised, but as steady state conditions existed during the large part of the experiment, most of the inherent extra sensitivity of the steady state will have been utilised.

There can still of course be apparent intensity anomalies due to nuclei with the spectrum having different T_1 values. Whereas all the transverse magnetisation may have decayed away during the time T_p not all the longitudinal magnetisation may have returned. Such variations are not due to the existence of spin echoes and consequently will not be effected by the use of random delays. They are the result of saturation (see section 5.7) and common to all forms of excitation. They can only be eliminated by lowering the mean energy of excitation.

5.3 THE EFFECT OF FINITE PULSE POWER

In the previous section it was implicitly assumed that all the nuclei within the spectrum of interest have been subjected to the same excitation. Now consider the more practical case where this is not so. There are two main factors to be considered; firstly, any spin i not at resonance in the rotating frame will precess about B_{eff} instead of B_1, and secondly, any pulse with a finite length B_1 is not constant throughout all the spectral region.

The problem can be approached from two viewpoints; either by considering only the magnitude of the r.f. field produced by a pulse, allowing the magnetisation to precess about B_{eff} (i.e. working in the time domain [1, 4], or by considering the phase of radio frequencies produced by a pulse and treating each component of the magnetisation separately (i.e. working in the frequency domain) [5].

Both methods produce effectively the same result. Consider the case of a 90° pulse; this is the longest pulse normally encountered in simple pulsed n.m.r., and, as the problems of finite pulse width obviously get more acute the further the pulse diverges from the ideal delta function, it is the most critical.

As was described in section 2.8 the magnitude of the effective field seen by a nucleus i in a reference frame rotating at the basic pulse frequency is given by

$$|B_{eff}| = [(\vartheta_c - \vartheta_0)^2/\gamma^2 + B_1^2]^{\frac{1}{2}} = [\vartheta_i^2/\gamma^2 + B_1^2]^{\frac{1}{2}} \tag{5.34}$$

The angle between B_{eff} and $B_1 \beta$ is given by [4]

$$\tan \beta = \frac{\vartheta_i}{\gamma B_1} \qquad |\beta| < 90° \tag{5.35}$$

During the pulse the magnetisation rotates about B_{eff} by an angle φ given by

$$\varphi = \gamma B_{eff} t_p \tag{5.36}$$

This angle can be related to the rotation about the x' axis (α) by

$$\varphi = \alpha \sec \beta \tag{5.37}$$

As B_1 increases β tends to zero and φ to α and we have the 'strong field' limit considered previously where all the nuclei rotate about x' by the same amount.

However when B_1 is not strong compared with spectral width (in field units) then each nucleus does not experience the same B_{eff}. As a resonance is offset from zero the direction of the effective field deviates from the x' axis towards the z axis. This effect alone would reduce the magnitude of magnetisation in the x'y' plane after a 90° pulse. However, there is a partial cancellation because, due to the increase in B_{eff} as ϑ_i increases (see eqn. 5.34), the rotation angle φ is also increased. The amplitude of the signal produced is thus relatively insensitive to the value of B_1, usually expressed in terms of $t_{90°}$ via eqn. (5.2), above a certain minimum value.

If the value of $t_{90°}$ is less than $(4\Delta)^{-1}$, where Δ is the spectral width of interest, β is the order of 30° at Δ Hz, and intensity errors are less than 2% [5]. Intensity errors of this magnitude are comparable with errors resulting from many other sources within the experiment. The effect of finite pulse power across a 10 kHz spectral width is shown in Fig. 5.3. From this illustration it can be seen that

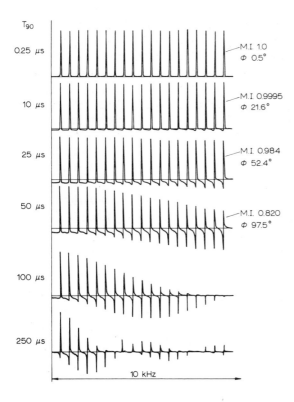

Fig. 5.3. The effect of finite pulse power on the phase and amplitude as a function of offset from 0 to 10 kHz. From P. Meakin and J.P. Jesson, J. Magn. Res., 10 (1973) 296.

the major effect on the spectrum is not seen in intensity changes but in phase changes. Fortunately these phase errors are an almost linear function of offset frequency and can be easily removed during the phase correction which must be carried out anyway to compensate for other instrument effects such as filters etc.

The phase and intensity distribution of signals resulting from a finite pulse can also be visualised by transforming into the frequency domain. If the pulse were indeed an impulse then, as can be seen in Table 3.1A, the B_1 exciting field would be uniform (the strong field case again). However, if the pulse were to have a finite duration and start at zero time, i.e. Table 3.1G and not F, then B_1 would have two components, one in phase, i.e. along x', whose amplitude would be

$\sin 2\pi\vartheta t_p/2\pi\vartheta t_p$, and one out of phase whose amplitude would be $(\sin \pi\vartheta t_p)^2/\pi\vartheta t_p$. At any offset frequency the signal produced corresponds to the magnitude of the various components of B_1 at that frequency. For example, at an offset of $\frac{1}{2} t_p$ the x' component of B_1 is zero and the y' component is $0.8 B_1$; the observed signal is therefore purely dispersive and reduced in amplitude. This model predicts behaviour shown in Fig. 5.3.

5.4 THE FREE INDUCTION DECAY

We have previously mentioned the free induction decay merely as an intermediate, whose Fourier transform gives as the desired n.m.r. spectrum. Considerable information can be gained from a detailed inspection of f.i.d. itself, and furthermore an understanding of the f.i.d. will be useful when considering practical and instrumental aspects of pulsed n.m.r.

First consider the case of a single line: at the end of a perfect $90°$ pulse $\underline{M}_0^+$ is along the y' axis. This magnetisation decays exponentially with a time constant T_2. The f.i.d. is simply $\underline{M}_0^+ \exp -t/T_2$ which on Fourier transformation (see Table 3.1C) yields a single Lorentzian line $1/\pi T_2$ Hz wide centred at zero frequency. Now if the single line is not on resonance the situation becomes more complex. The magnetisation vector still relaxes as $\exp -t/T_2$ but at the same time it precesses about z with its offset frequency ϑ_i, the tip of the vector, describing a decaying 'cork screw'. This situation is illustrated in Fig. 5.4. N.m.r. spectrometers normally only detect the y' components and from this viewpoint the f.i.d. appears as the projection of this complex motion onto the $y't$ plane. This projection is simply $M_0^+ \exp(-t/T_2) \cos 2\pi\vartheta_i t$. The signal detected along the x' axis is $M_0^+ \exp(-t/T_2) \sin 2\pi\vartheta_i t$. Examination of the table of Fourier transforms shows that the y' signal yields a Lorentzian line offset ϑ_i from zero and the x' signal the corresponding dispersion signal.

Now consider the f.i.d. which results from an 'imperfect pulse'. Such a pulse is too weak to rotate the signal purely about the x' axis, and the magnetisation after the pulse is located somewhere in the $x'y'$ plane. The magnetisation vector again precesses and relaxes; however, the x' and y' signals are now no longer purely sine and cosine functions and the resulting signals are mixtures of both phases, as are the lines after transformation. The spectrum has been subjected to a phase shift. A similar effect is produced if, for various

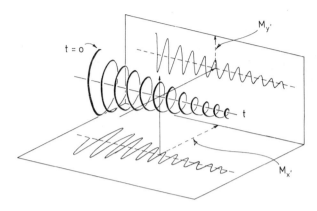

Fig. 5.4. The decay and precession of the nuclear magnetisation following a pulse showing the origin of the two components.

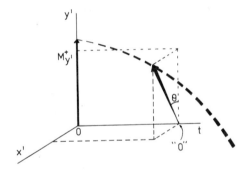

Fig. 5.5. Phase shifts resulting from the zero point for data collection being displaced from the true zero. The first data point is not the pure $M_{y'}^+$ that the calculation assumes it is.

instrumental reasons, we do not start examining the f.i.d. until sometime after it has started. This case is illustrated in Fig. 5.5.

Consideration of these effects points out the basic principle behind methods used to correct phase errors encountered in Fourier spectra. These errors, as already stated, can be thought of as arising because the experiment has not been started at zero time. If, for example, the motion of the magnetisation vectors in the case of the imperfect pulse were to be projected back in time at some point δ s, the vector would lie exactly on the y' axis and a pure absorption

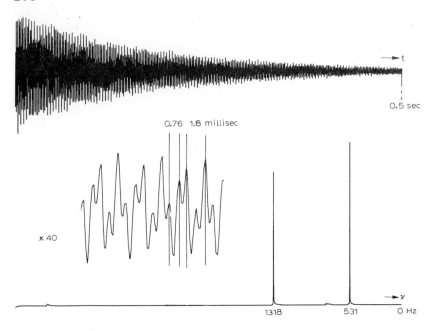

Fig. 5.6. The ^{13}C f.i.d. from ethanol. The inset shows a section of the f.i.d. magnified 40 times; the two repeating separations correspond to the offset frequencies of the two lines.

signal would result on transformation. The effects of an imperfect pulse and delaying the start of data acquisition are the inverse of each other. We will return to the practical details of this technique in chapter 6.

If the spectrum consists of more than one line, as is nearly always the case, the free induction decay consists of the sum of the decays of all individual lines Initially all these lines are in phase, but as they relax they get out of phase with each other. At a time later equal to ϑ^{-1} they are in phase again; they reinforce each other and a 'beat' is seen. Such effects can easily be seen for a simple case, e.g. the f.i.d. of ethanol shown in Fig. 5.6. Here beats due to the two frequencies can be identified and the appropriate frequency differences calculated.

A further aspect of the f.i.d. that should be considered is its amplitude at $t = 0$. It is a general rule of Fourier transforms that the value of a function at the origin in one domain is proportional to the area of that function in the co-domain. It follows therefore, and is illustrated in Table 3.1, that the area of an n.m.r. peak is equal to $M_0^+/2$ and the peak height is $M_0^+/2T_2$. This statement implies that M^+ occurs at zero time; however, as seen in the previous paragraph, the point considered to be the origin in the time domain is related to the signal phase. In other words, we only get a true value from the integral if we correctly locate the origin within the time domain, i.e. set the spectrum phase correctly. This condition is true also when working in the frequency domain (with a c.w. spectrometer) where it is necessary to set the spectrometer's phase correctly to produce a true absorption signal to result in a true integral. This point can be confirmed by reference to Fig. 2.9, noting that the dispersion mode signal has zero integral.

The realisation that the area of an n.m.r. signal obtained by Fourier transformation depends only on the value to f.i.d. at zero time has important consequences when considering a real experiment. Such an experiment involves digitisation of the time domain signal, as consequently there are limitations imposing on the magnitude and accuracy to which this value can be recorded, which in turn reflect on the accuracy of an integral. A final property of the free induction decay is its behaviour as the exciting power is increased. This, unlike the frequency domain spectrum, (section 2.7) is straight forward. The initial value for any line simply depends on the angle of rotation produced by the pulse (see eqn. 5.2), and has a maximum when this angle is $90°$ (or multiples thereof). Unlike c.w. n.m.r., increasing the value of B_1 has no effect on the line width.

5.5 THE GAIN IN SENSITIVITY USING PULSED EXCITATION

It was shown previously that as one excites all the nuclear spins in a spectrum simultaneously, an improvement in sensitivity per unit time in the order of (spectral width/typical line width)$^{1/2}$ would be achieved by the Fellgett principle. In this section we will briefly consider the pulsed experiment compared with the swept experiment from the view point of sensitivity and return to the question of optimisation of a pulse spectrometer in more detail later.

The response of the system $M_{y'}(t)$ is assumed to be accompanied by white random noise $s(t)$, originating from noise sources within the spectrometer. The sensitivity is measured by the so called signal to noise ratio S/N, defined as the ratio of the signal peak height to the root mean square value of the noise. The maximum achievable signal to noise ratio, assuming the use of a matched filter, is given by the ratio of the total signal energy to the power spectra density per Hz of the noise, W_{ss} thus

$$(S/N)^2_{max} = \tfrac{1}{2} \int_{-\infty}^{\infty} M_{y'}(t)^2 \, dt/W_{ss} \tag{5.38}$$

The factor $\tfrac{1}{2}$ must be included to take into account in pulsed n.m.r. the equal contributions of both positive and negative noise frequencies. In swept spectroscopy this is equivalent to contributions from both the upper and lower modulation side bands. The factor may be dropped if in the c.w. experiment single side band receivers are used and in pulsed n.m.r. if quadrature detection is used [6].

For frequency sweep excitation with a sweep rate of a Hz/s the v mode signal is given by

$$M_{y'}(t) = \frac{M_0 \gamma B_1 T_2}{1 + T_1 T_2 (\gamma B_1)^2 + (2\pi a t T_2)^2} \tag{5.39}$$

If Δ is the spectral width in question and T the total time available for the experiment, then

$$a = \Delta/T \tag{5.40}$$

In order to produce 'true' spectra, i.e. showing no transient effects, $2\pi a T_1 T_2^*$ must be less than one. Now it is known that the maximum signal to noise ratio for a swept experiment occurs when $T_1 T_2 (\gamma B_1)^2 = 1$. Substituting this value into eqn. (5.39), and then substituting that into eqn. (5.38), we obtain, assuming the normal side band detection, that [1]

$$^{c.w.}(S/N)^2_{max} = \frac{M_0^2 T}{12(3)^{1/2} T_1 W_{ss}} \tag{5.41}$$

Now consider pulsed excitation. In the total experiment time T we will add together T/T_p (assumed to be a whole number) of free induction decays. The total signal energy is thus that of one pulse multiplied by $(T/T_p)^{1/2}$. The maximum signal to noise ratio is thus given by [1]

$$^{P}(S/N)^2_{max} = \frac{T}{2T_p W_{ss}} \int_0^{T_p} M_{y'}(t)^2 \, dt \tag{5.42}$$

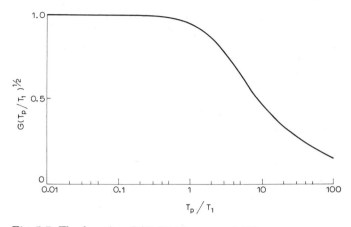

Fig. 5.7. The function $G(T_p/T_1)$ (see eqn. 5.45).

Substituting the value of $M_{y'}(t)$ from eqn. (5.17) one obtains

$$^P(S/N)^2_{max} = \frac{M_0^2 T_2 (1 - E_1) T}{4 T_p (1 + E_1) W_{ss}} \tag{5.43}$$

This itself has a maximum when $T_p \ll T_1$ where it becomes

$$^P(S/N)^2_{max} = \frac{M_0^2 T_2 T}{8 T_1 W_{ss}} \tag{5.44}$$

which is the same sensitivity attainable from stochastic excitation under the more practical limitation of using optimum r.f. power [7].

Finally, on dividing eqn. (5.44) by (5.41), we arrive at the gain in sensitivity per unit time of pulsed excitation with respect to continuous wave excitation [1].

$$\frac{^P(S/N)^2}{^{c.w.}(S/N)^2} = \frac{3(3)^{1/2}}{2} T_2 \Delta G \left(\frac{T_p}{T_1}\right) \tag{5.45}$$

where

$$G \left(\frac{T_p}{T_1}\right) = \frac{2(1 - E_1)}{(1 + E_1)} \cdot \left(\frac{T_p}{T_1}\right)^{-1} \cong 1 \text{ for } T_p \leqslant T_1$$

The function $G(T_p/T_1)$ is illustrated in Fig. 5.7. If we remember that the line width of half height is $1/\pi T_2$, then eqn. (5.45) is of the form predicted from the multichannel approach. Placing typical values in

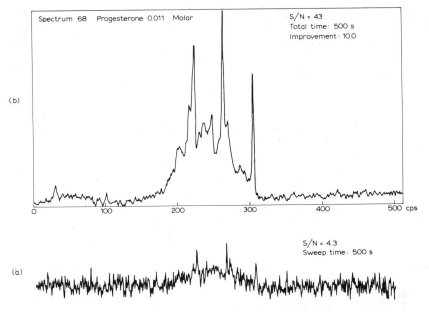

Fig. 5.8. Two spectra of $0.011\,M$ progesterone in fluorobenzene, both spectra were obtained in 500 s. (a) A single scan; (b) the spectrum produced by averaging 500 pulses 1 s apart and Fourier transforming the result. From R.R. Ernst and W.A. Anderson, Rev. Sci. Instrum., 37 (1966) 93.

eqn. (5.45) gives the gain in sensitivity achieved by the use of pulsed excitation as $10 \sim 20$ for protons and, due to the larger chemical shift range, about $60 \sim 80$ for ^{13}C. Fig. 5.8 shows the original illustration of this gain for the protons of progesterone by Ernst and Anderson in 1966 [1].

To achieve maximum sensitivity in the pulse mode requires that $G(T_p/T_1)$ be unity, which in turn require $T_p \ll T_1$. The dependence on sensitivity on the ratio of T_p to T_1 is, however, small up to $T_p = T_1$, where the sensitivity has dropped to 95% of the maximum possible. This is fortunate as the resolution achievable in a pulsed experiment is limited to $1/T_p$ Hz. If, in order to achieve increased sensitivity by pulsed excitation, T_p had to be made much less than T_1, the achievable resolution would be unacceptable for high resolution work.

In order to proceed further with this discussion we must broaden our concept of T_2. Previously it had been assumed that a free induction decay did in fact decay as the spin—spin relaxation time.

This is only true in a perfect magnet and in the absence of relaxation processes induced by the use of noise decouplers normally used in ^{13}C n.m.r. We therefore define an effective T_2^* as ($\pi \times$ line width at half height)$^{-1}$, which represents the sum of all the contributions to the rate of decay of the transverse magnetisation thus

$$1/T_2^* = 1/T_2 + 1/T_2' + 1/T_2^s \qquad (5.46)$$

where T_2' and T_2^s represent the contributions of magnet inhomogeneity and noise decouplers respectively (see section 9.8).

The achievable signal to noise ratio for both the time and frequency domain experiments depends on the ratio of T_2^*/T_1, and as T_2 cannot exceed T_1, both have a maximum efficiency when T_1 is equal to T_2. Deviation from the ideal conditions of $T_1 = T_2 = T_2^*$ have differing practical effects on the two techniques. For pulsed excitation the efficiency decreases as $(T_2^*/T_2)^{1/2}$. In the case of the sweep experiment a decrease in T_2^* can be used to advantage. As T_2^* decreases, the sweep rate may be increased without violating the limitation necessary to prevent transient effects. The loss in signal to noise ratio per unit time inherent in departure from the ideal case can thus be compensated for by speeding up the experiment. Much use was made of the approach in the early days of ^{13}C n.m.r. Under the conditions used to record these spectra T_2^* was very much shorter than T_1 and rapid scan rates (typically $100\,Hz/s$) would be used without noticeable signal distortion. This should be borne in mind when considering the gain in signal to noise achieved in practice by the use of pulse techniques. Putting typical values into eqn. (5.45) ($\Delta = 5000\,Hz$, $T_2 = 1\,s$) a gain of 70 or 80 is to be expected, whereas only factors of about 20 are achieved. The discrepancy arises because eqn. (5.45) gives the gain in signal to noise ratio per unit time when comparing pulsed excitation with a slow passage experiment. Typical c.w. ^{13}C spectra were anything but slow passage experiments.

In pulsed n.m.r. a shortening of T_2^* results in a lowering in sensitivity which cannot be as easily compensated for. The problem is illustrated in Fig. 5.9. The excitation pulse (assumed to be 90° for simplicity) rotates the magnetisation onto the y' axis (Fig. 5.9a). This magnetisation loses phase coherence as T_2^*. As long as there is magnetisation along the y' axis then a signal can be detected (Fig. 5.9b). Once a time longer than about $3T_2^*$ has elapsed no detectable signal remains (Fig. 5.9c) and the next pulse should be initiated. However, the magnetisation returns to its equilibrium value along z

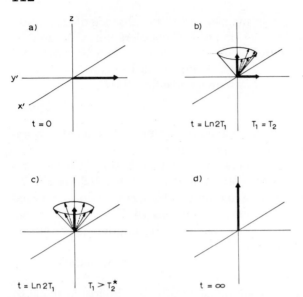

Fig. 5.9. Longitudinal and transverse relaxation following a pulse.

with a rate constant T_1. If T_1 is longer than T_2^*, as it generally is, then the result of initiating the second pulse at say $3T_2^*$ will result in smaller initial signal than after the first pulse. In order to utilize all the magnetisation for the second pulse a time of $3{-}5T_1$ must elapse between pulses; however, the detectable signal decays in $3{-}5T_2^*$ and so a compromise is necessary. These limitations also apply to c.w. spectroscopy but only if a line is scanned for a second time with $3{-}5T_1$ (a much less likely situation than in pulsed n.m.r.).

5.6 MULTIPULSE EXCITATION

(a) The liquid state

So far it has been assumed that all the pulses are of equal duration and equally spaced. From the earliest days of pulsed n.m.r. it has been realised that many interesting experiments can be performed with a repeating sequence of two or more different pulses. Multipulse techniques form the basis for relaxation time measurement (see chapter 10). They can also be used to increase the sensitivity of pulsed excitation.

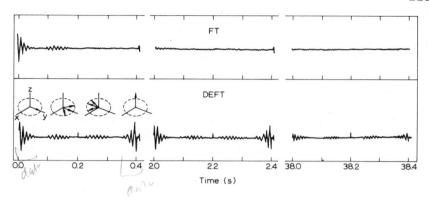

Fig. 5.10. Comparison of the signals obtained by repetitive use of 90° pulses and the d.e.f.t. sequence for the ^{13}C resonance of $CH_3 \, ^{13}COOH$. Each experiment consists of 20 repetitions, the 1st, 2nd and 20th of which are shown. From E.D. Beaker, J.A. Ferretti and T.C. Farrar, J. Amer. Chem. Soc., 91 (1969) 7784.

During the simple pulsed excitation the signal is lost due to decay processes caused by magnet inhomogeneity. If the sample is subjected to a 90° pulse and then τ s later a 180° pulse a spin echo appears, a second 180° pulse τ s later produces a further echo, etc. [8]. Details of the formation of these echoes is given in section 10.13a. The echoes can be considered as two back-to-back f.i.d.'s. Each echo has decreased with respect to the previous one by exp $(-\tau/T_2)$. If the echoes are correctly digitised and transformed they can in theory produce normal high resolution spectra. Since the signal lost due to inhomogeneity dephasing is constantly refocussed, a higher signal to noise ratio should result. This technique was called s.e.f.t. (spin echo Fourier transform) [9]. For reasons concerning pulse timing and spectrometer stability s.e.f.t. has never been successfully applied to high resolution spectroscopy [10].

A more successful approach to the recovery of the dephased magnetisation is taken in d.e.f.t. (driven equilibrium Fourier transformation) [11]. Here at the time of the first echo a 90° pulse is applied. This pulse restores the remaining transverse magnetisation to the z axis ready for the next sequence. D.e.f.t. is illustrated in Fig. 5.10 and can be written as:

$$[90° - T_A - 180° - T_A - 90° - T_d]$$

The time T_A between the pulses is fixed by the spectral width (and sampling theory, see section 6.10); for the case where T_1 equals T_2 optimum sensitivity is achieved when T_d is $4T_A$ [12]. Data is collected, as indicated, after the first $90°$ pulse; it may also be accumulated after the $180°$ pulse if the memory advance ramp is inverted as the signal is now increasing with time. When both sets of data are collected the experiment is sometimes called 'super d.e.f.t.'.

Initially there were hopes that d.e.f.t. could achieve sensitivities up to 20 times that of conventional pulsed excitation for ^{13}C, where T_1's are in general longish and homonuclear spin decoupling absent [11]. Such gains have never been realised in practice. Multi-pulsed experiments based on the Carr—Purcell sequence only refocus magnetisation lost due to magnet inhomogeneity effects; it follows therefore that d.e.f.t. will only be superior to normal f.t. when such losses are comparable to, or exceed the rate of, natural and non-refocussable relaxation processes, i.e. $(T_2')^{-1} < (T_2)^{-1} + (T_2^s)^{-1}$. initially T_2 was thought to be much longer ($\sim 20\,s$) than in practice it has been found to be ($1-10\,s$ for medium sized molecules) and the consequences of using proton noise decouplers (T_2^s) was not fully appreciated. It is not surprising therefore that when more detailed evaluations were carried out the gains were found to be nearer 2 than 20 [13].

Considering the added demands placed on a spectrometer, particularly with respect to pulsed power and accuracy, to perform d.e.f.t. successfully, its value for routine ^{13}C work is very low [13]. d.e.f.t. however has the advantage that it can be optimised without a knowledge of the T_1's involved. It is also useful where really long T_1's are involved e.g. ^{15}N and if either magnet homogeneity and/or computer memory limits resolution to $5\,Hz$ or below.

(b) The solid state

As was mentioned in section 2.6, high resolution spectra are normally only obtained from samples in the liquid and gaseous phase, where rapid molecular motion averages to zero the effects of direct dipole/dipole interaction. In the solid state such averaging does not occur and very broad lines result. The Hamiltonian for the dipolar interaction is

$$\mathcal{H}_D = \sum_{i<j} \sum 2r_{ij}^{-3} (1 - 3\cos^2\theta_{ij}) [\mu_i\mu_j - 3(\mu_z)_i(\mu_z)_j] \qquad (5.47)$$

where r_{ij} is the vector joining the nuclei i and j which is at an angle θ to B_0. This interaction will be zero under two conditions: (a) when θ is averaged over all values, as is the case in the liquid and gaseous phases and (b) when $\theta = 54°44'$ when $\cos^2 \theta = \frac{1}{3}$. The second situation can be successfully realised in the solid by mechanically spinning the sample very rapidly ($\sim 10\,\mathrm{kHz}$) about an axis which is itself at $54°44'$ with respect to the static magnetic field. The experimental problems are however extreme [14].

A second method is to apply a series of pulses so that in the rotating frame the nuclear spins precess about the 'magic angle' [15]. One such sequence is

$$[90°_{x'}, 2\tau, 90°_{-x'}, \tau, 90°_{y'}, 2\tau, 90°_{-y'}, \tau]$$

The first pulse rotates M_0 from its equilibrium position along the z axis to the y$'$ axis where it stays for 2τ. It is then returned to the z axis and τ s later placed along the y$'$ axis and finally returned to the z axis. The net magnetisation has thus spent equal periods of time along each of the axes and $\mathcal{H}_D$ is averaged to zero. The pulses must be short compared with τ and the whole sequence fast compared with T_2 i.e. $6\tau \ll T_2$. Since in solids T_2 is typically 10^{-4} s pulse widths of less than a microsecond must be used. The signal is detected during an appropriate part of the cycle, i.e. when M is along the x$'$ or y$'$ axes. The spectrum is then obtained by Fourier transforming the signal observed in successive cycles.

Scalar couplings are not affected by such a pulse sequence. Chemical shifts, however, are. M_0 behaves as if it is inclined at the "magnetic angle" to B_0; chemical shifts are reduced to the value of the projection of M_0 onto the z axis i.e. by a factor of $\cos 54°44' = 1/\sqrt{3}$. Any anisotropy in the chemical shift tensor, which is normally averaged to zero by molecular tumbling in the liquid state is also reduced by $1/\sqrt{3}$. The alternative technique of rapid sample rotation about the magic angle however does average chemical shift anisotropy effects. For simple molecules, where the more complex spectra resulting from presence of chemical shift anisotropy can be interpreted, considerable extra information can be obtained. However for large molecules the spectra are in general too complex to be analysed.

5.7 SATURATION EFFECTS IN PULSED N.M.R. [16]

As discussed earlier the area of an n.m.r. signal is proportional to the excitation power and the number of corresponding nuclei, provided saturation does not occur. Saturation is reached when the mean energy applied to the spin system is greater than the energy which can be dissipated into the lattice by relaxation effects. The populations in the ground and excited states becomes equal and the n.m.r. signal disappears. Saturating is an important effect both from the sensitivity view-point and also when, as is often the case, relative areas are of importance. The effects of the power used in a slow sweep experiment are illustrated in Fig. 2.10, each line being excited independently. In pulsed n.m.r. if a high mean power is applied, all lines in the spectrum are excited equally and simultaneously, but they are not necessarily saturated equally. This causes a redistribution of the spin populations among the different energy levels. and it is not immediately obvious how the relative intensities of the lines will be affected. The effects of any spin echoes which may be built up will not be considered.

For spectra which exhibit no spin—spin coupling there are no connected transitions, and therefore the saturation behaviour of the signals are the same whether c.w. or pulsed excitation is used. The most important case of this type is ^{13}C n.m.r. with the use of proton noise decoupling. For spectra where there is homonuclear coupling the problem is not quite as simple, and a density matrix theory is necessary in order to allow for all the possible relaxation processes. The full theory will not be given (see ref. 16), only the general conclusions, which, to a first approximation, also apply to stochastic excitation [7]; however, in order to explain saturation effects fully in this experiment, it may be necessary to include the higher terms of eqn. (4.23/24). More complex effects resulting in lines 'breaking up' occur with high powers [17].

The intensity of a peak in pulsed n.m.r. depends on the pulse interval and angle. The dependence on both these parameters is periodic and neither parameter gives an effect parallel to the saturation parameter $\gamma B_1 (T_1 T_2)^{1/2}$ used in swept n.m.r. Saturation is best studied using a constant α and (T_1/T_p) as the saturation parameter. Figure 5.11 shows saturation curves of an uncoupled nuclear spin with $T_1 = T_2$ for five different α values; also shown is the slow passage saturation curve. The similarity between the two types of saturation behaviour is obvious, and the parameter in pulsed saturation experiments which corresponds to power in swept experiments is $1/T_p$. Similar relationships exist when $T_1 \neq T_2$. If one uses $1/T_p$ as a saturation parameter, the line width also changes in a

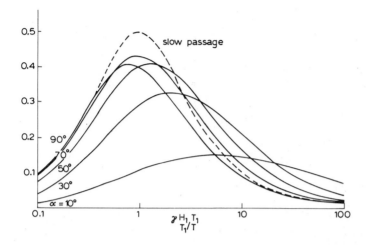

Fig. 5.11. Saturation curves in a pulse experiment for five different pulse angles. The peak height accumulated/unit time is shown as a function of T_1/T_p. It is assumed that $T_1 = T_2$ and that any residual transverse magnetisation is destroyed before the next pulse. The saturation curve for the slow passage experiment is also given. From R.R. Ernst and R.E. Morgan, Mol. Phys., 26 (1973) 49.

parallel manner in the two experiments, since the resolution of the experiment is also dependent on $1/T_A$. This in turn depends on T_p ($T_p = T_a + T_\alpha$). However, if a fixed acquisition time is used, there is no effect on line width as a function of power expressed in terms of the pulse angle. The above results only apply when echoes have been eliminated, as explained in section 5.2; under these conditions intensities of comparable accuracy, and hence integrals, can be obtained by both techniques, provided saturation is avoided. Pulsed excitation is naturally much more sensitive.

The saturation behaviour of strongly coupled spin systems is normally more complex than for uncoupled spins, especially when using swept rather than pulsed excitation. In the latter case, an initial state described by a Boltzman distribution is transformed into a new state effectively instantaneously and independent of the different relaxation rates of each energy level. In the slow passage experiment each line is selectively saturated, and its area therefore depends strongly on its particular relaxation ratio. Thus in a closely coupled spin system (e.g. a proton spectrum) where the spin lattice relaxation time of a multiplet is described by a series of coupled equations, the

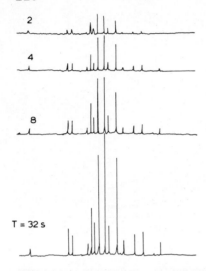

Fig. 5.12. Fourier spectra of a near degassed acrylonitrite for four different pulse spacings. The pulse angle was 90°. The single line on the left hand side is an impurity. From R.R. Ernst and R.E. Morgan, Mol. Phys., 26 (1973) 49.

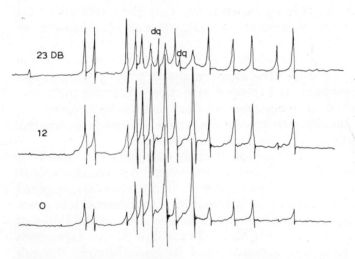

Fig. 5.13. Slow passage spectra of near degassed acrylonitrite for 3 r.f. field strengths, indicated in dB's relative to an arbitrary zero. Double quantum transitions are shown dq. From R.R. Ernst and R.E. Morgan, Mol. Phys., 26 (1973) 49.

relative signal intensities are less sensitive to saturation effects using pulsed excitation than swept excitation. For example, whenever external dipolar relaxation dominates and effects all nuclei equally, even strongly coupled spin systems saturate homogeneously with pulsed excitation, whereas they do not if swept excitation is used. These points are illustrated in Figs. 5.12 and 5.13 which show the saturation behaviour of the ABC spin system of acrylonitrile under both pulsed and swept excitation. The slow passage spectra show strong relative intensity variations and double quantum transitions, whereas the pulsed spectra show only slight intensity changes which can be attributed to contributions from intra-molecular dipolar relaxation. Intra-molecular effects are enhanced when the sample is diluted in an inert solvent. External effects are only attenuated if the sample is degassed, otherwise inter-molecular relaxation via oxygen is a significant source of relaxation. Pulsed excitation is therefore shown to be particularly suited to obtaining accurate intensity ratios in strongly coupled systems, and to be equal to other excitation techniques in the simpler cases.

REFERENCES

1 R.R. Ernst and W.A. Anderson, Rev. Sci. Instrum., 37, 93 (1966).
2 R. Freeman and H.D.W. Hill, J. Magn. Res., 4, (1971) 366.
3 E.L. Hahn, Phys. Rev., 80 (1950) 580.
4 D.E. Jones, J. Magn. Res., 6, (1972) 183.
5 P. Meakin and J.P. Jeason, J. Magn. Res., 10, (1973) 296.
6 J.D. Ellett, M.G. Gibby, U. Haeberlen, L.M. Huker, M. Mehring, A. Pines and J.S. Waugh, Advan. Magn. Res., 5, (1971) 117.
7 R.R. Ernst, J. Magn. Res., 3, (1970) 10.
8 H.Y. Carr and E.M. Purcell, Phys. Rev., 94, (1954) 630.
9 A. Allerhand and D.C. Cochran, J. Amer. Chem. Soc., 92, (1970) 4482.
10 A. Allerhand, Rev. Sci. Instrum., 41, (1970) 269.
11 E.D. Becker, J.A. Ferretti and T.C. Farrar, J. Amer. Chem. Soc., 91, (1969) 7784.
12 J.S. Waugh, J. Mol. Spectrosc., 35, (1970) 258.
13 R.R. Shoup, E.D. Becker and T.C. Farrar, J. Magn. Res. 8, (1972) 290.
14 E.R. Andrew, Prog. NMR Spectrosc., 8, (1971) 1.
15 U. Haeberlen and J.S. Waugh, Phys. Rev., 175, (1960) 453.
16 R.R. Ernst and R.E. Morgan, Mol. Phys., 26, (1973) 49.
17 H.D.W. Hill, private communication.

CHAPTER 6

INSTRUMENTATION

6.1 INTRODUCTION

The purpose of this chapter is to discuss the instrumentation required to perform pulsed high resolution n.m.r. experiments. The term high resolution is used to distinguish the spectrometers discussed in this section from the 'conventional' pulsed (wideline) spectrometers. Instruments of this type have been around since the early days of n.m.r. Their applications are mainly concerned with the measurement of relaxation times and they work in the solid state. They are characterised by the use of magnets with relatively low resolution. We will be discussing high resolution instruments capable of producing spectra containing coupling constant and chemical shift information, but using a pulsed mode of excitation, unlike the conventional swept experiments. The requirements of these instruments are not only different concerning the magnet, but they also have different criteria placed on the pulse power and parameters of this type. We will, where appropriate, contrast the performance of the two types of spectrometers, but as details of the 'pure' pulsed spectrometer are available elsewhere [1] we will not discuss the requirements of these systems in detail.

The high resolution pulsed spectrometer is shown diagrammatically in Fig. 6.1; it can be basically divided into three parts. The three parts of the systems are: (a) the spectrometer itself containing the magnet, probe, r.f. circuitry etc., (b) the data system containing the computer interface, the central processing unit itself, plus any peripheral devices which are attached to it, and (c) the software which controls the computer. As the spectrometer and the associated computer hardware become more and more of an integrated unit, the software associated with the computer becomes an increasingly integrated part of the n.m.r. spectrometer, and thus deserves discussion in this chapter on instrumentation.

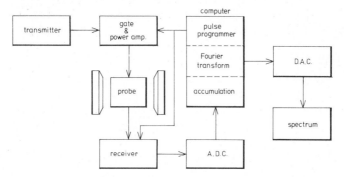

Fig. 6.1. Block diagram of a typical high resolution pulse spectrometer.

6.2 THE MAGNET

The heart of any magnetic resonance spectrometer is its magnet. This is especially true of high resolution spectrometers, which, as the name suggests, are designed to run at the highest possible resolution. There are three types of magnets used in high resolution work. They are the permanent, electro- and super-conducting (cryogenic) magnets. Each fulfils the basic requirement of providing a volume of highly homogeneous field into which the sample can be placed in order to perform an n.m.r. experiment, but they have different strengths and weaknesses and are accordingly used in different types of high resolution spectrometer.

The permanent magnet was originally used by Prof. Purcell's group and has been used in high resolution n.m.r. since its conception [2]. The initial advantage of a permanent magnet was that with a high degree of thermal stabilisation it produced a very stable magnetic field and hence it permitted the use of pre-calibrated charts without the need for any field frequency lock. Parallel with this stability in absolute field is the stability in homogeneity. The limitation of a permanent magnet is the rapid increase in size which occurs if the magnet is pushed to fields greater than about 1.4 Tesla or if a large gap is attempted, e.g. for the use of sample tubes over 5 mm. The modern application of the permanent magnet is the low cost, routine, multiple access type of system where its inherent stability and low running costs are invaluable assets.

The electro-magnet was the other original type of magnet system used in high resolution work, and is now probably the most common.

With an electro-magnetic, power has to be continuously applied from a stable power supply and excess heat removed by water cooling. The magnet is certainly not as simple or compact as a permanent magnet. Neither has it got very high inherent stability of either field or resolution. The advantages of the electro-magnet are its flexibility, being able to produce fields with a large volume of homogeneity. 2.3 Tesla (100 Hz) is about the upper field limit of an electro-magnet for n.m.r. work on protons. Above this field magnetic saturation of the iron core occurs and the power required to increase the field further is prohibitive.

The third source of a magnetic field used in n.m.r. is that of a superconducting magnet. Such a magnet consists of a solenoid, made of superconducting wire in a dewar containing liquid helium. Within the coil a large electric current circulates continuously. As long as the magnet stays at liquid helium temperatures the system is essentially a permanent magnet, and, as such, has the advantages of stability in both field and resolution outlined previously. However, the main advantage of the superconducting magnet is its ability to operate at very high fields. Originally, superconducting magnets were operated at a field which corresponds to 220 MHz for protons [3], but this field is continually being raised as technology improves. At the present moment the highest achievable field is 360 MHz. These high fields have the advantages of increasing the sensitivity of the basic spectrometer, and also increasing the chemical shift.

The main application of superconducting magnets is in proton studies of biological systems where both advantages are exploited to the full. They do, however, have the disadvantage of being expensive to manufacture.

None of the magnets described above is capable of producing an intrinsic field which is homogeneous enough for a high resolution spectrometer, and to achieve an improvement in this field homogeneity two techniques are employed. Firstly, the sample is spun in the probe about one axis in order to average out inhomogeneities in the perpendicular plane [4]. From this point of view there is an interesting distinction between superconducting magnets and the iron core magnets. In the latter the axis of rotation, which is conventionally termed the y axis, is at right angles to the basic magnetic field. However, in the superconducting magnet, which uses a solenoid design, the angle of rotation can only be parallel to the basic magnetic field which is the z axis. This distinction is of practical importance when considering liquid crystal studies (see section 8.7g). In order to achieve adequate averaging of the magnetic inhomogeneities the sample has to be rotated sufficiently fast that the following inequality is met:

Spinning rate (rps) $\geqslant$ γ (range of B_0 values encountered during one revolution)

From the above inequality you can see that the more inhomogeneous the base field, the faster the sample has to be rotated in order to average the inhomogeneities away. It is advantageous to keep this speed of rotation as slow as possible in order to minimise complications which can occur due to vortex formation in the top of the rotating sample. If this occurs it will itself introduce inhomogeneities which are probably more serious than those that the sample rotation is trying to eliminate. Problems associated with vortexing are more severe when handling large sample tubes, because not only will such tubes show a vortex at lower rotation speeds, but also the inhomogeneities seen in a revolution will, in general, be larger than those encountered when using a small sample tube. Rotation of the sample in an inhomogeneous magnetic field creates a modulation which produces 'spinning' side bands. These appear on each side of the peak separated from that peak by the rotation frequency. Typically such spinning side bands represent about 1% of the area of the centre band and their magnitude decreases as the speed of rotation is increased.

The second technique by which the homogeneity of the basic magnet can be improved is by the use of 'shim coils' [5]. These are a series of coils wound either on the probe or into the pole caps of the magnet, which generate small local field gradients. The principle is that by applying field gradients of equal magnitude and opposite sense to the residual field gradients in the basic magnetic field, the resolution of the parent magnet can be improved. Typical shim coils are illustrated in Fig. 6.2 and form the terms of a power series. All the first order terms, i.e. the x, y and z, are present, as are, usually, the second order terms involving the y axis. Higher order terms in y, the non-averaged gradient, are usually also present. The shim coils are adjusted to give the highest possible homogeneity by monitoring the height of a single peak, usually the lock signal: as the area of a signal is constant, the best homogeneity, i.e. minimum line width, corresponds to maximum peak height.

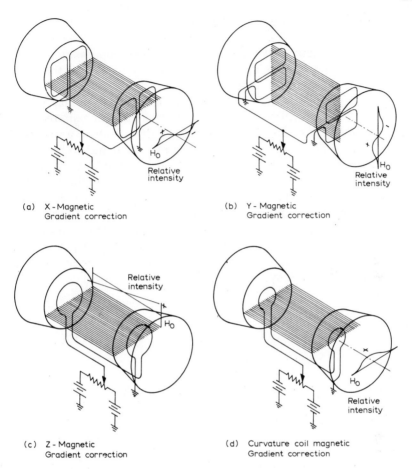

(a) X - Magnetic
Gradient correction

(b) Y - Magnetic
Gradient correction

(c) Z - Magnetic
Gradient correction

(d) Curvature coil magnetic
Gradient correction

Fig. 6.2. Field gradient correcting fields produced by shim coils, (a) x-gradient, (b) y-gradient, (c) z-gradient, (d) curvature (courtesy of Varian Associates).

In some experiments it is necessary to destroy the transverse magnetisation of the sample rapidly; such a case would be the elimination of any residual magnetisation after an imperfect $180°$ pulse. Since dephasing of the $x'y'$ signal occurs in an inhomogeneous field the desired result can be achieved by making the magnet field inhomogeneous. The inhomogeneity, which is only necessary for a few milliseconds, is generated by applying a large voltage to one of the shim coils (usually the z). This is termed a homogeneity spoiling pulse (h.s.p.).

6.3 FIELD/FREQUENCY LOCK

To obtain a spectrum it may at first sight seem necessary to have both the field and the exciting frequency stable to at least the same order as the interaction that is being measured. For most high resolution experiments this would require individual stability in the order of 0.01 ppm, an almost unobtainable goal, as can be appreciated if it is expressed in terms of time, where it corresponds to a variation of 0.3 s/year! Fortunately, it is only the ratio that has to be stable to this degree, the absolute value of either need only be stable to 0.01%. It is fairly easy to generate frequencies with the required relative stability by the techniques described in section 6.5. The desired experimental conditions can thus be achieved by using one of these frequencies to control, or lock, the magnet while another is used for excitation [6]. This control can be realised by performing a second n.m.r. experiment, as, for a specific resonance, field is proportional to frequency.

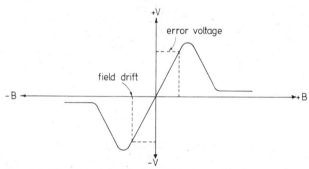

Fig. 6.3. The principle of a field/frequency lock.

The principle of the field/frequency lock is illustrated in Fig. 6.3. The receiver's phase sensitive detector is set to detect the dispersion mode signal. When measuring this signal, zero voltage is detected at the precise resonance condition of the reference compound. If there is a departure from this field/frequency ratio, either a positive or negative voltage will be detected depending on the sense of the change. The voltage after suitable amplification is converted into a correcting signal to complete a classical feed back loop. Where the signal is applied depends on the design of the spectrometer; it is usually added to the magnetic field, either directly, as in the case of a permanent magnet, or to the flux stabilizer when an electro-magnet is used. The reference signal can either be derived from resonances within the analytical sample, an 'internal lock', or from a small external sample situated as close as possible to the analytical sample [7]. External locks usually use a sample containing a strong proton or fluorine signal. They are very convenient as they function at all times, i.e. they do not switch off when the analytical sample is removed, nor do they place any chemical demands on it. Their draw-back is that they control the field, not at the sample under investigation, but 10 mm or so away. Even this small distance leads to a field/frequency drift over a period of time.

The internal lock does not suffer from this disadvantage, but it does necessitate a suitable line in the sample. Such a reference line can either come from a nucleus of the same type as that under investigation, in which case it is called a homo lock, or it can come from different nuclear species when it is called a hetero lock. The original type of internal lock was homonuclear as it was instrumentally the easiest. The two frequencies required for this type of lock were audio modulation side bands of a single basic radio frequency; their relative stability thus depended only on the audio oscillators used (and to achieve a stability of 0.1 Hz in an audio oscillator ($\sim$ 5 kHz) is straight forward.) Furthermore the probe only needed to be tuned to a single radio frequency. Heteronuclear locking on the other hand, requires the generation of two stable radio frequencies and the probe to be double tuned to accept both: a much more demanding task.

When using pulsed excitation, homonuclear locking becomes difficult but not impossible. The difficulties arise from two sources, firstly, a line intense enough to give an adequate lock signal, and to do so it must have a good signal to noise ratio in a single pulse, can cause both dynamic range and purity problems in a dilute sample. Secondly, the lock is, by nature of the excitation, intermittent, which leads to considerable difficulties when long acquisition or delay times are involved. For these reasons, whereas internal homo locks are still used extensively in c.w. proton spectrometers (where TMS at about 0.5% provides a convenient lock) they are hardly ever used in pulsed spectrometers.

The most commonly used locking system for pulsed spectrometers is the internal deuterium lock using time share modulation. Deuterium is not chosen because it is a very sensitive nucleus (in fact just the reverse is true (see Table 2.1) but because it is present to a very high concentration in the deuterated solvents commonly used in n.m.r. The high concentration compensates for the low sensitivity. Using deuterium as a locking nucleus permits noise decoupling of protons to be carried out while studying nuclei like ^{13}C. Such an experiment would not, of course, be possible if proton hetero locking (perhaps, at first sight, the obvious choice by reason of its sensitivity) were to be used.

6.4 THE PROBE

The probe is the key component of the spectrometer. Its functions are to contain the sample within the magnet gap, provide the necessary hardware to permit the samples temperature to be varied and, usually in the form of an air turbine, to spin the sample. The probe must also house the coils and associated radiofrequency electronics, in order to permit the excitation and detection of an n.m.r. signal. A typical probe is shown diagramatically in Fig. 6.4.

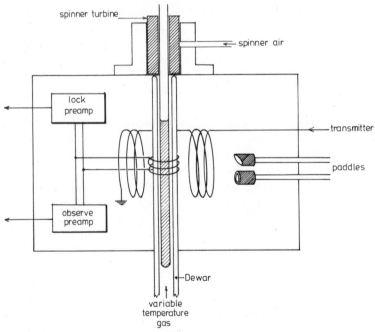

Fig. 6.4. Outline of a crossed coil probe.

Probes can be split into two general classes, depending on the geometry and number of coils used. The crossed coil probe has the transmitter and receiver coils as physically separate orthogonal coils, and the single coil probe, as its name suggests, uses a single coil for both functions.

In the crossed coil probe the transmitter coil is at right angles to the sample tube; the receiver coil is a helical coil with 2~8 turns wrapped around a glass former into which the sample tube fits. Ideally the r.f. field generated by the transmitter coil is orthogonal to the receiver coil and thus no direct coupling between the receiver coil and transmitter coil is possible. In practice such orthogonallity cannot be achieved and 'paddles' are placed in the probe, whose function is to induce leakage of an opposite sign and equal magnitude to that inherent from the basic coil arrangement [8]. In this way the probe is said to be balanced. Direct leakage between the transmitter coil and the receiver coil is very important when using the probe in a continuous wave spectrometer with field modulation where the receiver and transmitter are on simultaneously. Direct leakage, being coherent with the transmitter, appears as a d.c. bias in the spectrum. When using the probe in a pulse spectrometer the problem of leakage is not as critical, as the transmitter and receiver are never on at the same time. When a crossed coil probe is used for high resolution pulsed work, it still has paddles in it to minimise the direct leakage. The advantages of a crossed coil probe are that since the transmitter and receiver coils are separate they may be optimised independently, and the transmitter coil generates a large volume of highly homogeneous radio frequency field which is particularly advantageous when the spectrometer is used at the limits of its achievable resolution and sensitivity. R.f. homogeneity is important when measuring relaxation times. The main draw-back of the crossed coil probe is also connected with the large volume of homogeneity it produces, as a relatively large amount of power is consumed in order to generate a given effective field at the sample. As in pulsed spectrometers one is always striving to achieve the maximum effective field, this limitation is a fairly serious one.

The single coil probe uses what is the receiver coil of the crossed coil probe to perform both functions, and is by far the most common type of probe used in pulsed work, being mechanically simpler and electrically more efficient than the crossed coil probe. Such a probe is illustrated schematically in Fig. 6.5. The crossed diodes, which

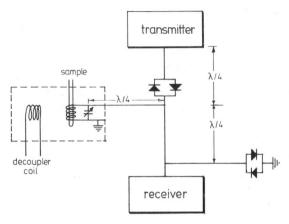

Fig. 6.5. Schematic showing the operation of a single coil probe [1].

under a high voltage effectively act as a zero resistance but under a low voltage as an infinite resistance, are used to optimise the performance of the single coil when performing both its functions, and to protect the receiver from an over-load. An excellent description of the requirements and operation of a single coil probe is given in the book by Farrar and Becker [1].

The receiver coil itself is the most critical component within the probe when considering sensitivity. Consequently a receiver coil is normally optimised for the detection of a specific nucleus or a fairly narrow range of nuclei. A separate probe is thus used for each nucleus or group of nuclei; in the case of the so called Universal probe, where the receiver coil is mounted on an interchangeable glass insert, separate inserts are used. In the latter method, one probe, with all the associated variable temperature equipment, spinner housing etc., can be utilized to cover the whole of the n.m.r. frequency range. If this approach is taken, various other key components which are frequency-specific, like the preamplifier, are also made in a modular form so that they may easily be interchanged. With improving electronic design the bandwidth of all parts of spectrometers is increasing [9]. In spectrometers which are dedicated to one specific nucleus, or, at the most, two, the method of the Universal probe is usually uneconomical and the specific probe for each nucleus is used.

As well as fulfilling the requirements for the observing channel, the probe must also contain a suitable transmitter coil for decoupling experiments. As this often has to carry relatively high voltages in order to generate the required fields, particularly for noise decoupling, and as it frequently operates at a very different frequency from the rest of the probe, it is usually a separate coil. Finally the probe must have a facility to handle a third frequency for locking. The heterolocking function is usually achieved by double tuning the main receiver coil to the lock frequency and using this as a single coil probe for the locking experiment.

6.5 FREQUENCY GENERATION

In principle the flexible n.m.r. spectrometer needs to generate three independent r.f. fields: one for observing, one for decoupling and one for locking. These three fields may be the same frequency or they may be different. In the early days of n.m.r. instrumentation the only way frequencies could be generated to the required degree of stability was to use one radio frequency oscillator and audio modulation techniques to produce side bands: an approach obviously limiting the spectrometer to a homonuclear operation with respect to both locking and decoupling [6]. In modern spectrometers, the three frequencies, even if they are only a few Hz apart, are independently synthesized from one parent frequency source, the so-called master oscillator of the spectrometer.

Frequency synthesis is usually carried out using one of two methods. The first uses phase locked loops shown in Fig. 6.6. In

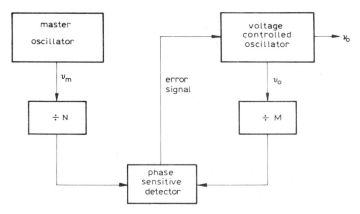

Fig. 6.6. The principle of the phase lock loop. The output frequency ϑ_0 is directly related to the master oscillator.

order to generate a frequency ϑ_0 a voltage controlled oscillator with ϑ_0 as its centre frequency is used. The output frequency of the v.c.o. is divided by a number M; the output from the master crystal is also divided only by a different number N. M and N are chosen such that after division the two frequencies are equal; these frequencies are then fed into a phase sensitive detector which produces a d.c. output proportional to the difference in phase between the two frequencies. This error signal is fed back into the v.c.o. and corrects the frequency of the v.c.o. In this way the voltage controlled oscillator is 'phase locked' to the master crystal. The process is repeated with different values of M and N for as many frequencies as are required within the spectrometer. The high frequency voltage controlled oscillator is used to produce the short term stability, whereas the long term stability of the frequency is governed by the master crystal itself. All frequencies within the spectrometer are thus governed by the master crystal and cannot drift relative to each other. As mentioned previously it is only the relative stability of these frequencies which is critical and not their absolute value. The second method uses electronic devices called mixers which either add or subtract two frequencies*. The function of a mixer is illustrated in Fig. 6.7; it is really a multiplier, the product of two sine waves being equivalent to the two frequencies corresponding to the sum and difference of the inputs. Using mixers, harmonics of the master frequency are added and subtracted to produce the desired frequenty. As all frequencies used are harmonics of the master frequency, relative stability of all the frequencies is ensured. Both methods have their strengths and weaknesses.

The methods of the synthesis outlined so far are only capable of producing a specific frequency. It will in general be necessary to

---

* There are three electronic circuits which accept two input frequencies and have as an output the sum end difference of these frequencies; the three are phase sensitive detectors, mixers and balanced modulators. The difference between them is in the relative magnitude of the input frequencies. In the phase sensitive detector the two frequencies are nearly equal, the sum is filtered out and only the difference between the frequencies appears at the output (see Fig. 3.5 and section 3.6). In a mixer both the sum and difference frequencies are produced and the unwanted frequency is filtered off. The balanced modulator is designed to work where one frequency is very small compared to the other; the low frequency is modulated onto the high frequency, and two 'side band' frequencies are produced.

Fig. 6.7. The operation of a mixer.

move this frequency or offset it to other values. This can be achieved in the following manner. An audio frequency is generated in a manner parallel to the one above using an audio frequency synthesizer working with a reference frequency which is itself derived from the master radio frequency oscillator. The audio frequency is added via the phase lock loop which generates the basic radio frequency or with the aid of a mixer. In this way the transmitter frequency can be offset within the range covered by the audio synthesizer, while still maintaining a direct relationship between the total frequency and the master clock. Frequency synthesis cannot produce a continuous range of frequencies; it is limited by the smallest locked step the audio synthesizer is capable of producing. For a pulsed spectrometer this is an acceptable limitation if the step is small enough. It is obviously not an acceptable limitation if the spectrometer has to operate in a frequency swept mode. The spectrometer must then have one free running audio oscillator whose frequency can be varied by spectrometer sweep circuity, usually in the recorder. In this experiment the sweep is normally generated by sweeping a

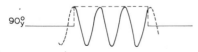

Fig. 6.8. 2 pulses 90° out of phase.

high level oscillator up to maybe 100 kHz in frequency and dividing that frequency down by a suitable number to produce the required frequency sweep. For example, non-linearities in the sweeping oscillator are minimised by the division process.

Finally a pulsed spectrometer may require two frequencies which are identical except that one is shifted in phase with respect to the other by a fixed amount, normally 90° and 180°. These frequencies are used to generate phase shifted pulses of the type shown in Fig. 6.8 which are used in the more complicated multipulsed experiments discussed elsewhere in the book. The phase shift can be achieved by routing one of the frequencies through an electronic device called a delay line which transmits a radio frequency at a different velocity to that of a simple coaxial cable. The characteristics of the delay line are chosen to delay the radio frequency being used by a time corresponding to the desired phase shift. Transformers can also be used, as can digital devices.

6.6 TRANSMITTER AND POWER AMPLIFIER

From the frequency generation module the required radio frequency is fed into the transmitter, whose function is to amplify the signal and to apply it to the transmitter coil of the probe. The transmitter can be thought of as consisting of two basic components: an r.f. switch, or gate, whose function is to switch on and off the radio frequency at the desired time, followed by a pulse amplifier which amplifies the signal from the relatively low levels produced by the frequency generation part of the spectrometer, to the fairly high values required at the probe. For electronic reasons the gate is placed before the power amplifier and consequently must be a very efficient switch, as any power that leaks through in the off position may be amplified and cause direct leakage at the probe. The power that is required depends on the specific probe characteristics and functions to which the spectrometer is being put. In general terms, for a single coil probe in the order of 50—100 W is adequate to produce 90° pulses in the order of 20 μs. However, for a crossed coil probe, four times more power is required to produce the same B_{eff}. It should be noted that the field generated is proportional to the square of the power used. Thus, to double the effective field, it is necessary to produce four times the power. If one wants to improve the effective B_1 field for a specific experiment, it is therefore much better to concentrate on improving the efficiency of the probe, where the dependence between B_{eff} is linear, not quadratic.

The characteristics of pulse amplifiers are slightly different from those used in continuous wave work. Firstly, they have to generate higher powers, but only for a very limited duration. If a pulse amplifier is run continually it can normally produce only about 10% of its 'rated' or peak power. Secondly, as the name implies, it has to amplify a pulse. As we have seen previously (see Fig. 3.1) to handle a pulse without distortion a range of frequencies outside the normal pulse frequency has to be amplified equally. A pulse amplifier must therefore, unlike a c.w. amplifier, have a large bandwidth.

The significance of these properties of a pulse amplifier can be illustrated by examining the shape of a real r.f. pulse as opposed to the 'ideal pulse.' The envelopes of an ideal and a real pulse are shown in Fig. 6.9. The ideal pulse, unlike a real pulse, goes from zero to its maximum value instantaneously, stays constant, then switches off instantaneously. A real pulse has a 'rise time' which is the time taken to reach its full amplitude. Reference to Figure 3.1 shows that the rise time will depend on the bandwidth of the amplifier; to have a sharp pulse, high frequencies must be handled. To produce a rise time in the order of a μs requires a bandwidth of a few MHz. Note that it is the bandwidth which matters, not the absolute frequency; problems of this type are therefore more severe at low frequencies

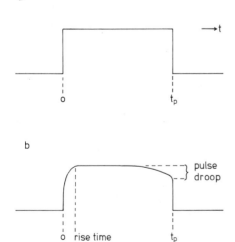

Fig. 6.9. Comparison of an ideal pulse (a) and a real pulse (b).

where the bandwidth forms a significant fraction of the center frequency. Long rise times lead to a non-uniform (non sinc ϑ) power distribution and transient effects which can cause phase errors. Pulse amplifiers usually draw their peak power from a capacitor bank which is recharged during the off period. Their power thus decays exponentially with time; the effect is that the pulse 'droops'. A correctly designed amplifier will not drop significantly during a normal pulse; if, however, it does, this can easily be detected, as a 360° pulse is more than four times a 90° pulse.

6.7 THE RECEIVER

The precession of the nuclei following excitation induces a voltage in the probe's receiver coil. The function of the receiver is to amplify the signal in a very linear manner from its initial microvolt level up to the ten volts or so required to derive the analogue to digital converter. It also must detect the signal, i.e. subtract the basic transmitter frequency leaving only audio or offset frequencies. These functions are achieved in the sequence given in Fig. 6.10.

The first part of the receiver is called the preamplifier and is situated as close to the receiver coil as possible, usually within the probe itself, see Fig. 6.4. The preamp's function is to provide the first critical amplification of the signal; its gain is usually only a factor of about ten but its signal to noise ratio is as high as possible. It is at this stage where the signal voltage is very weak that electronic noise is critical. After leaving the preamp the signal is routed into the main amplifier which is usually in the main spectrometer console. As

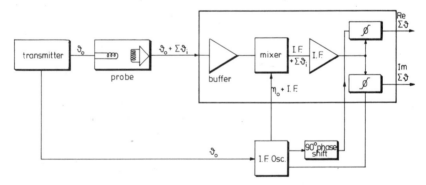

Fig. 6.10. The r.f. section of a Fourier spectrometer using quadrature detection.

the preamplifier is designed to work on the microvolt level it must be protected from the very high voltage present during the actual pulse. This protection is afforded either by switching the preamp off during the pulse or by the use of crossed diodes, as in Fig. 6.5. The switching off or gating is controlled by the pulse programmer.

The signal is next fed into the main amplifier. In a multinuclear spectrometer the frequency to be amplified may vary over a considerable range. To avoid duplicating the whole of the main amplifier for each frequency one of two alternative approaches is taken. Either the amplifier is constructed with an appropriately wide band, or the scheme shown in Fig. 6.10 is used. As a result of the fact that it is still difficult to construct wide band amplifiers which have as good noise figures as narrow band ones, the second approach is commonly used.

The transmitter frequency is fed into a mixer along with a second or local oscillator frequency, which is the sum of two frequencies, the transmitter frequency, and a third or intermediate frequency. The object of the design is that the output of the mixer is at the intermediate frequency, no matter what excitation frequency is used. All the main amplification within the spectrometer is thus carried out efficiently at one frequency, independent of the nucleus being studied. If single side band crystal filters are used they are normally placed in the i.f. stage of the receiver since here they are independent of transmitter frequency [11]. The crystal filters are constructed so that they will only pass frequencies from the i.f. frequency to either plus or minus their width. Their asymmetry greatly reduces the alaising of noise from the appropriate side of the centre frequency [12]. After amplification the signal is detected using either one or two phase sensitive detectors. If two are used (quadrature detection), a $90°$ phase shift is introduced between their reference frequencies. The advantage of quadrature detection is that frequencies of $\pm \vartheta_i$ can be distinguished due to the phase difference of $90°$. The pulse can thus be placed in the middle of the spectrum, which halves the data handling requirements and the receiver bandwidth, a spectrum now only containing frequencies between $\pm \frac{1}{2}\Delta$ [13]. Halving the bandwidth increases the signal to noise ratio by $(2)^{1/2}$ (see section 6.13). The crystal filters described above also halve the receiver bandwidth from $\pm \Delta$ to Δ, but not the data handling rate. Crystal filters and quadrature cannot, of course, be used together.

The receiver in a pulse spectrometer must fulfil certain requirements

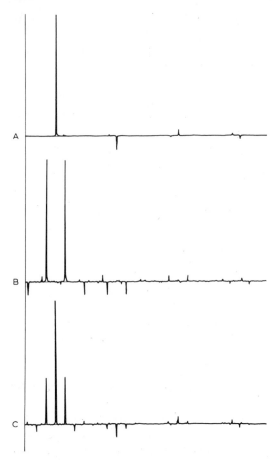

Fig. 6.11. The effect of non-linearity in the receiver on the spectrum obtained from (A) a single line, (B) a doublet and (C) a triplet [14].

not demanded, or at least not to the same extent, in a swept spectrometer. The first of these is linearity. If the receiver is not linear in its gain then, as illustrated in Fig. 6.11, side bands which are harmonically related to the offset frequency of a line will be produced [14]. The receivers of commercial spectrometers are usually linear and such problems only manifest themselves in the presence of a very large resonance, e.g. solvent resonance. The second requirement is that of bandwidth; a free induction decay is a range of frequencies, unlike the signal in swept experiments which has a discrete value, and the receiver must handle them equally. The equality must

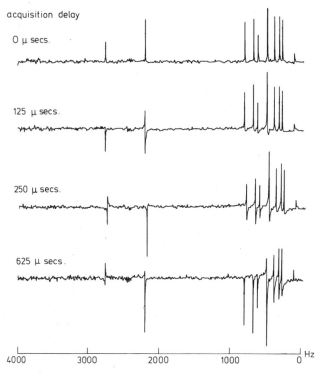

Fig. 6.12. The effect of delaying the start of data acquisition on ^{13}C spectrum of α-pinene.

obviously apply to gain but less obviously to velocity. If one frequency travels slower, and takes longer in the receiver than another, a relative phase shift will occur. Phase shifts are inherent in the components used and impossible to eliminate, but if they are small (less than 360° over the spectral width) and linear their results can be satisfactorily corrected by software. A final requirement is also concerned with the phase of the final spectrum and that is fast recovery from the effects of the pulse. Ideally the first sample on the f.i.d. should be taken at zero time; any delay (δ) in doing so will result in a frequency dependent phase shift of $(360\,\vartheta\delta)°$. Figure 6.12 shows the effect of delaying the start of data acquisition by varying times on the ^{13}C spectrum of α-pinene. The 'zero delay' spectrum was obtained using the minimum possible delay and then phase correction routines to extrapolate to 'zero time'. In the other spectra the phase correction used in the first experiment was retained.

6.8 THE PULSE PROGRAMMER

The pulse programmer is the part of the spectrometer which is responsible for timing the pulses and switching on and off the decoupler if necessary. In early spectrometers it existed as a piece of hardware in its own right; for reasons of flexibility it is now normally a part of the interface. Conceptively, however, it still exists as a functional part of the system. The programmer sends out control pulses to the transmitter to switch the various gates on and off; the timing of these pulses is the critical function. This timing is achieved by the use of a stable oscillator or 'clock' of known frequency. A register counts the number of cycles, in the form of pulses generated by the clock, and thus it measures time. The stability of the clock determines the accuracy of the timing, its frequency the precision. A 1 MHz clock can define a pulse to $1 \mu s$.

Two general types of programmer can be defined those controlled by hardware and software respectively. The hardware type has the pulse durations and intervals required set by switches and it contains its own clock. It is self sufficient but limited in flexibility. The software type on the other hand needs the presence of the computer at all times, but as it has almost unlimited flexibility, this type is universally used for high resolution work. The programmer must function on two different time scales, the first of which is the timing of the pulse width. Here the duration is small, a few hundred μs at the most, but the precision must be high, especially for flexible excitation. The clock used here is a high frequency oscillator (10 MHz). The pulse duration is defined by the computer entering an appropriate number into a register; at the beginning of the pulse the clock starts subtracting from this register, and the pulse is terminated when the counter reaches zero. The second group of timings required are the various delays between pulses. These are in the order of seconds and require a millisecond precision. To use the above oscillator would require unnecessarily large registers and so a slower clock is used, usually the basic clock of the central processing unit, one of the arithmetic registers being used for the timing.

6.9 THE SPIN DECOUPLER

The spin decoupler provides a second radio frequency source which is needed to perform spin decoupling experiments. First consider the simpler case of heteronuclear decoupling. The appropriate frequency is generated and referenced to the spectrometer's master crystal, as previously outlined. It must then be amplified and applied via the decoupler coil within the probe to the sample. In order to perform gated decoupling experiments it must be arranged that the output can be gated on and off under computer control at the desired part of the pulse train. If the spectrometer is required to undertake noise modulation, these facilities must be provided.

Noise modulation of a decoupler frequency is normally achieved by using a pseudo-random sequence generated in the shift register in the way outlined in section 4.6. It is usually arranged that the output from this shift register induces a 180° phase shift whenever the output is positive. The width over which the decoupling frequency is spread is quantified by the so called 'band width' of the noise decoupler. This is normally taken as the audio frequency fed into the shift register which generates the pseudo-random sequence, and it corresponds to half the difference between the zero points in the sinc ϑ power distribution. With the decoupling frequency in the middle of the spectrum of interest, the effective field generated at a frequency, which is offset from the midpoint by half the band width of the decoupler, is about 65% of that at the centre frequency.

The power output of a decoupler can be expressed in watts of r.f. power. It is more meaningful, however, to report the effective field generated by this power at the sample in the probe. In principle one cannot have too high a decoupling field. In practice the available power is limited by the performance of the power amplifier used in the decoupler and by the problems associated with the removal of the heat generated by the field within the well insulated confines of the probe. Heteronuclear decouplers typically produce 10 to 20 W of r.f. power with which they generate fields of $50 - 100\,\mu T$. Rather than consider the fields generated in terms of magnetic field units, it is conceptionally much easier to express these in terms of frequency units. For heteronuclear proton decoupling typical fields generated are 2—5 kHz at 100 mHz. As will be seen in section 10.7 the efficiency of the ^1H noise decoupler can be the limiting factor in determining the line widths of a proton noise decoupled carbon spectrum.

The situation concerning homonuclear decoupling in a pulsed spectrometer is not quite as simple as that of heteronuclear decoupling. The required frequency can easily be generated from a master crystal as usual. The complications arise due to the presence within the receiver's band width of a large decoupler signal which can cause problems in two areas. Firstly, if care is not taken, it will overload the receiver and cause distortion of the signal, introducing harmonic side bands into the output r.f. Problems of this type do not arise in the case of heteronuclear decoupling where the large decoupler signal has a frequency which is well outside the receiver band width and hence is not detected, nor do they occur to the same extent in c.w. spectrometers, where the receivers have a very narrow band width, and are only affected by the decoupler when it is examining the region very close to the point of irradiation.

Secondly, ϑ_3 will appear in the free induction decay as a large signal which can give rise to dynamic range problems at the digitisation stage, unless an adequate wordlength a.d.c. is used. The problem of receiver saturation can be countered in two ways; firstly by the use of single side band receivers, and secondly by using time share modulation [15]. The second approach, the use of time share, is the most common. Here it is arranged that the decoupler and the receiver are never on at the same time, thus minimising the direct interaction between them. A timing diagram is shown in Fig. 6.13. Following the excitation pulse, samples of the free induction decay are taken at the normal rate determined by the spectral width being

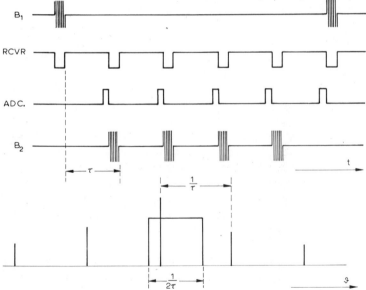

Fig. 6.13. Timing diagram for time-shared homonuclear decoupling.

studied, the interval between the samples being $(2 \times \text{spectral width})^{-1}$. This time interval is shared between the decoupler and the receiver as illustrated. The receiver is switched on as long as possible; it is then sampled by the a.d.c., and switched off. As soon as it is switched off, the decoupler field is switched on and left on for the required period of time depending on the decoupling power required. This switching is a modulation and as such produces side bands, but none of these fall within the spectral range being studied, as can be seen from the frequency spectrum illustrated at the bottom of Fig. 6.13. The receiver only detects the effective radio frequency component and not the unwanted side bands. The limitation of time share decoupling is that only a variable fraction of the total power from the decoupler is available at the required frequency. The rest is dissipated into the unwanted side bands. If the decoupler is on for τ s, then the fraction of the total decoupler power (B_2^2) available in the center band is given by the square of the area in the time domain spectrum i.e. $(B_2\tau)^2$. The amplitude of the B_2 field available for decoupling is only $B_2\tau$ [15]. As the maximum value τ can take is limited to a maximum of $\frac{1}{2}\Delta$, less the time required for the receiver to recover and the a.d.c. to sample the signal, the available decoupler power is rather ineffeciently used, particularly when dealing with large spectral widths where the time available for the decoupler becomes very small.

When discussing gated decoupling experiments which separate the nuclear Overhauser effect from its associated decoupling effect, we use as an explanation the observation that the nuclear Overhauser effect builds up after the application of a second if on the same scale T_1, whereas the decoupling effect is 'instantaneous' [16]. Having made this statement it might first be thought that the time share decoupling experiment outlined above will not work, because if the decoupling effect is generated and disappears instantaneously, or at least on a microsecond time scale, then what should be observed in the above experiment is a non-decoupled spectrum. The fallacy here is concerned with the two domains. We know that within the time domain the signal is being switched on and off. Within the frequency domain the consequence of this switching on and off is the production of side bands. Each of the frequency components then is effectively a continuous frequency. We have allowed for the discontinuity once, by introducing the concept of side bands; we must not allow for it twice by considering the frequency components that we have generated as being discontinuous. Within the frequency domain, therefore, the nuclear spins see a continuous series of r.f. side bands, one of which is at the correct frequency to perform a decoupling experiment.

6.10 THE INTERFACE

Up to now we have been dealing mainly with signals which have been continuous or analogue in nature as opposed to discrete or digital. In order to carry out any data processing using a computer the analogue data must be converted into a digital form. This is not theoretically necessary as analogue computers do exist and have even been applied to Fourier n.m.r. but their operation is so limited that they will not be considered further. To output data from a computer the reverse digital to analogue process must be undertaken. The devices which perform these and other functions concerning the interaction of the computer and the spectrometer are grouped together under the term 'interface'. It is assumed that the

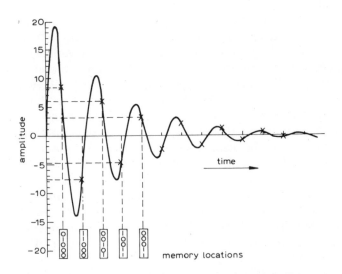

Fig. 6.14. The digitisation of a f.i.d. using a 4 bit and sign a.d.c.

reader has some familiarity with binary numbers, which form the basic concepts of digitisation and computer operation.

The first part of the interface to consider is the analogue to digital converter or a.d.c. The task of this is to produce a digital presentation of the f.i.d., which is achieved by measuring the instantaneous value of the f.i.d. at a series of equal time intervals as shown in Fig. 6.14. There are many types of a.d.c.; the most commonly used in pulsed n.m.r. is the 'successive approximations' type, whose operation is illustrated in Fig. 6.15. The first stage is a sample and hold circuit. On receiving the instruction to take a sample a capacitor is connected to the receiver; this quickly charges to the potential of the signal (a few μs) and is then disconnected from the input and holds the voltage it has acquired. It is only during this brief period that the a.d.c. samples the receiver, consequently it is only during this and any period required to achieve stability that the receiver need be on. It is thus this time which limits the decoupler duty cycle in a time share experiment. The voltage from the sample V_s is fed into a comparitor, and a reference voltage from a digital to analogue converter (d.a.c.) is subtracted. If the output is positive then a 1 is entered in the buffer register, if negative a zero. The d.a.c. then converts the new value of the register and the process is repeated until the register is full, at which time it contains a binary number which represents the sampled voltage. The binary number is then transferred to the computer. The speed of the conversion is controlled by the internal timer and is in the order of 1—1.5 μs/bit. A 10 bit a.d.c. operating

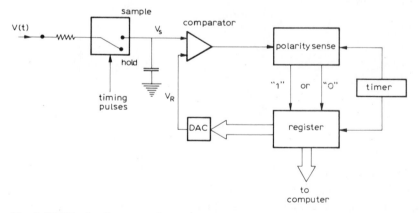

Fig. 6.15. Block diagram of a 'successive approximation' analogue to digital converter.

this way can convert a sample in $10-15\,\mu s$, i.e. has a data rate of $75-100\,\text{kHz}$.

How fast, precisely and accurately must the a.d.c. function? The first can be answered by consideration of sampling theory which tells us that to represent accurately a sinusoidal function, which an n.m.r. signal basically is, we must sample it at least twice per cycle. Or, in other words, the conversion rate must be at least twice the spectral width. This minimum sampling rate is known as the Nyquist frequency and as it requires the minimum number of data points is the sampling rate normally used. If there is a resonance in the spectrum which is outside the spectral width being considered, consequently for which the Nyquist theorem is not obeyed, then this line will be aliased or folded back and will appear in the spectrum. How this happens is illustrated in Fig. 6.16. The solid line represents the true frequency being digitised, the dotted line represents a frequency, outside the spectral width, which will also yield the same digitised values and is indistinguishable from the 'true frequency'. Any frequency from outside the spectral width will be digitised as if it has an 'apparent frequency' which is within the

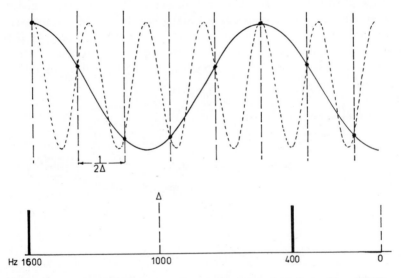

Fig. 6.16. Sampling at the 2000 samples/second defines the 400 Hz signal (solid line) however a 1600 Hz frequency (dotted line) would also give the same data points. The 1600 Hz frequence is 'folded over'.

spectral width. Inspection of Fig. 6.16 shows that

Apparent frequency $= 2 \cdot \Delta -$ true frequency

Not only signals from outside the spectral width, but also noise, which is a mixture of an infinite number of frequency components, can be folded in. To maximise signal to noise one therefore filters the signal prior to digitisation with a low pass filter usually of the Butterworth or Bessel type, see Fig. 6.17. Filters with a sharper cut-off (e.g. Tschebyscheff filters) are available; a 4 pole Butterworth filter typically only halves the power as the frequency is doubled (6 db/octane cut-off); however, the sharper filters induce extreme and non-linear phase shifts near the cut-off frequency which cannot be corrected later. It is more important, especially when dealing with

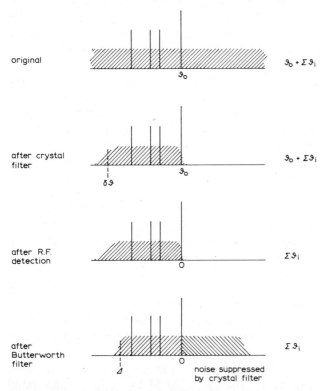

Fig. 6.17. The effect of Butterworth and crystal filters on the frequencies in the spectrum.

broad lines, to have low phase distortion than a sharp cut-off, and in this case Bessel filters are preferable. Figure 6.17 also illustrates the use of an r.f. or crystal filter which, if quadrature detection is not used, can halve the noise arriving at the a.d.c., thus increasing the sensitivity by $2^{1/2}$ (see section 6.7).

Now to the question of accuracy and precision. The accuracy that is required of an a.d.c. concerns the timing of the start of the sampling rate which on transformation governs the frequency accuracy. In view of the accuracy required, the timing pulses for the start of the sample and hold operation are normally obtained by suitable division of the spectrometer's master crystal frequency. The precision or dynamic range of the digitisation process is limited by the word length of the a.d.c.; a 10 bit a.d.c. can represent a signal to 1 part in 2^{10} i.e. 0.1%.

The question of dynamic range is a complex and important one which always arises since digital devices have a finite word length and can only represent a value to a fixed precision. Dynamic range considerations arise twice in a Fourier spectrometer, firstly at the a.d.c. stage, and secondly during the summation in the computer, and set a limit on the ratio of maximum to minimum signals which can be handled. Consider an a.d.c. with a 5 bit word length as represented in Fig. 6.14. The first or most significant bit represents the sign, the other bits represent the magnitude of the sample. If the input range is $\pm 10\,V$ then the minimum detectable signal is $0.625\,V$ $(1/2^4)$ and the maximum of course is $10\,V$. Any signal greater than $10\,V$ will be represented as $10\,V$ and below $0.625\,V$ as zero. Therefore, to time average a signal out of the noise, the peak to peak amplitude of the noise must be greater than $1.25\,V$.

Now if a strong signal and a weak signal are present together, problems can arise. In the example chosen, if the signal to noise ratio of the strong peak is greater than about $10/1$, it will not be possible to digitise accurately the noise and no amount of averaging will recover the small signal. The situation can be eased by increasing the word length of the a.d.c. but there are limits to this, set firstly by necessary conversion rate and secondly, as we will see, by the word length of the associated computer. Typical a.d.c. word lengths are 9—12 bit.

The complementary component to the a.d.c. is the digital to analogue converter or d.a.c. The principle of the d.a.c. is illustrated in Fig. 6.18. The binary number which is to be converted is fed into

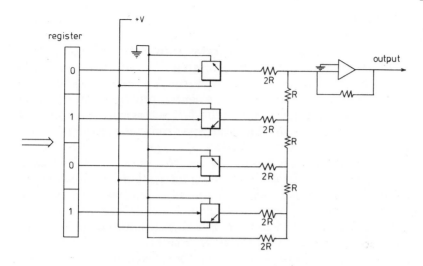

Fig. 6.18. The principle of a digital to analogue converter.

a register. Each bit, depending on whether it is '1' or '0' either opens or closes the appropriate gate. If the gate is closed, a voltage corresponding to the significance of that bit is added into the 'voltage ladder'. The total of these voltages is sent out via a buffer amplifier. In an interface two d.a.c.'s are required to drive the x and y axes of the display unit, usually a scope. The accuracy required is limited by the display unit to about 0.1%, i.e. 10 bits. When plotting a spectrum on a recorder it is normally only the vertical axis which is controlled via a d.a.c., the horizontal axis of the recorder being driven by a 'stepper motor', i.e. one which rotates by a fixed amount when subjected to a voltage pulse. Using a stepper motor enables the computer to control the recorder directly in a very simple manner.

The interface, apart from the major items dealt with above, also contains the appropriate hardware which enables the computer to sense the state of certain functions within the spectrometer, e.g. is the spectrometer field/frequency locked? Is the recorder set at the left hand limit switch? etc. The interface also contains a multiplexer, in effect a programmable switch which enables a single a.d.c. to be connected to various voltages within the spectrometer, not just to the output from the receiver, e.g. the voltage on the 'phase' knobs.

6.11 THE COMPUTER

The next consideration is generally known as 'the computer', whose function is data handling, computation etc. The computer, which is normally digital, although it can be analogue in design, can be subdivided into two parts, the central processing unit (c.p.u.) consisting of the arithmetic unit plus fast access core memory and the peripheral devices, of which the spectrometer via the interface is one. The computer is not necessarily an integral part of the spectrometer system. If the computer only has to Fourier transform then it can be very remote; however, as the sophistication of pulsed experiments and spectrometers increases, the situation has been reached where the two are no longer separable. When talking of the computer we will therefore assume it is a digital mini computer built into, and totally dedicated to, the spectrometer.

The c.p.u. is characterised by its speed, vocabulary, size and wordlength. Speed is expressed in terms of the core cycle time, i.e. the time taken to call and replace a word from core, which depending on the type of memory, is in the order of nanoseconds, 250 for MOS memory, and 900 for magnetic memory core. Most computers are more than fast enough for the functions required in an n.m.r. spectrometer; limitations due to computer speed can, however, arise when timing short pulses and calculating tailored excitation sequences during accumulation. Vocabulary is the number of words in the basic machine code used to programme the c.p.u. The longer the vocabulary, the fewer the number of instructions necessary to execute a specific function and consequently the faster it can be carried out. Size refers to the size of the rapid access core memory used. The final characteristic, and from an n.m.r. point of view maybe the most important, is the wordlength. This obviously refers to the number of bits in a computer word and governs the precision to which spectra can be recorded. Most mini computers used in n.m.r. systems have 16 or 20 bit words.

Connected to the computer are a number of external or peripheral devices whose function is to input and output data, and/or provide bulk data stage. Data input and output falls into two categories, high speed transfer to and from the spectrometer, which is handled directly via the interface, and slow speed communication with the operator. Communication with the operator is normally achieved via an alphanumeric keyboard and display. As well as its core memory the c.p.u. can have back-up storage in the form of magnetic tape or disc units; these can be used either to provide a permanent record of stored data, or, in more advanced systems using high speed discs, to act effectively as extra 'core', thus permitting the handling of very large data tables, for example a 32 K double precision data table can be successfully and rapidly manipulated in a 16 K core computer equipped with a disc.

6.12 SOFTWARE

Software is the name given to the programme etc. used in a computer. It is outside the scope of this book to consider how the software is written and how it functions in any great detail; what do concern us are the steps necessary to produce a spectrum. Programmes are normally written directly in machine code, as this represents the most efficient use of the available computer space [18]. In larger systems facilities may be provided to permit the user to write his own programmes in a high level language like Fortran and Basic.

(a) The Executive

The heart of the software is the 'Executive'; this programme oversees and co-ordinates all the other subroutines which are responsible for specific tasks within the spectrometer. A very simplified flow chart for the executive is given in Fig. 6.19. All parameters under software control are normally identified by two letter nemonics as two alphanumeric characters fill one 16 bit computer word. The required values are assigned to these parameters; a hardcopy list is generated if required, and the experiment started. The executive sets up all the timing routines and pulse programmer, and initiates the data acquisition.

(b) Data acquisition and dynamic range [19, 20]

The data acquisition subroutines initiate the a.d.c. and store the values in consecutive locations in the computer core until the data table is full. After any appropriate delay the sequence is repeated, the new values from the a.d.c. are added to the existing ones, and so on until the specified number of transients have been averaged.

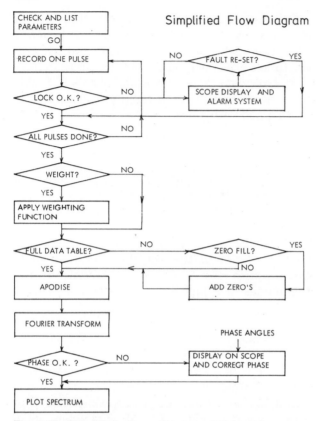

Fig. 6.19. Simplified flow diagram showing the progress of a pulse Fourier experiment.

Now return to the problem of dynamic range. If one averages data from an n bit a.d.c. using an N bit computer, then if one data point on each f.i.d. fills the a.d.c., after 2^{N-n} co-additions the computer word would be full and addition of the data from any further pulses would result in overflow, i.e. it would need a $(N + 1)$ bit in the computer word to represent the data accurately. Since no such bit is available the data will be inaccurate. This cannot be allowed to occur since in the time domain every point contains information about every point in the frequency domain; one incorrect time domain point would affect the whole spectrum. Either data acquisition must be stopped at this point or the existing data must be scaled down. If the existing data is scaled, so must be the incoming data in order that each transient has the same significance.

The simplest scaling procedure is to shift each bit in the word one place to the right, which is equivalent to dividing by two.

| 1 | 0 | 0 | = 4 |
| $\rightarrow$ | 1 | 0 | = 2 |

Scaling cannot go on indefinitely, or valuable information would be shifted out of the word and information lost. There is therefore a maximum signal to noise ratio which can be achieved. This will depend on the wordlength of both the a.d.c. and the computer, and also on the S/N of the incoming signal (2^x). If 'a' scans are added together, the S/N will be increased by a factor of $2^{a/2}$ as discussed previously; the final S/N will be given by

$$(S/N)_a = 2^x \times 2^{a/2} = 2^{(2x+a)/2} \tag{6.1}$$

The maximum achievable signal to noise therefore depends on the maximum value of 'a' which in turn is dependent on n and N.

Consider first the case where the signal to noise ratio is less than unity; as the noise dominates, the signal in the computer memory doubles every four scans. A scaling process will therefore potentially be necessary every 4, 16, 64 pulses i.e

2^a pulses require $2^{a/2}$ scalings

Scaling can continue until $(N - n)$ shifts have been accomplished, after which point any further scaling will result in the loss of the least significant bit of the incoming data. At this stage of the experiment, which is reached after 2^{N-x} pulses, data acquisition should be stopped unless some method of recovery of the lost information is used. The limiting signal to noise ratio is

$$(S/N)_{max} = 2^{(N+x)/2} \tag{6.2}$$

provided less than 2^{2n} pulses are required. In excess of this number, n shifts will have been made and no further data will reach the computer.

If the signal to noise ratio of the incoming signal is above unity, scaling will be necessary more frequently as the signal will double every 2 pulses. The maximum allowable number of scalings will therefore be reached much sooner. Fortunately, software techniques are available which enable data from a.d.c. bits shifted outside the computer wordlength to be recovered by suitable round-off techniques (see ref. 20 for further details). If data scaling routines are not used then eqn. (6.2) holds; however, using these techniques, acquisition can continue for up to 2^N pulses resulting in a signal to noise ratio given by

$$(S/N)_{max} = 2^{(N+2x)/2} \leqslant 2^N \tag{6.3}$$

The limit of 2^N is set by the dynamic range of the computer word.

(c) Removal of systematic noise (leakage)

During data acquisition, r.f. from the transmitter can, despite every precaution, leak directly into the receiver; if this happens it will manifest itself as an unwanted d.c. output from the phase sensitive detector which after digitisation will appear in the accumulated data. Leakage can arise from two sources, pulse breakthrough where the transmitter coil is still radiating energy when the first data sample is taken, and/or direct radiation into the receiver from within the spectrometer's electronics. The first case lasts a few microseconds and results in only the first few points being heavily biased. The f.i.d. can be considered as having been multiplied by a constant voltage for a short period (in effect a pulse), on transformation; the frequency spectrum is convoluted with a sinc ϑ function, and a rolling base line results. Pulse breakthrough can be corrected for either by discarding the first few offending points or by delaying the start of data acquisition, both of which will induce phase shifts (see Fig. 6.12).

The direct leakage gives rise to a constant bias in the f.i.d. which, if left uncorrected, causes a spike at zero frequency on transformation (see Table 3.1A and Fig. 6.20). There are two remedies for direct leakage; to subtract it away or to use a phase alternating pulse sequence. In the first case, the mean value of the f.i.d. is subtracted from each data location, and the correction is normally made prior to transformation. In the second case, after each pulse or group of pulses, the phase of the excitation is changed by $180°$. The resultant f.i.d. is inverted in sign and is therefore subtracted from the accumulated data. N.m.r. signals are, to the first approximation, unaffected by the phase alternation, except for the change in sign, and averaging proceeds normally. Unwanted d.c. biases do not change sign and so average to zero.

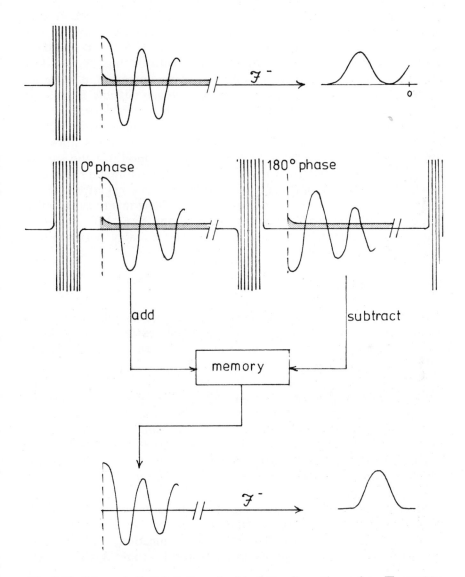

Fig. 6.20. Removal of systematic noise by phase alternation pulses. The systematic noise is represented by a.d.c. bias (the shaded area under the f.i.d.) which if uncorrected leads to a spike at zero frequency.

(d) Data routing with quadrature detection [19]

The merits of quadrature detection have been enumerated previously; however, unless the two phase detectors are identical and are using reference frequencies which are *exactly* 90° out of phase with each other, reflections appear, as do distortions at zero frequency. It is possible to use components whose precision is suitable, but it is much easier to use software to average out any imbalance in the detectors and error in the 90° phase shift. The desired averaging may be achieved using pulses whose phase is successively shifted by 90°. Initially the output of detector A (the v mode) is stored in the first half of the data table, that from B (the u mode) in the second half. After the second pulse the output of A is the u mode signal and is added into the second half; the output of B is a negative v mode signal; this is subtracted from the first half. Using this routing, pulse phase shifting averages away any error in the fixed 90° shift and since both detectors contribute to both modes, imbalance here and in any filters is also removed.

(e) Apodisation

When performing a Fourier transform, although the data is assumed to extend to infinity, it does not do so; it only extends to the end of the acquisition time. At this time there is a discontinuity; the data is truncated. This is important if there is still detectable *signal* present at that time i.e. $T_a < 3T_2^*$; if this is the case the f.i.d. is effectively multiplied by a window function which is unity from $t = 0$ to T_a and zero elsewhere. On transformation the spectrum is convoluted (broadened) by the sinc function resulting from transformation of the window as is illustrated in Fig. 6.21a. The effect of truncation can be reduced by multiplying the end of the f.i.d. by a function which reduces it to zero. A crude function is a triangle and its effect is shown in Table 3.1. A cosine (Hanning) function (see Fig. 6.21b) is normally used. The process is called apodisation (which 'translated' means 'cutting off the feet'). The larger the percentage of the f.i.d. which is apodised the less the effects of truncation, but the more distorted the resulting line shape and intensities are. See Fig. 7.10.

(f) Weighting

The next stage is the multiplication of the f.i.d. by an exponential function $\exp(t/TC)$ to optimise either the signal to noise ratio or the

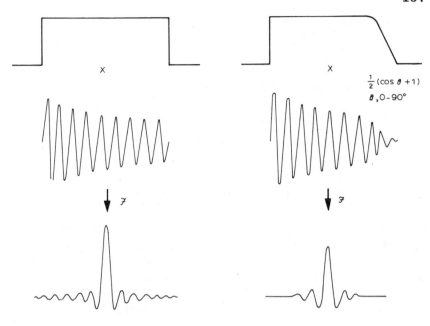

Fig. 6.21. The application of an apodisation function to the end of the f.i.d. reduces the consequences of truncation.

resolution of the spectrum. Prior to applying the weighting function, any d.c. bias in the signal should be removed by subtracting from each location the mean value of all the data points. Since f.i.d. is essentially symmetrical its mean value is zero.

(g) The Fourier transform

The next step is the actual Fourier transformation itself (see ref. 18). The transform is accomplished using the Cooley Tukey [21] or fast Fourier transform (f.f.t.) algorithm, rather than an algorithm devised simply from eqn. (3.40). A straight forward algorithm for N data points would require the computer to perform N^2 multiplications whereas the f.f.t. method requires only $2N \log_2 N$ multiplications; the latter is therefore much faster. As an example, for 4096 data points the saving in time is about 190 and the saving is even more significant when larger data tables are used, e.g. nearly 1500 when 32,768 data points are used.

If we rewrite eqn. (3.39) and (3.40) in a single matrix form they become

$$M(n) = 1/N \, [W_n^{kn}] \quad [m(k)] \tag{6.4}$$

where $W_n = \exp 2\pi i n/N$, which for $N = 4$ is

$$
\begin{bmatrix} M_0 \\ M_1 \\ M_2 \\ M_3 \end{bmatrix} = 1/N
\begin{bmatrix}
W_1^0 & W_2^0 & W_3^0 & W_4^0 \\
W_1^0 & W_2^{-1} & W_3^{-2} & W_4^{-3} \\
W_1^0 & W_2^{-2} & W_3^{-4} & W_4^{-6} \\
W_1^0 & W_2^{-3} & W_3^{-6} & W_4^{-9}
\end{bmatrix}
\begin{bmatrix} m_0 \\ m_1 \\ m_2 \\ m_3 \end{bmatrix}
\tag{6.5}
$$

By considering the solution to eqn. (6.5) it becomes apparent that each time domain point contributes to each frequency domain point. The chief task in the transform is solving the W matrix. A time saving is made in the Cooley-Tukey approach by limiting N to a value of 2^z where z can be any number i.e. N can equal 1024 but not 1023. With this constraint imposed and n and k rewritten as binary numbers e.g.

$$k = (k_2 \cdot 4) + (k_1 \cdot 2) + k_0$$

where k_2 etc. can now be only 0 or 1, it is possible to factorise the W matrix and use symmetry properties to achieve the increase in speed outlined previously.

The fast Fourier transform is carried out 'in place'; the final frequency spectrum occupies the same area of the computer memory as did the original data. Figure 6.22 shows the progress of the fast Fourier transform. Such an efficient use of computer memory has obvious economic advantages, but has the disadvantage that the original data is lost. If, after transformation, it is found that more pulses would be necessary or another weighting function would be advantageous, a copy of the original f.i.d. would be necessary, either in another part of the memory or with the aid of a back-up store, e.g. a magnetic tape or disc. Such a precaution is well worth taking especially as the data has taken days to collect! An inverse transform to regenerate the original spectrum is not possible with the programmes normally in n.m.r. spectrometers and would not be a good substitute since the transformation process itself would add 'noise' to the spectrum.

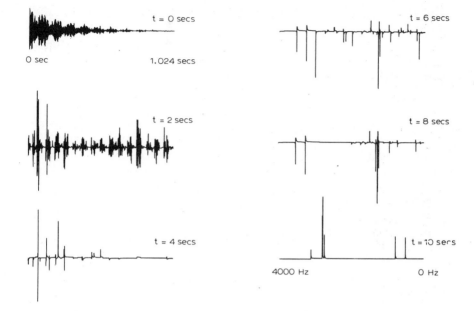

Fig. 6.22. The progress of Fourier transformation using the Cooley-Tukey algorithm. The calculation was stopped at various times and the content of the memory plotted.

(h) Display and phase setting

The final stage is the display and plotting of the spectrum. If the spectrometer were perfect, the values which correspond to absorption and the dispersion spectrum would be obtained simply by displaying the corresponding data locations. For various reasons we have already mentioned, spectrometers are not perfect and phase shifts will have been introduced. If the phase shift is θ then the real and imaginary data are related to the absorption and dispersion modes thus

$$r_i = v_i \cos \theta_i + u_i \sin \theta_i \qquad\qquad , \ (6.6)$$

$$l_i = v_i \sin \theta_i - u_i \cos \theta_i \qquad\qquad (6.7)$$

The pure absorption spectrum can be obtained by taking combinations of the real and imaginary values thus

$$v_i = l_i \sin \theta_i + r_i \cos \theta_i \qquad\qquad (6.8)$$

The only problem left is how to chose the correct value of θ. The phase shift, as one would expect, is not always constant through the spectrum; fortunately its change can be approximated to a linear variation with frequency, and consequently can be defined by two parameters thus

$$\theta_i = \theta_0 + \theta_1 \vartheta_i \qquad (6.9)$$

The constants θ_0 and θ_1 are usually set by the operator while viewing the spectrum on the scope to produce the absorption spectrum. The parameters are set using two potentiometers which produce a voltage that is digitised via the a.d.c. and used by the computer as the values of θ_0 and θ_1 during the display (see Fig. 6.23).

It is possible to by-pass the phase setting problem if necessary by displaying either the magnitude or the power spectrum, neither of which has a phase dependence. Both of these displays produce distorted spectra as can be seen. Magnitude spectra have a non-Lorentzian shape, cannot be integrated (the integral is divergent)

TABLE 6.1

Methods of displaying n.m.r. spectra

$M(\nu) = u + iv$

Name	Mode	Form	Shape	S/N	Comments
Adsorption	v	$\dfrac{K}{1+\nu^2}$	Lorentzian	1	normally used for spectra
Dispersion[a]	u	$\dfrac{K\nu}{1+\nu^2}$	Dispersive	1	used for locking
Magnitude	$(u^2+v^2)^{\frac{1}{2}}$	$\left(K\,\dfrac{1}{1+\nu^2}\right)^{\frac{1}{2}}$	—	$(2)^{\frac{1}{2}}$	used when phase is not required. Integral does not converge
Power	(u^2+v^2)	$\dfrac{K^2}{1+\nu^2}$	Lorentzian	2	does not involve phase, signal prop. to square of number of nuclei

[a]This mode should not be confused with the derivative mode e.g. as used in e.s.r. which is $\nu/(1+\nu^2)^2$.

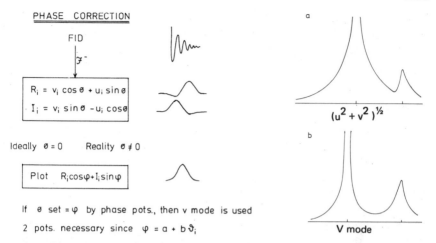

PHASE CORRECTION

FID

$\mathscr{F}^-$

$$R_i = v_i \cos\theta + u_i \sin\theta$$
$$I_i = v_i \sin\theta - u_i \cos\theta$$

Ideally $\theta = 0$ Reality $\theta \neq 0$

Plot $R_i\cos\varphi + I_i\sin\varphi$

If θ set $= \varphi$ by phase pots., then v mode is used

2 pots. necessary since $\varphi = a + b\vartheta_i$

$(u^2 + v^2)^{1/2}$

a

b

V mode

Fig. 6.23. The principle of phase correction in Fourier n.m.r. spectra. Two potentiometers in the spectrometer are used to define a phase angle φ which is equal to the unwanted phase shift introduced by the spectrometer.

Fig. 6.24. The appearance of two overlapping lines when using the magnitude (a) and v mode (b) presentation.

and are non-additive in areas of overlap (see Fig. 6.24). Power spectra are Lorentzian but their area is proportional to the square of the number of nuclei to which they correspond. The possible methods of displaying n.m.r. spectra are summarised in Table 6.1.

(i) Data reduction

In the pulsed experiment the collection and plotting of data are two separate functions. This enables one to treat data output from a very different point of view from that taken in swept n.m.r. As the data is stored in a digital computer, it can be continually examined on an oscilloscope screen, the phase can be corrected etc., and any specific required area can be plotted onto a recorder with virtually an unlimited choice of parameters. It must, however, be remembered that spectral expansions obtained in this way differ from those obtained by using a narrow sweep width in the swept experiment, as in the pulsed case a different display of the same data does not represent a new experiment. Having the data in a digital form, a computer memory also enables one to do away with the tedium of measuring line positions and intensities. Software in the system can very easily measure chemical shift with respect to either the transmitter frequency or any line identified as zero. In this way a list can be printed of all the chemical shifts in a particular spectrum referred to, say tetramethyl silane. If this is done, the plotted spectrum becomes purely a visual aid.

(j) Miscellaneous

The areas of software we have discussed previously are just the major groups of sub-routines used in a high resolution spectrometer. Within the executive there are programmes to do many other functions far too numerous to mention in detail, particularly as they are to a certain extent specific to the spectrometer in question. In the more software controlled spectrometers it is possible by means of the software for the operator to control such parameters as the offset of the spectrometer's observing frequency, the decoupling frequency, the mode of decoupling, the a.d.c. filter used, and many others.

6.13 FACTORS AFFECTING THE SIGNAL TO NOISE RATIO OF A SPECTROMETER [22, 23]

N.m.r. being a basically insensitive technique, the sensitivity a spectrometer can achieve is arguably its most important characteristic. So far we have discussed the many functional parts which make up a spectrometer; now consider how they contribute to the overall sensitivity of the system.

The sensitivity of an n.m.r. spectrometer is normally defined in terms of the ratio of the peak height of a specific line in a standard sample to twice the root mean square noise level present in the spectrum. The full definition of the latter term is not always clearly stated or understood. Peak to peak noise is defined as the distance between the maximum positive and negative deflections of the trace in a peak free region of the base line where the pen has crossed the mean base line one hundred times. In noise theory peak to peak noise has no significance, whereas the root mean square noise or noise power does. Unfortunately, there is no analytical relationship between the two and an arbitrary conversion factor has to be used; in n.m.r. the factor normally chosen is 2.5, as statistical noise theory shows that the probability of the peak to peak noise equalling or exceeding 2.5 times the r.m.s. noise is only 1%. The so called signal to noise ratio is hence defined by

$$\text{Signal to noise ratio} = \frac{2.5 \times \text{signal height}}{\text{peak to peak noise}} \qquad (6.10)$$

There are obviously two ways of improving the performance of a spectrometer; reduce the noise and increase the signal. Noise originates from two areas, firstly the basic spin system over which one has no control, and secondly from within the spectrometer itself. However carefully the circuitry is designed and built, it will produce some noise, which will be most significant where the signal is weakest, i.e. in the very first stages of the preamplifier. In a well designed system negligible noise should be added later.

First examine the n.m.r. signal. The signal originates from the macroscopic magnetic mount of the sample M_0. As shown in Chapter 2, eqns. (2.28—30), the moment is related to the physical properties of the nucleus thus

$$M_0 = \frac{N\gamma^2 h^2 I(I+1)}{3kT^0} B_0 \qquad (6.11)$$

or replacing B_0 by ϑ/γ

$$M_0 = \frac{N\gamma h^2 \vartheta I (I+1)}{3kT^0} \qquad (6.12)$$

This equation shows us that from the n.m.r. viewpoint we can increase the sensitivity in three ways.

    (i) by increasing the magnetic field ($\propto B$ or ϑ)

    (ii) by lowering the temperature ($\propto 1/T$)

    (iii) by increasing the number of samples in the active volume ($\propto N$ = number of nuclei/unit volume)

Following the application of a 90° pulse the magnetisation precesses and decays in a time constant T_2^*. The flux through the receiver coil generated by this precession is given by

$$\Phi = \int_0^A M_0 \exp\left[-(t/T_2^*) - i2\pi\vartheta_0 t\right] \, da \qquad (6.13)$$

where the integration is carried out over a cross section A of the receiver coil. This approximates to the situation of the crossed coil probe in which the transmitter coil has a much larger volume than the receiver coil. The flux through the receiver coil in practice may be written as

$$\Phi = \xi M_0 \exp\left[-(t/T_2^*) - i2\pi\vartheta_0 t\right] V_c/a \tag{6.14}$$

where ξ = the coil 'filling factor'. V_c is the enclosed volume of the receiver coil and a is its length.

If the magnetisation is not uniform throughout an infinite cylinder then eqn. (6.14) can still be used, but the expression for the filling factor must be modified. For example, in a single coil probe the excitation decreases rapidly beyond the ends of the coil, resulting in a smaller detected signal than for the ideal cross coil probe. Detailed analysis of this situation shows that the voltage across the receiver coil generated by this magnetic flux can be given in terms of certain physical properties of the receiver coil by the following equation.

$$V_s(t) = -K[\exp - (t/T_2^*)] \cos 2\pi\vartheta_0 t \tag{6.15}$$

where $K = n\rho\xi Q M_0 \vartheta_0 V_c/a$

The appropriate physical properties of the receiver coil are respectively its filling factor volume and length mentioned above, the number of turns n, it's quality factor (Q), and ρ, which is the ratio of the inductance of the coil to the total inductance of the tune circuit, of which the receiver coil is part (i.e. including the inductance of the coil leads).

The tuned circuit within which the nuclear signal generates a voltage is in itself a source of noise. A detailed analysis shows that the inherent noise within a tuned circuit can be given in terms of the parameters introduced above and the additional ones, of which L is the inductance of the circuit and $\Delta\vartheta$ is the bandwidth (or total spectral width) of the experiment in question by

$$V_N(t) = (8\pi k T \vartheta_0 L \rho Q(\Delta\vartheta))^{\frac{1}{2}} \tag{6.16}$$

The sum of the voltages arising from both the signal and the noise generated within the tuned circuit containing the receiver coil is then amplified by the spectrometer and detected with respect to the reference frequency to produce the output frequency ϑ_i. The amplification and detection stage is itself a further source of noise. This is represented in terms of 'a noise figure' F. Thus the output of the spectrometer, the signal voltage, is given by

$$S(t) = AK \exp(-t/E_2^*) \cos 2\pi\vartheta_0 t \tag{6.17}$$

where A represents the spectrometer gain and the output noise voltage is given by

$$N(t) = A(16\pi kT\vartheta_0 L\rho Q(\Delta\vartheta)F)^{\frac{1}{2}} \tag{6.18}$$

The additional factor of $2^{1/2}$ has been included assuming that single phase detection and not quadrature has been used, i.e. the spectrometer is sensitive to noise components at $\pm\vartheta_i$.

To achieve optimum sensitivity the spectrometer output (f.i.d.) is normally multiplied by a weighting function of the form

$$W(t) = \exp(-t/TC) \tag{6.19}$$

Since the Fourier transform is linear, a transformation of the signal and noise contributions can be considered independently. We will firstly consider the signal. In the frequency domain this is given by the Fourier transform of the produce of eqn. (6.17) and (6.19) which results in

$$S(\vartheta) = \frac{AK\beta}{2}\left[\frac{1 + 2\pi i(\vartheta_0 - \vartheta)}{1 + 4\pi^2(\vartheta_0 - \vartheta)^2 B^2}\right] \tag{6.20}$$

where $\beta = \dfrac{T_2^* \cdot TC}{(T_2^* + TC)}$

This equation, as expected, is of Lorentzian form which has a maximum when $\vartheta = \vartheta_0$ given by

$$S_p(\vartheta) = \frac{AK\beta}{4} \tag{6.21}$$

The simplest way to transform the noise into the frequency domain is to use the fact that the noise energy must be the same in both domains. This can be expressed in terms of Plancherel's theorum as follows

$$\frac{1}{\pi}\int_{-\vartheta}^{+\vartheta} N^2(\vartheta)\,d\vartheta = \int_{-\infty}^{\infty} N^2(t)\exp(-2t/TC)\,dt \tag{6.22}$$

If we assume that the noise we are dealing with is white noise, i.e. it is independent of time and frequency within the limits of the experiments considered, then eqn. (6.22) has a solution which is given by

$$N(\vartheta) = A(2\pi kT\vartheta_0 L\rho Q(TC)F)^{\frac{1}{2}} \tag{6.23}$$

This equation represents the r.m.s. noise voltage hence we have the necessary information to present an analytical equation for the sensitivity of n.m.r. spectrometers thus

$$S/N = \frac{\beta K}{4N(\vartheta)} \qquad (6.24)$$

It is convenient in order to simplify the final equation to express the inductance of the receiver coil L in terms of other parameters of the coil by

$$L = \lambda\mu_0 V_c (n/a)^2 \qquad (6.25)$$

where (n/a) is the number of terms per unit length of the coil and μ_0 is the permeability of free space. The factor λ represents Nagaoka's correction for finite length of the coil which is a function of the ratio of the coil diameter to the coil length.

Finally, assuming the use of a matched filter, i.e. $TC = T_2^*$, we obtain from eqns. (6.23—25), an expression for the signal to noise ratio of an n.m.r. spectrometer in terms of the various factors involved given by

$$S/N = \left[\frac{N(\vartheta_0)^{3/2}(T_2^*)^{\frac{1}{2}}}{T^{\circ 3/2}}\right] \left(\frac{\xi(\rho Q V_c)^{\frac{1}{2}}}{(\lambda\Delta\vartheta)^{\frac{1}{2}}}\right) \left(\frac{\gamma h^2 I(I+1)}{24 k^{3/2}\mu_0^{1/2}}\right) \qquad (6.26)$$

The first bracket groups together all the physical properties associated directly with the nuclear spins. The second bracket includes all the instrumental parameters associated with the receiver coil and the early stages of the receiver. The third bracket comprises the fundamental physical constant of the nucleus under study.

Having derived this expression for the signal to noise ratio we will discuss the terms in it and the form in which they occur in order to see how best to optimise the performance of the spectrometer.

N

The number of nuclei per unit volume. Sensitivity is directly proportional to N, hence every effort to maximise it should be made. There are two basic ways to maximise this term, either in the sample-limited case to concentrate the sample and place it in the smallest sample tube the spectrometer can accept, or in the solubility case, to use the largest sample tube which the spectrometer can accept.

ϑ_0

Operating radio frequency (or magnetic field since they are proportional). This enters into our calculation in three ways, (i) in the expression for the magnetic moment, eqn. (6.12), (ii) as the frequency at which the magnetic field fluctuates in the receiver coil, eqn. (6.13), (iii) as $\vartheta_0^{-1/2}$ in the frequency dependency of the noise, eqn. (6.17). The net result is that the sensitivity of an n.m.r. spectrometer increases as $\vartheta_0^{3/2}$. Consequently n.m.r. spectrometers operate at the highest available magnetic field compatible with other practical requirements such as cost. The operating frequency also indirectly affects such factors as the Q of the receiver coil and the noise figure of the receiver.

T°

Temperature. Sensitivity of the experiment is proportional to $T^{-3/2}$. The magnetic moment increases as the temperature decreases, since the lower energy levels become more populated. There is an additional factor of $T^{-1/2}$ resulting from the decreasing receiver noise at low temperatures (see later). In practice one is not always free to make use of this parameter in order to increase the signal to noise ratio as cooling the sample may cause solubility problems and induce line broadening resulting from increasing viscosity.

T_2^*

The longer the free induction persists the stronger the signal that is detected. In this treatment it is assumed that an optimum exponential weighting function has been used to multiply the free induction decay. The total energy of the noise components is therefore proportional to $(T_2^*)^{1/2}$, making the sensitivity also proportional to $(T_2^*)^{1/2}$. Improvement in magnetic field homogeneity can increase the signal to noise ratio of the spectrometer provided that the decay in the free induction decay is dominated by this inhomogeneity rather than natural relaxation processes.

ρ

The ratio of the receiver coil inductance to the total coil inductance of the tuned circuit. The inductance in the leads to the receiver coil is largely wasted as negligible magnetic flux is induced in them. At higher radio frequencies (100 MHz and above) this loss in efficiency can be quite important.

Q

The quality factor of the receiver coil. This appears to the power of $\frac{1}{2}$, since the detected signal increases proportionally to Q, whereas the noise is proportional to the square root of Q. The Q factor of a coil increases with its size since the energy stored is approximately proportional to the volume, while the power dissipated is porportional to the length of the wire involved in the coil.

F

Receiver noise figure. This is defined as the ratio of the signal to noise power at the input of the receiver to that at the output. Consequently sensitivity decreases with the square root of F. The noise figure of the receiver is a function of the receiver coil temperature (see eqn. 6.18); decreasing the latter will therefore increase the signal to noise ratio. It should be noted that it is the receiver coil temperature which is involved in here, not the sample temperature. In the analysis it has been implicitly assumed that these temperatures are the same, but this is not necessarily the case. One can envisage the situation, particularly using a superconducting spectrometer, where the receiver coil is at a much lower temperature than the sample.

V_c

The enclosed volume of the receiver coil. In association with the filling factor the enclosed volume determines the effective sample volume, and hence the strength of the signal. It also determines the inductance of the coil and thus the noise power, see eqn. (6.25). As a result sensitivity increases as $V_c^{1/2}$. There are also some second order effects of increasing V, larger coils mean a higher Q, larger filling factors and in many cases a somewhat increased ρ factor. Increasing the receiver coil size rapidly increases sensitivity, provided there is sufficient material to fill the coil.

ξ

The receiver coil filling factor. This can be used to correct for two different effects: firstly, the practical consideration that not all the volume within the receiver coil is in fact sample. This correction is significant with small sample diameters where a considerable percentage of the active volume is taken up by glassware or air, where a 5 mm tube, for example, is placed in a 10 mm coil. The second effect which can be expressed in terms of the filling factor is r.f. inhomogeneity. In a cross coil probe, with a large transmitter coil the r.f. field may be considered as uniform and all the nuclei within the sample are excited equally. However, for a single coil probe the exciting field generated will be far from homogeneous, decreasing rapidly outside the coil. As a result molecules initially outside the coil, but which may, by diffusion enter the active volume of the receiver coil within T_2^*, have been excited by a lower field than those initially in the centre of the coil which have diffused outward. The net effect is to lower the detectable signal by up to 60%. Effects of this type can be accounted for by reducing the filling factor by the appropriate amount.

λ

Nagaoka's constant expresses the actual inductance of the receiver coil in terms of the inductance of a section of equal length from an infinitely long solenoid (where the magnetic flux lines are parallel). This constant favours short coils ranging from a value of 0.99 for a coil where the diameter/length ratio is 0.02 to the value of 0.52 for a coil having a diameter to length ratio of 2.

A parallel analysis to the above for the continuous wave situation yields the same result as given in eqn. (6.26) provided that the receiver bandwidth in the latter case is made equal to $(T_2^*)^{-1}$. The receiver bandwidth decreases in the c.w. experiment because here, where only one line is observed at one time, the bandwidth of the receiver can be restricted to that of a typical line width. In a Fourier experiment on the other hand, where all the resonant nuclei are to be studied at the same time, the receiver bandwidth must be at least $\Delta\vartheta$ to accept all frequencies present in the f.i.d. The equations derived in this section calculate the achievable signal to noise ratio and do not take into account the time to obtain the signal. It is therefore to be expected that the continuous wave and Fourier transform expressions are the same.

REFERENCES

1 T.C. Farrar and E.D. Becker, Pulse and Fourier Transform N.M.R. Academic Press, New York, 1973, Ch. 3.
2 E.M. Purcell, H.C. Torrey and R.V. Pound, Phys. Rev., 69, (1946) 37.
3 F.A. Nelson and H.E. Weaver, Science, 146, (1964) 223.
4 F. Bloch, Phys. Rev., 94, (1954) 496.
5 W.A. Anderson, Rev. Sci. Instrum., 32, (1961) 241.

6 R. Freeman and D.H. Whiffen, Mol. Phys., 4, (1961) 321.

7 W.A. Anderson, Rev. Sci. Instrum., 33, (1962) 1160.

8 F. Bloch, W.W. Hansen and M.G. Packard, Phys. Rev., 70, (1946) 474.

9 D.D. Traficante, J.A. Simms and M. Malcay, J. Magn. Res., 15, (1974) 484.

10 Hewlett Packard Journal, June 1970.

11 A.G. Redfield and R.K. Gupta, Advan. Magn. Res., 5, (1971) 82.

12 A. Allerhand, R.F. Childes and E. Oldfield, J. Magn. Res., 11, (1973) 272.

13 E.O. Stejskai and J. Schaefer, J. Magn. Res., 14, (1974) 160.

14 P. Meakin and J.P. Jesson, J. Magn. Res., 10, (1973) 290.

15 J.P. Jesson, P. Meakin and G. Kneissel, J. Am. Chem. Soc., 95, (1973) 618.

16 J. Feeney, P. Pauwells and D. Shaw, Chem. Commun., (1970) 554.

17 See F. Gruenberger and D. Babcock, Computing with Minicomputers, Melridle
   Publishing Co., Los Angeles, 1973.

18 O. Ziessov, On-line Rechner in der Chemie, De Gruyter, Berlin, 1973.

19 E.O. Stejskal and I. Schaefer, J. Magn. Res., 14 (1974) 173.

20 R. Freeman, Varian Res. Rept., (1974) 107.

21 J.W. Cooley and J.W. Tukey, Math. Comput., 19, (1965) 297.

22 R. Codrington, H.D.W. Hill and R. Freeman, Varian Res. Rept., (1974) 105.

23 R.R. Ernst, Advan. Magn. Res., 2 (1967) 1 and references therein.

CHAPTER 7

EXPERIMENTAL TECHNIQUES

7.1 INTRODUCTION

The aim of this chapter is to provide the reader with a set of guide lines for preparing a sample and selecting the conditions and parameters necessary to produce a satisfactory n.m.r. spectrum. As appropriate to a book on Fourier n.m.r. the criteria used will be those belonging to a spectrometer operating in the pulsed mode. Whereas the information given is of a purely general nature, it will be heavily biased towards the recording of proton and ^{13}C spectra, these being the two nuclei most commonly studied when using Fourier transform techniques.

7.2. SAMPLE PREPARATION

In order to obtain a high resolution spectrum the sample must, in general, be in solution. It is possible, however, to record spectra in a gaseous state where molecular motion is also capable of averaging away the unwanted direct dipole—dipole broadening. Nevertheless, the gaseous state requires special experimental techniques and will not be dealt with in this general chapter. Also of a more specialist nature is the measurement of high resolution spectra in the solid state by the use of multiphase techniques.

Assuming that the sample is a solid it must be dissolved in a solvent. Table 7.1 shows the chemical shift of the resonance of typical solvents used in proton and ^{13}C n.m.r.

When choosing a solvent certain guide lines must be followed.

(a) The solvent should not contain any resonance lines which will directly overlap any signals anticipated from the solute.

(b) The solvent must contain suitable locking material, e.g. where a heteronuclear deuterium lock is used the solvent should be deuterated. If homonuclear locking is used, it should contain a

TABLE 7.1

Chemical shifts of common solvents (ppm)

Compound	Proton chemical shift	Carbon chemical shift[a]	
		protio	per deutero
TMS	0.0	0	—
Acetone	2.17	30.4	29.2
DMSO	2.62	40.5	39.6
Dioxane	3.70	67.4	—
Chloroform	7.27	77.2	76.9
Benzene	7.37	128.5	128.0
Water	4.75	—	—
Carbon tetrachloride	—	96.0	—

[a] From Levey and Nelson, ^{13}C N.M.R. for Organic Chemists, Wiley, New York, 1972.

suitable reference line, either originating from the solvent itself, or, preferably, to avoid complications due to solvent effects, from tetra-methylsilane dissolved in the solvent.

(c) The solvent itself should not give rise to a large signal as this will result in dynamic range problems at the digitisation stage. This problem is particularly important in proton n.m.r. where, if at all possible, deuterated solvents should be used.

Having chosen a solvent the next choice which must be made is that of the size of sample tube. In modern spectrometers the normal choice of tube diameters ranges from 1 mm to as large as 25 mm. If, for various reasons, it is desired to work at the lowest possible concentration, the choice is easy; use the largest sample tube available. If, as is more often the case, sample volume must be optimised for maximum signal to noise ratio, the choice is more complex. To achieve this end the maximum number of nuclei must be placed in the active volume of the receiver coil. As was outlined in section 6.4, typical receiver coils are broad flat coils; for such coils the active region extends outside the actual volume in a very non-linear manner. If the coil has a radius r then the signal detected along the axis of the coil (S_y) is related to that at the centre (S_0) by [1]

$$S_y = S_0 r^3 (r^2 + y^2)^{-3/2} \tag{7.1}$$

thus if $y = 2$ the signal per unit volume is only 10% that at the

TABLE 7.2

Effect of tube size on achievable signal to noise ratio

N.B. Constant resolution on all sample sizes is assumed. This is questionable for tubes over about 10—12 mm with nuclei giving narrow lines, particularly protons.

Size of tube	S/N/unit sample	S/N/unit concentration
1 mm	1	1
5 mm	0.040	25
8 mm	0.015	64
10 mm	0.010	100
25 mm	0.002	625

centre. Based on such arguments the optimum sample length would be only about 10 mm. However, because of the difference in susceptibility of the sample and the surrounding air, the magnetic field is distorted at the ends of the sample, and the resolution of the spectrometer is adversely affected. About a 20 mm length of sample is normally required to overcome end effects, although less than 50% of the volume is actually used. If a spherical sample is used, all the active volume of the receiver coil can be utilised and no end effects occur, but the coil filling factor is lower [2]. A good rule of thumb to maximise the achievable signal to noise ratio is therefore to dissolve the maximum amount of sample which is available in the minimum volume of the chosen solvent, and then use the largest sample tube the spectrometer can accommodate in which the solution yields a column at least 20 mm in length. Making the solution as concentrated as possible also minimises unwanted effects due to the signal of the solvent and any impurities in it. If the volume of solution obtained is not sufficient to provide a 20 mm length in the smallest sample tube then it is possible to use microcells to contain the sample. Microcells are normally small spherical containers which fit inside the conventional cylindrical tube; being spherical they make the most efficient use of the receiver coil's active volume. Microcells are less efficient than using a smaller sample tube and much less convenient to handle [2].

In the above discussion, when considering the 'smallest tube available' it is assumed that the spectrometer has a probe (or insert) optimised to accommodate such a tube. The filling factor, and consequently the sensitivity of, for example, a 5 mm tube in a 10 mm probe, is very much lower than for a 5 mm tube in a 5 mm probe (or insert). The situation with regard to the efficiency of various sample tubes is summarised in Table 7.2.

In order to measure the chemical shifts of the various resonances within the spectrum it is necessary to have a reference substance. This may be contained either in the solution, which is termed an internal reference, or externally in a separate small capillary within the n.m.r. sample tube, known as external referencing. The use of an external reference has the advantage that there can be no displacement of the reference peak to to intermolecular effects between the reference and the solute or solvent. It has, however, the decided disadvantage that the solution and the reference do not experience the same magnetic field, owing to their different magnetic susceptibilities. When the reference compound and the solution are contained in long cylindrical tubes, which is normally the case in high resolution n.m.r., the chemical shift measured with respect to the external reference may be corrected for susceptibility errors using the following formula.

$$\delta_{corr} = \delta_{obs} + \frac{2\pi}{3} \left(\chi_{ref} - \chi_{solution} \right) \tag{7.2}$$

This susceptibility shift can sometimes be used the other way around and be used to determine the susceptibility of the solution [3].

Using internal referencing, no bulk susceptibility corrections are necessary. One must, however, be aware of any possible effects of molecular association between the reference and the sample. In the case of tetramethylsilane (TMS), the reference compound normally used in both proton and carbom n.m.r., such effects will be minimal as it is a ball-like, isotropic and chemically alert molecule. Such a generalisation cannot be made for reference compounds used when studying other nuclei, e.g. $CFCl_3$ in fluorine n.m.r. TMS has two major drawbacks, firstly its high volatility makes it inconvenient when working at high temperatures, and secondly it is immiscible with certain solvents, particularly water. The first problem can be solved by the use of hexamethyldisiloxane (HMDS) which has a much higher boiling point and a chemical shift of only 0.06 ppm with respect to TMS. The second situation can be overcome by the use of the water soluble sodium salt 2, 2-dimethyl-2-silipentane-5-sulphonate (DDS) [4]. The methyl protons of this compound give a suitable strong single line, but unfortunately the methylene protons give appreciable signals of high gain. In carbon n.m.r., because the much larger total chemical shift range errors due to solvent—solute interactions are less significant, one of the solvents listed in Table 7.1 may be used as the reference material.

7.3 SPECTRAL WIDTH AND OFFSET

The spectral width (in ppm) used to record a spectrum will obviously depend primarily on the nucleus under investigation, and secondarily on the type of compound being analysed. A typical width for a proton spectrum is 10 ppm and for ^{13}C, 200 ppm; the typical range for other common nuclei can be seen by reference to Fig. 1.1. At the same time as deciding on a spectral width, one must decide where to place the frequency of the r.f. excitation, which will in turn depend on the locking material used in the spectrometer and the reference compound. After detection in the receiver all n.m.r. lines appear as offset frequencies from the frequency of the pulse, which thus becomes the zero point of the spectrum. Because of the problems associated with alaising or fold-over, it is simpler when recording a Fourier spectrum initially to set the spectral width and offset so that the spectral width covers the total expected chemical shift range, and the carrier is positioned either in the middle of the spectrum or at one edge, depending on whether quadrature phase detection is being used or not. In this way one can be certain of the true positions of all the lines within the spectrum. If at a later date it is necessary to expand the spectrum by decreasing the spectral width and changing the pulse frequency appropriately, then one is aware of the chemical shifts of all the major features within the spectrum, and can recognise any lines which have been alaised.

If it is required to look in more detail at a specific part of the spectrum in order to improve the digital resolution, it is necessary to decrease the spectral width. Great care must be taken with such experiments to place the pulse or zero frequency in the correct position to minimise the effects of fold-over. The origin of alaising due to digitisation has already been discussed in chapter 6 and is illustrated in Fig. 6.18. A second type of fold-over can occur if quadrature detection is not used, algebraic fold-over. It arises due to the inability of the spectrometer's electronics to distinguish positive and negative frequencies (with respect to the carrier). Effects of both types of fold-over are illustrated in Fig. 7.1. Figure 7.1b illustrates

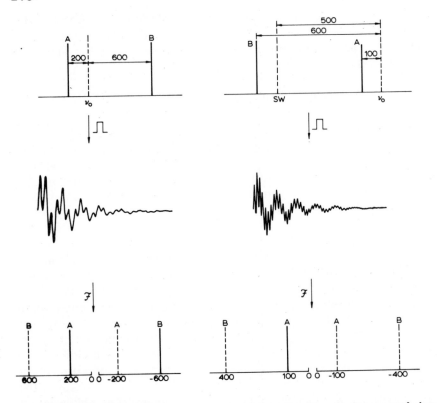

Fig. 7.1. Fold-over in n.m.r. (a) Originating due to the sign of a frequency being undetermined, (b) originating due to an inadequate rate of digitisation. Solid lines show the true peak positions, dotted lines apparent position.

the problem of fold-over due to an inadequate digitisation rate. The line labelled B has a true frequency with respect to the carrier of 600 Hz, but due to fold-over about the spectral width, appears in the spectrum as a line at 400 Hz, with respect to zero, and so, if alaising is not recognised, one would deduce the incorrect frequency. Figure 7.1a illustrates the problem when the carrier is placed in the middle of the spectrum, and quadrature detection is not used. The line labelled B, whose true frequency is 600 Hz higher than the carrier, occurs at |600 Hz|; the apparent chemical shift with respect to line A is thus only 400 Hz instead of the correct value of 800 Hz. Even when using quadrature detection, 'algebraic' fold-over will still occur, due to instrumental limitations. The folded peak will be less than

0.1% of its 'true' intensity, and consequently only significant in spectra having a high dynamic range.

Alaising also differs in effect when using single or dual detection. The case of single detection is shown in Fig. 7.1b; the ghost line from B appears at 400 Hz with a 90° phase shift. Using quadrature detection this phase shift results in the line appearing at 100 Hz, i.e. the axis of fold-over is at the opposite end of the spectrum.

The choice of the correct offset to record a partial spectrum can be quite a difficult problem, and in general should not be attempted unless the total spectrum has first been recorded.

If it is necessary to use a very narrow spectral width remember one gains nothing by decreasing the spectral width to such an extent that the digital resolution far exceeds the experimental line width; this in fact, may cause unnecessary complications. If, for example, one is interested in a certain line in a proton aromatic multiplet it may be much easier to record the whole of the aromatic multiplet than it would be to try and selectively record the line of interest.

One final spectrometer function connected with spectral width is that of the pre-a.d.c. filter. This filter should ideally be set equal to the spectral width for single detection and to $\frac{1}{2}$ the spectral width for quadrature detection in order to minimise the fold-over of noise and any unwanted peaks from outside the spectral width of interest (see Fig. 6.17). A spectrometer only has a limited number of filters with discrete band widths. In order to avoid severe phase shifts at the extremes of the spectrum, the narrowest filter with a *bandpass* exceeding the spectral width should be selected.

7.4 ACQUISITION TIME

Having decided on the spectral width (Δ), this is translated within the spectrometer into an acquisition time (T_a). As we have seen, sampling must take place at the Niquist frequency, which is twice that of the highest frequency. If the spectrometer has N locations available for data storage, the acquisition time is given by $T_a = N/2\Delta$, and determines the digital resolution obtainable. Consider two frequencies ϑ Hz apart, and let them both have zero value at zero time. These frequencies can only be resolved or distinguished when they have been out of phase by one wave length, which will occur after $1/\vartheta$ seconds. Two frequencies (lines) in a f.i.d. can therefore only be resolved if their frequency difference is larger than $1/T_a$. The frequency resolution obtainable is thus limited to $1/T_a$ Hz; therefore from the point of view of resolution, the longer the acquisition time the better.

In practice the acquisition time is limited by two factors, the available data storage, and the typical line widths being studied. To obtain high resolution, long acquisition times are necessary. Or to put it another way, large data tables are necessary. The free induction decay decreases as $\exp(-t/T_2^*)$. Therefore after about $3T_2^*$, the signal has decreased to 5% of its maximum value, and further data collection is of little value. If data acquisition is terminated at this point, digital resolution $(1/3T_2^*)$ effectively equals the observable line width $(1/\pi T_2^*)$. For proton work, where typical line widths may be in the order of 1/10th of a Hz, an acquisition time in the order of seconds is required to reproduce the true line width. For ^{13}C however, where, at least in the noise decoupling experiment, line widths are in the Hz region, acquisition time in the order of tenths of a second are required. Whether a desired acquisition time can be achieved or not depends on the size of the available data storage and the spectral width being studied. Table 7.3 gives values of T_a and N for typical spectral widths. 8K is the typical data table size used in routing spectrometers; 32K data tables normally require the use of a disc back-up store, especially if double precision arithmetic is used. If the desired resolution cannot be achieved over the whole of the spectral region of interest because of the limitations in available data storage, spectral expansion has to take place using the methods outlined in the previous section.

Should excess data points be available, or should sensitivity be

TABLE 7.3

Effect of data table size on Fourier spectra

Spectra width (Hz)	4096 data points		8192 data points		32,768 data points	
	Acquisition time (s)	Resolution (Hz)	Acquisition time (s)	Resolution (Hz)	Acquisition time (s)	Resolution (Hz)
50	40.960	0.02	81.920	0.01	327.680	0.003
100	20.480	0.05	40.960	0.02	163.840	0.006
500	4.096	0.24	8.192	0.12	32.768	0.03
1000	2.048	0.49	4.096	0.24	16.384	0.06
5000	0.410	2.44	0.819	1.22	3.277	0.30
10000	0.205	4.88	0.410	2.44	1.638	0.61

the prime aim of the experiment, then the resolution criteria considered above should not be used to decide on the acquisition time. Under these conditions the most efficient method of using the available data locations is to fill half with data and set the other half to zero. After transforming the total data table, i.e. $N/2$ data points plus $N/2$ zero's, as shown previously (Ch.3) as a result of the principle of causality, the sine and cosine series each contain all the available information.

7.5 CHOICE OF PULSE WIDTH

Chosing the correction pulse width is probably conceptionally the most difficult operation in optimising a Fourier transform experiment. It is also a very critical one because not only does it directly affect the signal to noise ratio that is achieved in a given time, but also can it have a very large influence on the relative intensities of the resulting peaks. Optimisation of the pulse width is the equivalent of optimising r.f. power in the swept experiment; indeed, the mean power is the same in both experiments. The difference lies in the fact that for swept excitation one can in general see *a* signal in a single scan; the correct power can thus be formed by a quick iterative

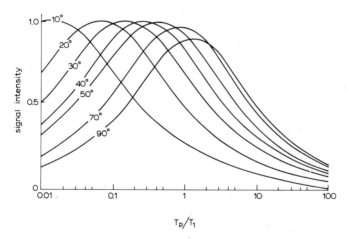

Fig. 7.2. The fractional intensity of a resonance obtained in a repetitive pulse experiment as a function of the interval between the pulses for various pulse angles.

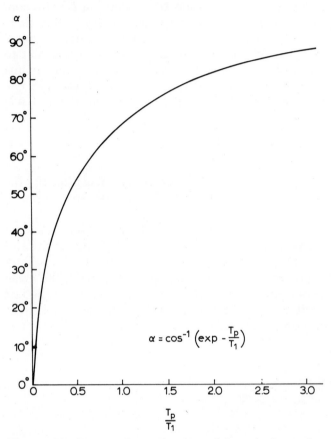

Fig. 7.3. The Ernst angle as a function of the pulse inverval.

process using the visible line. As pulsed experiments usually involve a significant degree of time averaging, such an *iterative* process is not so convenient and a prior decision has to be made. As was seen in Ch. 5, it is very difficult, if not impossible, to derive a criterion which will give the optimum pulse width for all lines within a given spectrum, even assuming that they all have the same spin lattice relaxation time. Figure 7.2 illustrates the achievable sensitivity in a Fourier experiment as a function of the pulse width used, expressed as an angle α°, and the ratio of the interval between the pulses and T_1. The optimum angle is the one which joins the tips of the family of lines in Fig. 7.2, and has hitherto been referred to as the Ernst angle (α_E) [5], thus:

$$\cos \alpha_E = \exp \left[-(T_p/T_1)\right] \qquad (7.3)$$

Equation (7.3) gives the best criterion to decide on the optimum pulse angle to use in a given Fourier experiment and is shown graphically in Fig. 7.3; this choice ignores any of the effects which may result from spin echoes built up by a steady state established within the experiment. If such steady states do exist a longer pulse width could profitably be used. Although the error in the pulse width can be as large as a factor of 2, even this, as Fig. 7.2 demonstrates, can still recover up to 90% of the available signal.

Having accepted that the choice of the Ernst angle is the best way of optimising the pulse width used in our experiment, we have still two degrees of freedom left, namely the spin lattice time we chose and the interval between the pulses. If we base our estimate of pulse power on the nuclei in the spectrum which have the longest T_1's then, whereas we will reproduce all the intensities within the spectrum as accurately as possible, we will be applying well below the optimum power to the lines with a shorter T_1, i.e. their signal to noise ratio will be lower than that which it is possible to achieve. Figure 7.2 illustrates the penalty paid. If on the other hand we use the nucleus with the shortest T_1 as a basis for our calculation, we will achieve the maximum signal to noise ratio at the nucleus. Applying this power, a longer T_1 will, however, saturate the lines from the other nuclei, which will be diminished in intensity compared with their true value, and may in some cases disappear completely from the spectrum. Again Fig. 7.2 gives the decrease in intensity. One must therefore decide what is the longest T_1 anticipated from a line whose accurate intensity is required in the final specctrum. In ^{13}C, quite often, there will be a nucleus, say a carbonyl, which has a very long T_1; if this is to be represented accurately in the final spectrum, the whole experiment must be slowed down to keep pace with the relaxation of this line. If, however, the resonance from the carbonyl is of no interest, then it could be ignored and the experiment optimised for the faster relaxing carbons of the aliphatic part of the molecule. Figure 7.4 illustrates the effect of using various pulse widths on the ^{13}C spectrum of α-pinene. The interval between the pulses in all cases is T_a. The spectra are plotted out with a constant gain and result from a constant number of pulses. When a low pulse angle, e.g. $10°$, is used, all the lines show their 'correct'

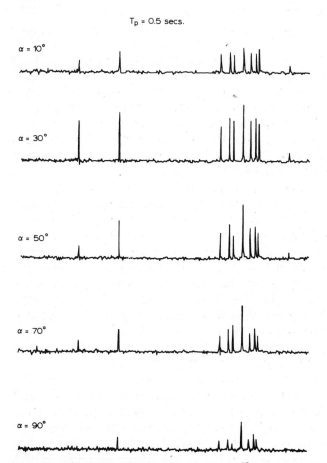

Fig. 7.4. The effect of pulse width on the ^{13}C spectrum of α-pinene. All spectra result from 100 pulses 0.5 s apart.

intensities. When the pulse angle is increased to approximately 30°, all the lines increase in intensity. However, when the pulse angle is increased beyond this up to about 50°, whereas the aliphatic carbons increase in intensity, the quatenary carbons show a slight decrease in intensity. By the time the pulse angle has been increased to a value of 90°, all lines are showing a decrease in intensity, and the quaternary carbons have decreased to such an extent that they are almost absent from the spectrum. It should be noted that, unlike swept n.m.r., no line broadening accompanies saturation using pulsed excitation.

Having decided on the value of T_1, to be used as a base for the optimisation of the pulse width, one still has two degree's of freedom left, the pulse width itself and the interval between the pulses. The latter is made up of two parts, the time required for data acquisition and any delay between the end of data acquisition and the start of the next pulse. A delay is necessary, for example, in a gated decoupling experiment (see section 7.6) to allow the nuclear Overhauser effect either to grow or decay. From the point of view of sensitivity it is best to use no delay unless necessary, i.e. set the pulse interval equal to the acquisition time dictated by the required spectral width and resolution and to adjust the pulse angle appropriately.

The advantages of using minimum pulse interval and angle, as opposed to the alternative approach of always using a 90° pulse optimising conditions by varying the pulse delay, are threefold. Firstly, as can be seen in Fig. 5.7, where the use of minimum pulse interval corresponds to a minimum ratio of T_p / T_1, the sensitivity achievable is higher than that obtainable using a 90° pulse (i.e. $T_p / T_1 = 1.25$) by up to 20%. Secondly, the use of minimum pulse width minimises the non-uniformity in power distribution across the spectrum resulting from finite pulse length. Finally, when studying systems where non-Boltzman populations are involved, e.g. CIDNP, where the Fourier relationship between the f.i.d. and the spectrum is strained, the strain is less when using small pulse angles (see section 4.11). There are, however, two instrumental criticisms of always using the minimum pulse width and interval, neither of which should be valid given good spectrometer design. They are, firstly, that the above rule leads to co-adding the largest number of f.i.d.'s which, if digitisation is itself a source of noise, would lower the sensitivity. Secondly it may be impossible to time accurately the short pulse width required.

The criteria disucssed above give an optimum pulse for the experiment in terms of an angle of rotation. If one wishes to set the spectrometer, one needs to convert this angle into a pulse width. The conversion can only be done knowing the effective field generated by the spectrometer. The most convenient way to measure this is by determining a specific pulse for the spectrometer. The determination can be made using either the 90°, 180° or 360° pulse. This value can be used via eqn. (5.2) to calculate the effective field of the spectrometer, or, more simply, to provide a calibration which enables pulse angles to be converted directly to pulse widths.

184

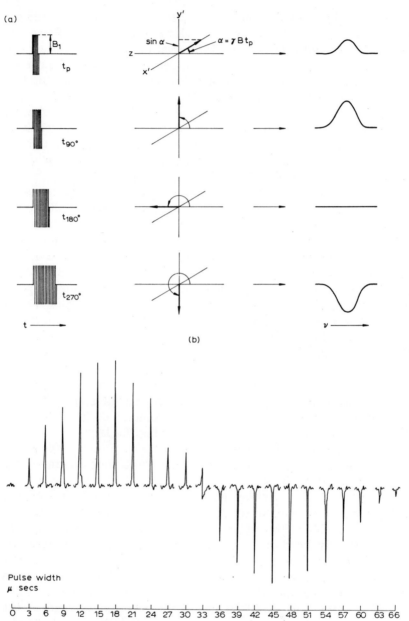

Fig. 7.5.(a) The effect of increasing pulse width on an n.m.r. line. (b) The calibration of pulse power using the ^{13}C resonance of benzene.

Figure 7.5a illustrates the theoretical behaviour of an n.m.r. signal as the duration of the pulse is increased and defines what is meant by a 90°, 180° pulse, etc. Figure 7.5b shows a real example, a single line, in this case the ^{13}C resonance of benzene, is recorded and the pulse width is progressively increased. The peak height increases and goes through a maximum, then crosses over, becomes negative and decreases to a minimum. The tip of the first maximum corresponds to the 90° pulse and the zero to the 180° pulse. If the pulse width is further increased, the next zero signal will correspond to a 360° pulse. Although in principle any of these points will be satisfactory, in practice it is much easier to determine the zero corresponding to a 180° or 360° pulse rather than to determine the maximum corresponding to a 90° pulse. In the case of a single coil probe, where the exciting field is not very homogeneous, in general it is almost impossible to get a zero corresponding to the 180° pulse, as a small residual out-of-phase signal will always be generated. If this is the case then the 360° pulse should be used as the calibration point. Here such effects are refocussed and a true zero is obtained. The value of the 90° pulse should be one quarter that of the 360° pulse. If this is found not to be the case, the 360° pulse being longer than four times the 90° pulse, this indicates that either the power amplifier on the spectrometer is 'drooping' i.e. it is not keeping up a constant r.f. power throughout the whole duration of the pulse (see section 6.6), or there is some incremental error in the pulse timing.

7.6 SPIN DECOUPLING

Before starting data accumulation any decoupling conditions required must be set up. Heteronuclear decoupling will now be considered from this point of view, taking proton decoupling whilst observing ^{13}C as an example. If straightforward noise decoupling is required, the decoupler frequency should be placed in the centre of the proton spectrum, the noise band width set to the width of the proton spectrum anticipated for the compound under investigation, and the full power used. Should selective decoupling be required then of course no noise modulation should be used and the decoupler frequency should be offset to the correct part of the proton spectrum. The power used depends on the type of decoupling being undertaken. The detail as to how the frequency of the decoupler can be correctly positioned depends on the type of spectrometer being used; calibration of the B_2 frequency with a known compound may possibly be required.

In order that selective heteronuclear and homonuclear decoupling experiments may be carried out effectively, it is necessary to know the field strength which is being generated by the spin decoupler. The strength of the decoupling field can be determined by measuring the residual splitting within a multiplet when the decoupler is offset by a known amount from the true resonance position of the other half of the spin multiplet. This is illustrated by Fig. 9.9 for the case of a proton decoupler being used in a ^{13}C experiment. As the decoupler frequency is incremented through the proton frequency, the doublet splitting in the carbon spectrum progressively decreases until, precisely at resonance, as expected, the ^{13}C spectrum is a single line. The multiplicity reappears as the offset from the true resonance is increased. This behaviour is described by the following equation provided that the offset frequency $\delta\vartheta$ is not too large (see section 9.2) [6].

$$J_R = \frac{J_{CH} \cdot \delta\vartheta}{\gamma B_2} \tag{7.4}$$

where J_{CH} is the CH coupling constant, $\delta\vartheta$ is the frequency difference between the correct decoupling frequency and the off-resonance frequency used, and the J_R is the residual splitting in Hz. From a series of such experiments the value of B_2 can be calculated. Armed with the knowledge of the magnitude of the decoupler field it is possible to set the power required in an off resonance experiment. If a gated decoupling experiment is to be performed either to maintain or to remove the nuclear Overhauser effect (see section 9.7), a delay must be introduced between the end of the data acquisition time and the next pulse to permit the growth or decay of the enhancement. Figure 7.6 shows approximately the n.o.e. present in the steady state for the two experiments.

The best optimisation procedure for quantitative results is probably to use a $90°$ exciting pulse, estimate T_p from Fig. 7.3, using the T_1 value of the slowest relaxing carbon of interest. The degree of n.o.e. remaining is then given by Fig. 7.6a; should this be unacceptably high for the accuracy required, then T_d must either be increased with a loss of sensitivity, or T_a decreased, also with a loss in sensitivity. Optimisation for sensitivity is not as easy; the logic outlined above is satisfactory except that since increasing T_d

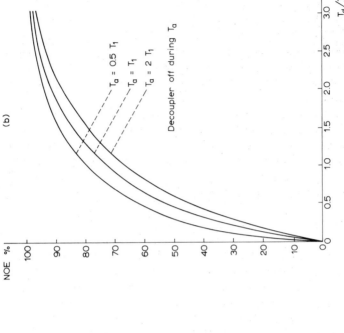

(a)

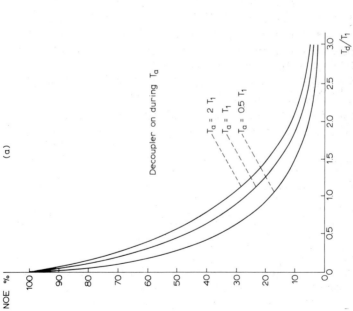

(b)

Fig. 7.6. The nuclear Overhauser effect present in the steady state of a gated decoupling experiment. (a) With the decoupler on during data acquisition; (b) with the decoupler off during data acquisition.

increases the n.o.e. but lowers the total sensitivity. Unless the contribution from the n.o.e. is known, optimisation can go no further than the value of T_d given by the Ernst angle formula.

In homonuclear decoupling experiments, the frequency of the decoupler is set to the appropriate value and the power adjusted to that required for the experiment (see Table 9.11). The detail as to how this is achieved depends on the type of decoupler used in the spectrometer.

For homonuclear decoupling a power in the order of 2—3 times the spin coupling constant to be 'removed' is required in order to achieve successful decoupling. Using this rule the power can be set, although in general it is not possible to set the precise frequency to achieve optimum decoupling because of the presence of Bloch-Siegert effects (see section 9.3).

7.7 NUMBER OF TRANSIENTS

Prior to commencing the data acquisition some estimate as to the number of free induction decays to be co-added prior to Fourier transform needs to be made. This choice is dictated either by the signal to noise ratio one wishes to achieve, or by the total time that is available to perform the experiment. It should be remembered that the signal to noise ratio only increases as the square root of the time spent; therefore, in order to double the signal to noise ratio, the experiment would need to run for four times as long, etc. Table 7.4 illustrates the effect of the available time of the achievable signal to noise ratio and gives the time required to co-add various numbers of free induction decays, assuming a data acquisition time of 1 s.

As a rule, it is impossible to judge from the time domain spectrum what the final signal to noise ratio of the frequency domain spectrum

TABLE 7.4

Effect of time averaging on sensitivity		Time required to co-added 1 s f.i.d.'s	
Duration of experiments	Sensitivity	Number of transients	Total time
1 second	1	1	1 second
1 minute	7.7	100	1.67 minute
1 hour	60	1000	16.67 minute
1 night (12 hours)	207	10000	2.78 hours
1 weekend	440	100000	1.16 days

is going to be without carrying out a Fourier transform. In some simple cases it is possible, but not in more complex ones, and it is impossible if a large solvent signal is present which totally dominates the free induction decay. If the spectrometer has the facility to keep a copy of the free induction decay either within the core of the computer or in a peripheral device, then from time to time the free induction decay may be copied; the original can then be transformed into a frequency domain spectrum, and a more accurate estimate of the time required to achieve the desired signal to noise ratio can be made. The copy is then reloaded into the spectrometer's data area and the data accumulation resumed. When this facility is not available the number of free induction decays to be co-added prior to transformation must be assessed via experience and a knowledge of the performance of the spectrometer being used.

7.8 SETTING THE GAIN

Having determined all the required parameters, data acquisition can now commence. There is, however, one final adjustment which

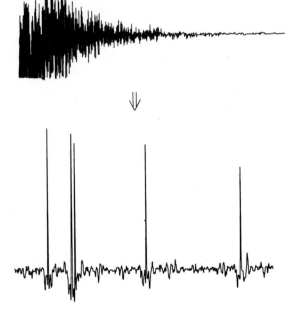

Fig. 7.7. The effect of a.d.c. overload.

must be made at this stage, i.e. setting the gain of the receiver so that it provides the optimum signal into the a.d.c. The gain setting can only be achieved while the correct pulse sequence is being applied to the sample, and hence should be done during a short preliminary run. It is desirable, in order to achieve the maximum dynamic range, to provide the analogue to digital converter with the largest signal it can conveniently handle. If the gain of the receiver is set too low, only a small part of the possible dynamic range of the analogue to digital converter will be used, and the various points on the free induction decay will only be specified by a few bits. Consequently, the accuracy to which they are represented is not the maximum available, and the experiment will suffer. On the other hand, if the gain of the receiver is such that the a.d.c. is overloaded, then a much worse situation results. The points at the beginning of the free induction decay will be represented inaccurately. Figure 7.7 illustrates such a situation. Here the voltage provided to the a.d.c. is such that approximately the first tenth of the free induction decay is distorted. The Fourier transform of this is also illustrated. Despite the fact that only the first 10% is distorted, because each point in the free induction decay contains information about every point in the frequency domain spectrum, the whole of the spectrum is distorted. The resultant distortion can easily be explained by considering the distorted f.i.d. as a product of the true f.i.d. and a square window function. On Fourier transformation the output is the convolution of the Fourier transforms of these two components. The Fourier transform of the true free induction decay is the required spectrum. The Fourier transform of the square window function is a sinc ϑ function and what is observed is the convolution of these two spectra. The true Lorentzian line is broadened to $1/t$ where t is the width of the window function, and shows 'ringing' before and after the peaks, with a frequency of $1/t$. Once such a distortion has taken place there is no way of recovering the information; a.d.c. overloading should therefore be avoided at all costs.

The first pulse in a sequence is not necessarily typical since it occurs before any steady state has had time to build up and before any consequences of 'saturation' have become apparent. The first few pulses of a sequence often give a larger free induction decay then the following ones. It is important, therefore, when setting the receiver gain to do it after the first few pulses have been initiated, and where conditions typical of those under which the majority of the data is to be collected have been established. If only a few transients are being stored, the first few typical values will contribute significantly to the time averaged data, and a steady state should be allowed to build up before the data acquisition starts. This is particularly true when using saturation and multipulse techniques to eliminate solvents (see section 7.10).

7.9 THE USE OF WEIGHTING FUNCTIONS

Having completed the data acquisition successfully, prior to transforming into the frequency domain, it is usual to multiply the free induction decay by a weighting function, to optimise either the signal to noise ratio or the resolution, depending on the required result. As we have seen, these weighting functions in n.m.r. are simply either positive or negative exponential functions. They can be

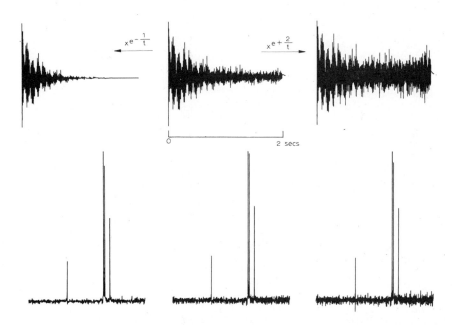

Fig. 7.8. The figure shows the effect on the ^{13}C spectrum of ethyl benzene of different weighting functions. The central spectrum is obtained from the unweighted free induction decay ($T_a = 2$ s, $T_2^* = 0.7$ s). The left hand spectrum is obtained by transforming the f.i.d. after multiplying it with a negative weighting function, the right hand spectrum after multiplying by a positive function.

described as exp (t/TC), where TC, the time constant, is a parameter that has the dimensions of time. The application of weighting functions with both positive and negative time constants are illustrated in Fig. 7.8. When chosing a time constant which will maximise the signal to noise ratio of the spectrum, it is necessary to decide what is the largest line width that is present in the spectrum or will be tolerated. On being multiplied by a negative weighting function, all lines of the spectrum will be broadened by $1/\pi\ TC$, as can be seen from Table 3.1.

The resultant linewidth is given by

$$\delta\vartheta \ = \ \frac{T_2^* - TC}{\pi T_2^* TC} \tag{7.5}$$

If the linewidth is known, TC should be set at $-\ T_2^*$: this results in a doubling of the linewidth. Doubling of the linewidth to achieve

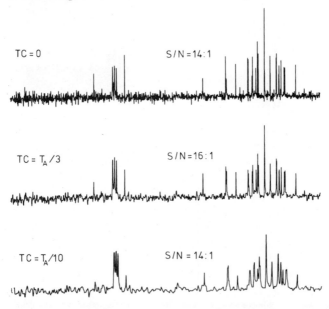

Fig. 7.9. The effect on signal to noise ratio of the ^{13}C spectrum of α-pinene induced by multiplying the free induction decay by different weighting functions (exp $-\ t/TC$). The middle spectrum results from the use of the matched filter, the other two spectra have the same signal to noise ratio but result from the use of incorrect filter functions. Note the excessive line broadening induced by using too small a time constant.

the optimum signal to noise ratio also occurs in swept n.m.r. where the use of a saturation field ($S = 1$ in eqn. 2.61) also doubles the linewidth [7]. If, as is generally the case, the linewidth is unknown prior to transformation, TC can be estimated from the f.i.d.; it should be equal to the time at which the signal has decayed to one e^{th} (~ 0.36) of its initial value. An even simpler criterion, though less accurate, is to set TC equal to $-1/3T_a$; this assumes that the linewidth of the final spectrum is limited by the acquisition time. Figure 7.9 shows the signal to noise ratio achievable by applying differing weighting functions; note the increasing linewidth and lack of resolution as TC is decreased.

The use of resolution enhancement functions (positive values of TC) requires a little more care. The main danger is using a time constant which is too small, which results, after weighting, in a free induction decay where the signal is larger at the end of the data acquisition than at the beginning. Transformation of such a spectrum is at the best misleading, and at the worst impossible. The optimum time constant for resolution enhancement, like that for sensitivity, is equal to the effective spin—spin relaxation time for the line being studied, but is positive in sign. The use of such a time constant theoretically produces a free induction decay which is constant with time. In practice, however, unless the free induction decay were to have an exceptionally good signal to noise ratio, the use of such a time constant would produce a prohibitively noisy spectrum and, unless severe apodisation were used, a very distorted one. In practice, time constants in the order of 2 or 3 times the effective spin—spin relaxation time are appropriate.

When experimenting with a choice of weighting functions it should be remembered that since the time constant appears in the exponent, the effect of multiplying the free induction decay by a weighting function three times is the same as multiplying it once by a weighting function equal to three times the value. It should also be remembered that weighting functions can be reversed, i.e. multiplying with a weighting function of -1 can be corrected by multiplying with a weighting function with a time constant of $+1$ (ignoring arithmatic errors). If facilities are available to make a copy of the free induction decay it may be advantageous to read out the spectrum using more than one value for the weighting function. Thus one region of the spectrum can be treated in such a way as to increase the resolution, whereas another area can be read out under conditions of high signal to noise ratio at the sacrifice of resolution, e.g. to detect a weak peak.

Severe problems with resolution enhancement can be met if the spectrum of interest comes from a small amount of solute in a magnetically active solvent, for example proton spectra recorded in D_2O. If the free induction decays resulting from this type of sample is multiplied by even a modest positive exponential, the signal resulting from the slowly relaxing HOD line very soon shows severe truncation effects, and any further resolution enhancement simply leads to a spectrum which is so distorted as to be unusable. One way round this problem is to use some of the subtraction properties of free induction decays rather than the multiplicative properties used so far [8]. To do so, one copy of the free induction decay is multiplied by a time constant TC_1. On transformation this will give a Lorentzian line with a line width of $(\pi T_A)^{-1}$ where $1/T_A = 1/T_2^* + 1/TC_1$. Similar treatment of the f.i.d. with an exponential function whose time constant is TC_2 leads to a line characterised by T_B. If the second is multiplied by a constant K and subtracted from the first spectrum, a third and different spectrum is generated, which has all the frequencies of the original spectrum, but is no longer a Lorentzian line shape. The new line shape is given by the following equation [8].

$$L(\vartheta) \propto \frac{T_A - KT_B + (2\pi\vartheta_i)^2 (T_A T_B^2 - KT_B T_A^2)}{1 + (2\pi\vartheta_i)^2 (T_A^2 + T_B^2) + (2\pi\vartheta_i)^4 T_A^2 T_B^2}$$

K is normally adjusted empirically to give the best line shape. The process is equivalent to multiplying the f.i.d. by $\exp - (t/TC_1 + K \exp - t/TC_2)$. The new line shape has the following properties.

(1) It reduces rapidly to zero due to the ϑ^4 appearing in the denominator.

(2) The intensity is reduced by $T_2^*/(T_A - T_B^2/T_A)$.

(3) It has a line width which is reduced to $\pi(T_A^2 + T_B^2)^{-1/2}$.

Spectra resulting from this technique, often called 'convolution difference', show resolution enhancement (by property 3) but with a lower signal to noise ratio than the original (by property 2) and some line shape distortion (property 1). The technique is of particular application in the case of protein solutions where the line shape is of less importance than the resolution of small separations, particularly when removing line broadening resulting from the addition of paramagentic ions. The convolution difference technique can also be used to remove broad lines. This is achieved if TC_1 and TC_2 are very much greater than T_2^*; their normal decay is virtually unaffected by the multiplication, and on subtraction the lines disappear from the final spectrum.

7.10 APODISATION

Apodisation functions are applied to minimise the consequences of inevitable discontinuity when data acquisition is terminated. These are only serious when there is still a finite signal present at the end of data acquisition. Normally, as is illustrated in Fig. 6.21 only the last part of data is apodised. However, in experiments where the discontinuity is extreme, e.g. where a resolution enhancement function has been used, a larger fraction of data must be apodised. Figure 7.10 illustrates the effect of increasing the degree of apodisation on a free induction decay which has been subjected to resolution enhancement. The distortions obviously present when no apodisation is used are progressively removed as the degree of apodisation is increased. Excessive apodisation can lead to the disappearance of weak lines.

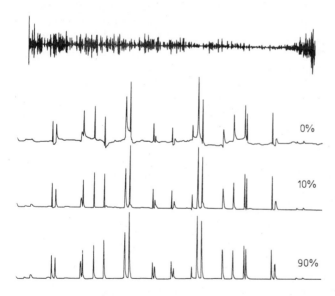

Fig. 7.10. The effect of increasing the percentage of apodisation on the resolution enhanced free induction decay from the protons in *ortho*-dichlorobenzene.

7.11 PROBLEMS ENCOUNTERED IN THE PRESENCE OF LARGE
SOLVENT PEAKS [9]

In proton Fourier transform work, particularly when dealing
with biochemical applications (where the majority of experiments
are performed in aqueous media), one is often confronted with the
problem of detecting dilute resonances ($10^{-3} - 10^{-5}$ molar) in the
presence of a large residual HOD line. As discussed earlier the situation poses a problem of dynamic range both for the analogue to
digital converter and for the word length of the computer memory.
The ways in which the instrument can best handle the dynamic range
problem have already been discussed. This section is concerned with
experimental techniques which can be used to prevent the large
signals appearing at the digitisation stage.

The problem of the large solvent signal can be tackled chemically
by the replacement of the nuclei in the solvent with a different
isotope which does not give resonances, in the area of interest. The
classical case of this is deuteration of solvents in proton n.m.r.
Deuteration, even to a very high extent, typically 99.9%, is possible,
but does not solve this problem completely as the ratio between
solvent and solute molecules can be very high (10,000:1). In ^{13}C
n.m.r. isotopic substitution can eliminate the problem; here ^{12}C
enriched solvents are available (or maybe ^{13}C depleted is a better way
of putting it). Because of the lower sensitivity of this technique,
concentrations which give solvent to solute ratios greater than a
few thousand to one are impractical because of the time it would
take to perform the experiment. In this case, sufficient enrichment
is possible. Basically four instrumental methods can be adopted.

(1) Use of a two pulse sequence to discriminate against the solvent
by virtue of its relaxation time [10].

(2) The 'saturation' of the solvent line by double irradiation [11].

(3) The partial saturation using a sequence of $90°$ pulses to establish
a steady state regime.

(4) The use of tailored excitation in such a way as not to excite
the unwanted solvent line [12].

The first method of eliminating the solvent signal takes advantage
of the difference in spin—lattice relaxation times between the nuclei
of sample and those of the solvent. Provided the solute nuclei relax
faster than those of the solvent, it is possible to apply a $(180° - \tau$
$90° - T_p)_n$ pulse sequence in order to eliminate the solvent magnetisation. If the pulse interval τ is chosen such that the $90°$ observing

pulse is applied at the instant when the solvent magnetisation passes through zero, the solvent line is eliminated from the spectrum. The experiment is greatly improved if a homogeneity spoiling pulse is introduced immediately following the $180°$ pulse to eliminate any transverse magnetisation remaining due to the imperfect nature of this pulse. In a simple experiment it is essential that T_p is greater than or equal to five times the T_1 of the solvent. Since the spin—lattice relaxation time of the proton in HOD can be as long as 20 s, the efficiency of the experiment with regard to signal to noise ratio is not very high. Increased efficiency can be achieved by running the experiment under steady state conditions. The values of τ and T_p are now adjusted in such a way that in the steady state the M component of the magnetisation due to the solvent resonance is zero immediately preceding the observing $90°$ pulse. The steady state condition can be described by the following equation [13]

$$\exp\left(-(T_{1s}\ln 2 + T_p)/T_{1s}\right) + \exp\left(-(T_{1s}\ln 2 - \tau)/T_{1s}\right) = 1 \qquad (7.6)$$

Equation (7.6) allows one to calculate τ for a chosen value of T_p, usually taken to be the acquisition time. Figure 7.11 is a graph of eqn. (7.6). A prerequisite for the application of this technique is a precise knowledge of the solvent spin—lattice relaxation time. For

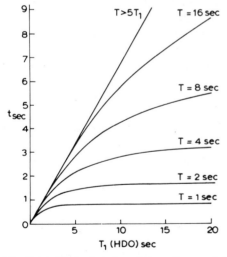

Fig. 7.11. Plots showing the optimum delay between the $180°$ and $90°$ pulses in solvent elimination experiments as a function of T_{1s} for various pulse delays.

aqueous solutions, which are the principal application of solvent elimination techniques, a T_{1s} measurement requires roughly 15 min. A steady state where there is a dynamic balance between the effect of the pulses and relaxation is usually obtained after the first three or four passes through the sequence, and data should not be collected until this steady state is established. For its success the two pulse sequence relies on the spin—lattice relaxation time of the solvent being longer than that of the sample. For this reason care must be taken to minimise the presence of any paramagnetic agents which may shorten the relaxation times drastically, and it is advisable if possible to de-gas the sample to remove dissolved oxygen.

The method is equally applicable to ^{13}C n.m.r. where occasionally the signal from deuterated solvents can cause problems. The situation here is more favourable than in proton n.m.r. because ^{13}C spin—lattice relaxation times of per-deuterated molecules are relatively long due to the unavailability of ^{13}CH dipolar interactions to provide a relaxation mechanism.

The second method of attenuating the solvent signal consists of irradiating it selectively. By this method, the populations of the two spin states of the solvent protons are equilibrated, and any signal resulting from it is removed from the spectrum. In order to avoid beat notes in the spectrum it is important to use as little decoupling power as possible. Practically, the optimum B_2 level is best achieved by observing the free induction decay of the solvent resonance on an oscilloscope. The decoupler frequency is then carefully adjusted until a minimum response is obtained; the gain of the spectrometer is then adjusted in order to make maximum use of the dynamic range of the a.d.c. The limitation of the double resonance method lies in the inability to retrieve signals near the solvent resonance, as is possible with the $180° - \tau - 90°$ technique. On the other hand it is entirely independent of the ratio of the T_1's of the solvent and solute.

The third technique, the use of $90°$ pulses, is the least powerful from the point of view of solvent suppression, but it is probably the most efficient with regard to the signal to noise ratio which can be achieved. As with the inversion recovery procedure the technique relies on the difference in spin—lattice relaxation times between the solvent and solute nuclei. When a series of $90°$ pulses are applied to the spin system and when the pulse interval is chosen to be relatively short compared with T_{1s}, the magnetisation cannot recover completely to

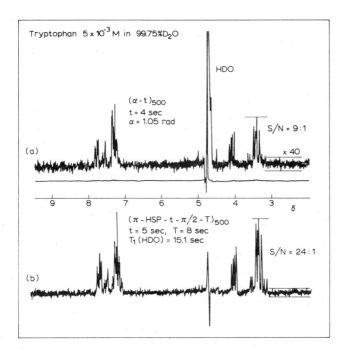

Fig. 7.12. Solvent elimination by the $180° - \tau - 90°$ method for $5 \times 10^{-3} M$ tryptophan in 99.75% D_2O [9].

its thermal equilibrium between the pulses, hence the resulting signals are of lower amplitude. The signals with the longer relaxation times, i.e. the solvent, are preferentially affected and the solvent signal is suppressed with respect to the solute. Again, in order to allow a steady state to be build up, data should not be acquired until the pulse sequence has been run through three or four times. The $(180° - \tau - 90°)$ approach is illustrated in Fig. 7.12.

The final approach to the elimination of a large unwanted solvent peak is simply not to irradiate it at all. To this end one uses a tailored excitation frequency as discussed in section 4.8. A notch is cut in the excitation spectrum at the solvent frequency. The relative efficiency of this technique again depends on T_1 properties. The application of tailored excitation to solvent elimination in ^1H n.m.r. is illustrated in Fig. 7.13, which compares the signal arising from tyrosine dissolved in D_2O, when conventional stochastic excitation is applied, with that arising from the use of flexible excitation with a notch in the

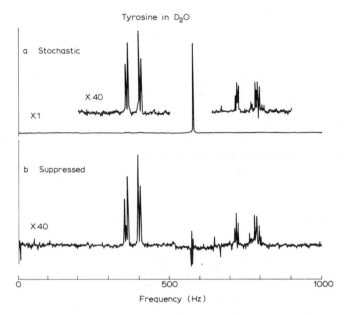

Fig. 7.13. Solvent elimination by tailored excitation for 12 mg of tyrosine in 1 ml of 99.7% D_2O. From B.L. Tomlinson and H.D.W. Hill, J. Chem. Phys., 59 (1973) 1775.

excitation function at the solvent position. The solvent suppression obtained here is in the order of 150:1 which compares favourably with the previous more conventional techniques. With tailored excitation one can also mimic the saturation method of eliminating solvents; rather than putting a notch in the excitation spectrum in order to excite the solvent, one puts a spike in the power spectrum to saturate the solvent signal. In practice it is quite often found that the second method is more efficient than the first.

7.12 SPECTRAL PRESENTATION

Having transformed the free induction decay into the frequency domain, the final aspect of the experiment is to present the data in the most informative way. It is initially displayed on an oscilloscope so that it may be phased and scaled etc. and then a hard copy is made using a suitable recorder. Various regions of the spectrum may be read out with different vertical and horizontal scales in order that the spectroscopist may extract all the available information. It must be remembered that all the 'spectra' obtained in this way are only different presentations of the original spectrum; they contain the same information.

The most important property of a line in an. n.m.r. spectrum is its frequency or chemical shift. The computer of a Fourier spectrometer is capable of calculating this parameter with respect to either the carrier frequency or a specific line identified as the reference. A line is detected by the computer as a point where the plot has a maximum value; noise, of course, has many thousands of such values. A threshold value must be defined, above which a maximum is a peak and below which it is noise. Setting a threshold is an easy task, given a good or reasonable signal to noise ratio and sharp lines; problems can arise when these conditions are not satisifed, especially when broad lines are present. The accuracy to which frequencies can be determined by the computer depends on the sophistication of the software; if the value taken is simply that of the data location of the maximum value, the accuracy is $\pm \Delta/N$; greater accuracy can be achieved if interpolation routines are used.

After the chemical shift a peak's area is probably the most important characteristic. Such areas are determined by integration using standard digital integration techniques on the stored data; a separate experiment is not required. If there is any slope or roll in the base line, it should be removed prior to integration. The phase should also be carefully set, as the presence of any u mode component will alter the area. Error in phase setting is probably the limiting factor in the accuracy of an integral. As discussed earlier, the integral depends on the magnitude of the magnetisation at zero time; phase setting is equivalent to adjusting the zero time point and consequently has a significant effect on the integration. In order to obtain accurate integrals rather than merely reproducible integrals, great attention must be paid to the setting of a pulse angle and any delays involved in the experiment in order to avoid saturation effects.

If it is required to compare the absolute value of an integral from two different samples it is necessary to ensure that the software keeps a correct account of any scaling operations it has performed during the data acquisition process and allows for them when outputting the integral. Using a suitable reference line as a calibration mark improves the accuracy of the procedure.

This chapter has discussed the various experimental parameters which have to be optimised and specified in order to perform a successful experiment. The sequence and method used to set up these parameters will obviously depend on the specific spectrometer being used; the logic, however, does not. A general, though by no means unique approach, for setting up a pulse Fourier transform experiment is given in Appendix 2 in the form of a flow chart.

REFERENCES

1 F.A. Nelson, N.M.R. and E.P.R. Spectroscopy, Pergamon, London, 1960.
2 B. Erwine, Sensitivity Enhancement for Analytical N.M.R. Spectroscopy, Varian Application Note, NMR-75-1, 1975.
3 D.F. Evans, J. Chem. Soc., (1958) 2003.
4 G.V.D. Tiers and R.I. Coon, J. Org. Chem., 26, (1960) 2097.
5 R.R. Ernst and W.A. Anderson, Rev. Sci. Instrum., 27, (1966) 93.
6 R.R. Ernst, J. Chem. Phys., 45, (1966) 3845.
7 R.R. Ernst, Advan. Magn. Res., 2, (1957) 1.
8 I.D. Campbell, C.M. Dobson, R.J.P. Williams and A.V. Xavier, J. Magn. Res., 11, (1973) 172.
9 F.W. Wehrli, Solvent Suppression Technique in Pulsed N.M.R., Varian Application Note NMR-75-2, 1975.
10 S.L. Pratt and D.B. Sykes, J. Chem. Phys., 56, (1972) 3182.
11 J.P. Jesson, P. Meakin and G. Kneissel, J. Amer. Chem. Soc., 95, (1973) 618.
12 B.L. Tomlinson and H.D.W. Hill, J. Chem. Phys., 59, (1973) 1775.
13 F.W. Berg, J. Feeney and G.C.K. Roberts, J. Magn. Res., 8, (1972) 114.

CHAPTER 8

THE N.M.R. SPECTRUM

8.1 INTRODUCTION

Up to this point both the theoretical and practical aspects of n.m.r. spectra have been dealt with. The end product however has been just $M(\vartheta)$. This chapter is concerned with the end product in its own right.

As was briefly outlined in the introduction, n.m.r. spectra consist of lines whose positions are described in terms of chemical shifts and coupling constants. We will initially concern ourselves with the origin of these parameters and how they interact, and then with how to assign the lines within a spectrum to specific nuclei within a molecule, and briefly how to apply this information to problems of chemical interest. Except in a few cases, the spectrum obtained is independent of the method used to obtain it. As a consequence, most of the material on these topics already published in other books is relevant. This chapter therefore does not aim to be rigorous and complete, but it does aim to give a sufficient account of the areas outlined above so that the reader can understand and use n.m.r. spectra in a general way. Particular attention will be drawn to areas where Fourier n.m.r. has developed or permitted new applications. References will be given to key works (in the form of books and review articles rather than the original papers) where further, more detailed, information can be found.

8.2 THE CHEMICAL SHIFT [1, 2]

Nuclei absorb energy and give an n.m.r. signal when the Larmor equation is satisfied. If every nucleus of the same species was subject to the same magnetic field, that of the spectrometer's magnet, then n.m.r. would be a dull and not very useful technique. Physicists could measure the nuclear magnetogyric ratio with even greater accuracy but not much else. Fortunately for the chemist not all the nuclei of the same species are subject to the same local magnetic field; the local chemical environment very slightly modifies the basic field so that

$$B_{loc} = B_0(1 - \sigma) \tag{8.1}$$

where σ is called the screening constant. Molecular screening is not isotopic as it differs along various axes within the molecule; σ is therefore a tensor. However in the gaseous and liquid phases, due to the rapid molecular motion, a nucleus is subject only to an average value of σ; the trace of the tensor. The individual elements of σ can be significant when studying samples where rapid isotropic motion is impossible, for example in liquid crystals and solids. The degree of anisotropy within the chemical shift is also important when discussing nuclear relaxation (see section 10.6). The resonance frequencies for two different nuclei of the same species are given by

$$\vartheta_A = \gamma(1 - \sigma_A)B_0 \tag{8.2}$$

$$\vartheta_X = \gamma(1 - \sigma_X)B_0 \tag{8.3}$$

and their relative, or chemical, shift is given by the difference of these equations

$$\vartheta_A - \vartheta_X = (\sigma_X - \sigma_A)\gamma B_0 = (\sigma_X - \sigma_A)\vartheta_0 \tag{8.4}$$

The first point to note is that the chemical shift is field dependent. Now if both ϑ and B_0 could be measured with sufficient accuracy, screening constants could be measured directly. It is however impossible to measure magnetic fields sufficiently accurately by any technique other than n.m.r. A circular argument ensues. The circle is broken by defining a reference signal, e.g. that of nucleus A in our example, and measuring all chemical shifts with respect to it, i.e. setting $\sigma_A = 0$. Chemical shifts measured this way are given the symbol delta (δ), and in order to avoid chemical shift values depending on the magnetic field being used in a particular spectrometer, they are quoted as a ratio between the frequency difference and the frequency of the reference compound. The values so obtained are in the order of parts per million, the ratio is therefore multiplied by 10^6 and given the units of parts per million or ppm. Thus

$$\delta_x = \frac{\vartheta_x - \vartheta_{Ref}}{\vartheta_{Ref}} \times 10^6 \text{ ppm} \tag{8.5}$$

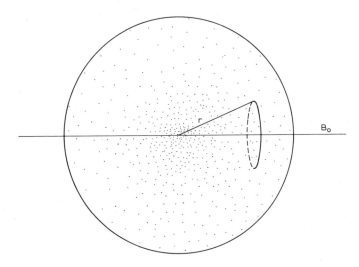

Fig. 8.1. The diamagnetic circulation of the electronic charge cloud under the influence of a magnetic field.

a positive value of δ_x means that the nuclei of x resonate at a higher frequency (lower field) than the reference compound.

In order to explain the origin of the screening constant, first consider a nucleus surrounded by a spherical electron cloud, i.e. in an 'S' state as illustrated in Fig. 8.1. In a magnetic field the entire electron system will precess at the appropriate Larmor frequency which is

$$\vartheta = \frac{e}{4\pi M_e} B_0 \qquad (8.6)$$

where M_e and e are the mass and charge of the electron respectively. This circulating current will generate a magnetic field opposed to B_0 which will screen the nucleus at its centre. Summing up all the possible currents gives a value for the screening constant, which for an 'S' atom of charge Z is

$$\sigma_d = \frac{e}{3M_e} \int_0^\infty r\rho(r)\mathrm{d}r \qquad (8.7)$$

where $\rho(r)$ describes the electron density as a function of the distance r from the nucleus. Calculation of σ_d from eqn. (8.7) gives a positive contribution to the screening which is called the Lamb or diamagnetic term [3]. The value given by (8.7) is 17.8 ppm for hydrogen and it increases rapidly as the number of electrons in the atom increases, being in the order of 260 ppm for ^{13}C and 10^{-2} for heavy metals [4]. However in n.m.r. it is only the differences between screening constants in different molecules or parts of a molecule which are important. These differences are only a small fraction of the total diamagnetic term. Decreasing the electron density round a nucleus decreases the diamagnetic screening constant and induces a shift to low field.

Molecules normally lack the symmetry of atoms; the electronic cloud is constrained by the localisation of specific nuclei and chemical bonds. Another term must therefore be introduced to describe the hindering of electronic motion by the electric field of the other nuclei present with the molecule. This second term, also due to Ramsey, is, as one might expect, more complex than the first term; it cannot be adequately described in classical terms [15]. The magnetic field is considered, using quantum theory, as causing a mixing of the ground and excited state molecular wave functions. This mixing results in a shift low field and is consequently referred to as the paramagnetic term (σ_p) to distinguish it from the Lamb or diamagnetic term (σ_d). Ramsey theory gives the following value for the screening constant

$$\sigma = \sigma_d + \sigma_p \tag{8.8}$$

with

$$\sigma_d = (e^2 \Xi / 12\pi M_e)(\overline{1/r}) \tag{8.9}$$

and

$$\sigma_p = -(e^2 \Xi_0 / 12\pi M_e^2) \sum_n \langle \psi_0 | \hat{I} | \psi_n \rangle \langle \psi_n | r^{-3} \hat{I} | \psi_0 \rangle / E_N - E_0 \tag{8.10}$$

Since in real molecules neither the excited state energies (E_n) nor their wave functions (ψ_n) are known, calculation of the screening constant is impossible.

The problem of ab initio calculation of chemical shifts is made more difficult by the two terms given by eqns. (8.9) and (8.10) having a comparable magnitude but of opposite sign and being dependent on the choice of gauge. Various methods, e.g. the one due to Flygare [16], have been tried to factorise the terms into forms where calculations are possible. A more or less empirical approach to understanding chemical shifts is to divide the contributions to the total shift into more manageable sub-groups. For a specific nucleus six major sub-groups can be used,

    (a) Local diamagnetic effects σ_d

    (b) Local paramagnetic effects σ_p

    (c) Anisotropic effects from local groups σ_a

    (d) Electric field effects σ_e

    (e) Unpaired electron shifts σ_k

    (f) Solvent effects

These do not form a complete or even a well behaved set, but a convenient set nevertheless.

(a) Diamagnetic shielding

Local diamagnetic effects originate in the symmetrical electronic shielding provided at the nucleus. The presence of an electron-donating or withdrawing group in the molecule will therefore have an effect on the chemical shift of nearby nuclei. An example of this effect is the proton chemical shifts of the methyl halides; these increase from iodine to fluorine as the increasing electronegativity of the halogen withdraws more electron density from the hydrogen. This contribution to the total chemical shift, which is normally in the order of a few ppm, is important in proton chemical shifts, as can be seen by examining a correlation chart of proton shifts (Fig. 8.9), but less significant for other nuclei.

(b) Paramagnetic shielding

The form of the paramagnetic term in chemical screening is given in eqn. (8.10) and, as mentioned previously, a detailed calculation is difficult; however some general trends can be usefully discussed. Firstly, s electrons cannot contribute to σ_p since they have zero orbital angular momentum, being spherically symmetric. Conversely, p and d electrons contribute strongly to paramagnetic screening as they are anisotropic. Secondly, as can be seen from eqn. (8.10), the σ_p term is inversely proportional to the excitation energy from the ground to the first excited state of the molecule, and hence is most significant when this is small. σ_p is larger than σ_d, being in the order of $100 \sim 10\,000$ ppm.

For the reasons given above σ_p does not contribute to proton chemical shifts, where there are no low lying p and d electrons, but for other nuclei it is the dominant effect, e.g. in ^{19}F and ^{13}C etc. For these nuclei the same diamagnetic effects which dominate proton n.m.r. are present, often larger due to the larger number of electrons involved, but they usually pale into insignificance compared to the paramagnetic contributions. Possibly the best example of paramagnetic shift is to be found in ^{59}Co n.m.r. Here, as in many transition metal complexes, the excitation energy ΔE is small, and consequently paramagnetic screening is efficient. In a series of d^6 cobalt complexes a correlation is found between the ^{59}Co shift and ΔE^{-1}[7], the latter having been obtained from u.v./visible spectroscopy. As predicted by ligand-field theory, the shielding decreases as the ligand-field strength decreases.

It must be remembered that despite its misleading name a paramagnetic contribution to chemical shift does *not* imply the presence of an unpaired electron. The shift induced by an unpaired electron in a paramagnetic molecule is referred to in this book as an unpaired electron shift and is dealt with in section 8.2e.

(c) Anisotropy effects

Anisotropy in neighbouring groups in the molecule will generate local magnetic field effects, and they in turn will induce chemical shifts which do not average to zero with molecular motion. The classic case of an effect of this type is the ring current shift in benzene. Shifts induced by local anisotropies are typically in the order of a few ppm, and are therefore very significant in proton n.m.r. but less so in the spectra of other nuclei where shifts due to the two previous effects are much larger.

If the electron cloud on a neighbouring atom or group is spherically symmetrical, then its local field at the nucleus under consideration will average to zero due to the random molecular motion in the solution. However, if the electronic cloud is not spherically symmetric, i.e. the group is magnetically anisotropic, then the local magnetic field does not average to zero and a chemical shift results. Consider the case of cylindrical symmetry; two suceptibilities are necessary to describe the problem, $\chi_{/\!/}$ and $\chi_{\perp}$ (in the case of lower symmetry more would of course be required). If r is the length of the vector between the centre of the anisotropic group and the nucleus in question and θ is the angle between this vector and axis of symmetry then the anisotropic shift is given by [8]

$$\sigma_a = (\chi_{/\!/} - \chi_\perp)(1 - 3\cos^2\theta)/12\pi r^3 \qquad (8.11)$$

This equation shows that the effect of σ_a is cone shaped (see Fig. 8.2), the conical angle being 54.7°. Inside the cone the effect is either to shield or deshield a nucleus, depending on the relative sign of the two susceptibilities; the effect will decrease to zero at the surface of the cone and then have the opposite behaviour outside the cone.

Consider benzene, which is magnetically anisotropic with cylindrical symmetry on account of the two π electron rings. In a magnetic field the electrons in the ring precess about the field and thus behave like an electric current flowing in a loop, generating a magnetic field. This field opposes B_0 in the centre of the loop and consequently enhances it outside the loop. Any nuclei situated outside the ring and also in the plane of the ring will be subjected to a negative

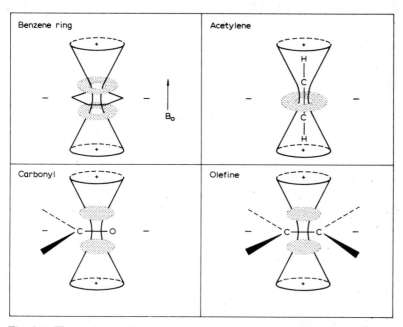

Fig. 8.2. The regions of magnetic shielding and de-shielding generated by some common anisotropic groups. A + sign indicates a shift to high field, a − sign to low field.

screening and a low field shift. Thus, the protons in benzene have a shift of 7.2 ppm compared with 1.5 for cyclohexane. For the case of the benzene ring the effect of the ring current on proton shifts is well documented and predicts that up-field shift should also be possible for nuclei inside or above the ring. The latter effect is found in many biomolecules, e.g. lysozyme, where protons appear high field of TMS due to their position with respect to benzene rings in other parts of the protein. It can also be illustrated by studying the shifts of the ring methylenes in 1, 4-polymethylenebenzenes [9]. Here the central methylene protons of the 1—4 bridging aliphatic chain are in the shielding part of the cone and do show high field shifts. Whereas it is not possible to get nuclei 'inside' a simple benzene ring, the NH protons in porphyrins lying inside their aromatic ring current do indeed show high field shifts, being 4 ppm high field from TMS, while the methine protons, in the same plane but outside the ring, are shifted to low field (7.5 ppm) [10].

Other common groups are also magnetically anisotropic and have similar σ_a effects; two of the most common are illustrated in Fig. 8.2 and these are the olefinic and carbonyl bonds. Even the σ C—C bond has a small degree of anisotropy. An example of the anisotropy effect of the carbonyl group is the extreme low field shifts found for aldehyde protons (~ 10 ppm).

Anisotropy in the chemical shielding of a group can lead to its effect on a neighbouring group being opposite to that on a nucleus within a group. An example is the high field shifts found in transition metal hydrides, e.g. $+ 21.7$ ppm for cis-IrHCl$_2$ (PEt$_3$)$_3$. These large positive shifts, which are sufficiently unique to characterise such protons and provide a good example of the usefulness of n.m.r. in functional group analysis, originate in the electronic distributions of the central metal. At the metal nucleus itself these electrons generate a low field paramagnetic shift, but at the proton induce a high field shift.

(d) Electric field effects

If there is present in a molecule a strongly polar group, such a group will influence the molecular electron dispersion and hence affect the chemical shifts. For nuclei in an axially symmetric environment the effect can be represented by [11]

$$\sigma_e = - AE_z - BE^2 \tag{8.12}$$

the z axis being along the chemical bond to the polar grouping. The first term simply allows for the electron drift in the polar bond against (hence the $-$ sign) the inherent electric field. The second term takes account of the asymmetry induced by this shift, which in turn will result in a paramagnetic deshielding.

The values of A and B differ from nucleus to nucleus. As is to be expected for protons, $A > B$ and both positive and negative shifts can result. For example in nitrobenzenes σ_e can be up to 0.8 ppm.

Electric field effects are more significant in ^{19}F shifts where $B > A$, and allowance needs to be taken for the effects originating in electric dipole effects mutually induced between an X—Y bond and the bond to the fluorine being considered. These induced dipoles have zero average for E_z but a non-zero average for E^2. The magnitude of this effect (which has essentially the same origins as the van der Waal's shift) depends on the polarisibility of the X—Y bond and the inverse sixth power of the distance between the nucleus and the X—Y bond. An example of σ_e is to be found in ^{19}F spectra of fluorinated benzenes, where a large low field shift is induced in a fluorine nucleus by replacing a fluorine *ortho* to it by the much more polarisible chlorine (the '*ortho* effect') [12].

(e) Unpaired electron shifts

The electron has a very large magnetic moment; therefore if an unpaired electron is present in the sample, e.g. in the form of a free radical or paramagnetic ion, it will have a dramatic effect on the n.m.r. spectrum. Firstly the nuclei and the electron will undergo scalar coupling of magnitude A (this is the hyperfine coupling constant observed in e.s.r. spectra). Secondly there will be a dipole—dipole interaction which may also provide an efficient relaxation mechanism for the nucleus (see section 10.9). The consequences of these two effects depend initially on the electron relaxation time (T_1^e). If this is comparable with A, $(T_1^e)^{-1} \sim A \sim$ MHz, as is normally the case in free radicals, then the n.m.r. spectrum is so broadened that it is unobservable. If, as is quite often the case in paramagnetic complexes $(T_1^e)^{-1} \gg A$ and or $(\tau_e)^{-1} \gg A$ where τ_e is an exchange lifetime, then sharp lines are observed in the n.m.r. spectrum which show large shifts. The first shifts due to unpaired electrons were observed in metals by W.D. Knight in 1949, the year before the first 'normal' chemical shift [13]. The paramagnetism in metals arises from the electrons in the conduction band.

The use of the term unpaired electron shifts distinguishes such effects from the so-called paramagnetic term in the nuclear screen. This is induced in diamagnetic molecules by the mixing of excited states into the ground state (see section 8.2b). The term unpaired electron shifts also does not imply any mechanism, as do the terms contact and pseudo-contact, often used to describe shifts of scalar and dipolar origin.

Firstly consider the case where there is scalar coupling between the proton and the electron. It is known from e.s.r. spectra that there will be coupling in the order of 20 MHz between an unpaired electron on a carbon atom, e.g. in an aromatic system, and a bonded proton. If T_2^e were to be long enough for such coupling to be resolved then the proton spectrum would consist of two lines 1.9 Tesla apart (see Fig. 8.3). If on the other hand T_2^e is very short compared with A^{-1} then only a single line will be seen. This line will, as usual, be at the weighted mean between the two lines of the original doublet. But unlike the case of a nuclear—nuclear coupling, the weighted mean will not be at the mid point (i.e. the chemical shift of the proton without the coupling); it will be shifted. In the case of

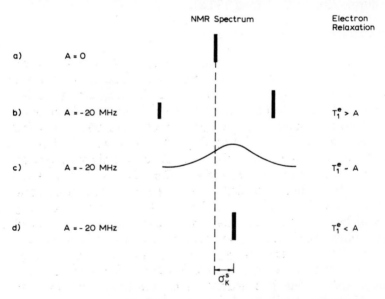

Fig. 8.3. The origin of the scalar unpaired electron shift. (a) The n.m.r. spectrum when there is no scalar nuclear-electron coupling; (b)—(d) the n.m.r. spectrum when there is coupling for various electron relaxation times.

J coupling the energy difference is small (a few Hz) and the population difference between the two components of a doublet is small. For A coupling, where the lines are MHz apart, the populations are very different and consequently, as can be seen in Fig. 8.3, the resulting spectrum shows a shift. The shift depends on the total electron spin quantum number S; the *sign* and magnitude of A thus [14]

$$\sigma_K^s = -Ag_e\beta_e S(S+1)/3\gamma_N\hbar kT \tag{8.13}$$

The hyperfine coupling constant is itself proportional to the unpaired spin density ρ and a constant (Q) dependent on the particular system

$$A = \rho Q S^{-1} \tag{8.14}$$

For aromatic C—H interactions $Q = -2.5$ Tesla but for methyl groups Q is of a similar magnitude but positive.

In a molecule containing an unpaired electron the electron is normally delocalised via aromatic systems but some delocalisation via σ electrons has been postulated. An elegant example of the study of scalar shifts is the work of Eaton and Philips on the nickel aminotroponeiminates, where the unpaired electron present in the tetrahedral form is delocalised into the aromatic ring. In these compounds the electron relaxation time is sufficiently short that excellent proton spectra are observed [15]. The unpaired electron shifts were related directly to the unpaired spin densities in the aromatic system measured by e.s.r. and also calculated from m.o. studies. It is impossible to obtain the sign of A directly from the e.s.r. experiment, whereas its value is very obvious from the n.m.r. spectrum.

The dipole—dipole interaction between the electron and the nucleus will also lead to chemical shift when T_2^e is short, unless the electron distribution is isotropic. The techniques used for averaging anisotropic motion are varied and the detail of the solution depends on the assumptions made with respect to the origin of electronic anisotropy, i.e. whether zero field splitting effects are included or not. If axial symmetry is assumed, two g values defined and zero field splitting effects ignored, it can be shown that [14]

$$\sigma_K^D = -\beta^2 S(S+1)(3\cos^2\theta-1)(\chi_{/\!/}+\chi_h)(\chi_{/\!/}-\chi_h)/9r^3kT \tag{8.15}$$

The dipolar or pseudo contact shift thus has the behaviour expected of a dipolar interaction, i.e. $a(3\cos^2\theta-1)/r^3$ and can give information of a stereochemical nature.

In 1969 C.C. Hinkley found that certain β-diketone lanthanide complexes could induce shifts in proton spectra without inducing significant line broadening [16]. The compounds now called 'shift reagents' have had a dramatic effect on n.m.r. In solution they form loose complexes with any polar group present in the solute molecule, and induce chemical shifts which depend mainly on the interse cube of the metal/nuclear vector and the angle between this vector and the principal axis. Some lanthanides, e.g. europium, produce low field shift and some, e.g. praseodymium, high field shifts. In their simplest use shift reagents are used to increase internal chemical shift and permit easier analysis (often first order) of complex molecules. In their most sophisticated applications they provide structural information in solution comparable with X-ray data.

It is important to distinguish between the requirements of a 'shift reagent' and a 'relaxation reagent'; both are paramagnetic complexes. The shift reagent should be anisotropic and have a very fast electronic relaxation (10^{-13} s) in order to induce little line broadening (see eqn. 10.38). It must also be complex, if only in a very loose way, with the solute. The relaxation reagent should be isotropic (hence producing no dipolar shift), relax slower (10^{-6} s) and not form any complex with the solute. The former reagents are normally co-ordinatively unsaturated lanthanide complexes, and the latter stable octahedral transition metal complexes.

If structural information is to be derived from shift reagents, the shifts studied must be purely dipolar in origin. Scalar shifts depend on unpaired spin density, not geometrical factors, and are rarely completely absent, especially for ^{13}C n.m.r. Saturation of the two effects is difficult, but it can be achieved by using a series of shift reagents containing different lanthanide ions, when the dipolar shift changes markedly and predictably with the different anisotropies of the central metal ion. A further way round the problem has been proposed by Fuller using tris-dipivalomethanatogadolinium which bonds specifically to polar groups without causing any shift; it does however act as a relaxation reagent, having a simple geometrical dependence of metal nuclear distance (r^{-6}) and angle [17].

(f) Solvent effects [18]

The solvent used in an n.m.r. experiment will itself affect the chemical shift of the solute, and also the reference. Even an isotropic molecule such as TMS will be affected by the solvent, a factor which must be borne in mind when measuring solvent effects.

Solvents can induce shifts in three major ways; firstly by the effect of their magnetic anisotropy. The most common example of this type is the shifts induced in benzene solution by the π-electron could. The ring currents can induce shifts either to high or low field depending on orientation (see section 8.2c). Due to molecular tumbling an average value is seen by the solute. Since benzene is not a symmetrical molecule, a non-zero average results and a solvent shift, usually to low field, results.

A second solvent effect can occur with polar solutes in polar solvents which is similar to the electric field (see section 8.2d). In such systems 'preferred' orientations of the solvent molecules around the solute occur which enhance any diamagnetic shielding caused by the basic electric dipole within the solute.

Thirdly, shifts can be induced if there are any specific interactions between solvent and solute. A common example of this is inter-molecular hydrogen bonding. The effect of hydrogen bonding on chemical shifts is complex and not well understood; it most probably results in many effects including those of electric fields. The effect of hydrogen bonding is large and to low field; for example, the chemical shift of the hydroxyl proton in methanol increases by about 10 ppm, i.e. is more deshielded in going from high dilution in CCl_4 (where the spectrum is effectively that of an isolated molecule) to the strongly hydrogen-bonded environment of the pure liquid. Intramolecular hydrogen bonds produce the largest chemical shifts, e.g. salicylaldehyde, where $\delta(OH) = 11.0$ ppm.

Hydrogen bonding normally occurs in systems where chemical exchange can occur, e.g. OH, NH, NH_2 etc. If the lifetime of the proton in and out of a hydrogen bonded site is long compared with the difference in chemical shifts between the two sites (measured in Hertz) then two separate resonance lines will occur, one corresponding to the chemical shift of each environment. If the lifetimes are short, then a single resonance at their mean chemical shift will result (see section 8.5). For intermediate lifetimes line broadening will occur; consequently the shift and width of a resonance from a proton involved in hydrogen bonding is very marked with changes in concentration and temperature.

8.3 SPIN—SPIN COUPLING [19]

The second interaction which characterises an n.m.r. spectrum is internuclear scalar coupling. This interaction was first detected by Proctor and Yu [20] while studying $NaSbF_6$. They found that the ^{121}Sb spectrum, despite careful purification, consisted of 5 lines. (Gutowsky and McCall [21] later pointed out two extra lines giving a 1:6:15:20:15:6:1 pattern). A systematic study by Gutowsky and McCall showed that fine structure occurred in the spectrum of molecules containing two or more nuclei which resonate at different fields. The nuclei could be either chemically shifted nuclei of the same species (homonuclear coupling) or nuclei of another species (heteronuclear coupling).

The number of lines in the spectrum of A in an AX_n molecule is given by $2nI_x + 1$ where I_x is the spin quantum of X, and if this is $\frac{1}{2}$ then the intensitives are given by the coefficients of the binomial series. For example, the ^{13}C spectrum of CH_4 is a quartet of equally spaced lines whose intensities are 1:3:3:1. The interaction is characterised by the spin coupling constant which is measured in Hz and is the separation between the two consecutive lines of the multiplet. In the methane example it is given the symbol $^1J(^{13}C-^1H)$ (the superscript refers to the number of chemical bonds between the nuclei, and the bracket contains the nuclei involved).

(a) Coupling mechanisms

Consider two nuclei A and $X(I = \frac{1}{2})$. Since they possess magnetic moments they interact with each other as magnetic dipoles, in a way which is proportional to their magnetic moments, the inverse cube of their internuclear distance and an angular term $(3 \cos^2\theta - 1)$, where θ is the angle between their internuclear vector and the magnetic field. In solution all angles θ are possible and their dipole fields average to zero, $3 \cos^2\theta = 1$, (see section 10.4) hence such interactions cannot explain the coupling described above.

The above analysis of the interaction of magnetic dipoles does not account for spin coupling. A second, indirect dipole—dipole interaction is possible between chemically bonded nuclei using the bonding electrons and orbital momentum as conductors. Since it involves chemical bonds molecular motion is not capable of averaging an interaction of this type to zero. However a classical analysis of the indirect interaction indicates it is too small to account for the observed couplings. The classical analysis assumes that the particles involved are point dipoles, which is certainly not a valid approximation at this level. Quantum mechanics covers this situation and introduces an electron nuclear interaction, the Fermi contact term, which allows for the finite probability of a bonding electron being found at the nucleus [22]. The Hamiltonian for the contact inter-action is

$$\hat{\mathcal{H}}_c = \tfrac{2}{3}\Xi_0 g_s \gamma \hbar \sum_e \delta(r_{eN}) \hat{I}_N \hat{S}_e \qquad (8.16)$$

where $\delta(r_{eN})$ is a Dirac delta function ($= 1$ if r_{eN} is zero and $= 0$ for all other values). A simple model of spin coupling can now be built up using the contact interaction. The model is a gross over-simplifi-cation but none the less predicts several important properties of nuclear spin—spin couplings. The first property is that they proceed via chemical bonds, and hence they are quite often referred to as scalar couplings, to distinguish them from the dipole interactions described previously.

Figure 8.4 illustrates the most stable, lowest energy state for two nuclei with positive magnetogyric ratios joined by a single bond. The nuclei and the electrons interact via the Fermi contact effect and spin pair: the bonding electrons themselves also spin pair, resulting in a lowest energy state where the two nuclei are anti-parallel to each other. Such an interaction is defined as a positive coupling constant. If one of the nuclei has a negative magnetogyric ratio then the lowest energy state for that nucleus would have the the magnetic moment of the nuclei parallel with that of the inter-acting electron. In this case, in the lowest energy state the two

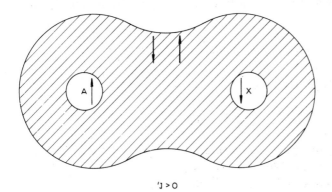

Fig. 8.4. The lower energy state for two nuclei coupled by the Fermi contact mechanism (J is positive).

coupled nuclei have parallel orientation. This is a negative coupling constant.

For a two bond coupling the above model results in a negative coupling contant (for γ positive) as can be seen by considering the CH_2 fragment where the lowest energy state is

$$\begin{array}{ccc} H & C & H \\ \downarrow \; \uparrow\downarrow & & \downarrow\uparrow \; \downarrow \end{array}$$

The 'central electrons' have a parallel relationship since they occupy equivalent orbitals on the carbon atom (Hund's rule).

If perfect pairing of the electron spins were rigorously required, only the molecular states outlined above would exist and the resonance lines would not be split. Spin coupling must therefore also involve contributions from excited triplet states having impaired electron spins. The situation can best be interpreted in terms of molecular orbital theory. All the electrons in the molecule are supposed to move independently in a set of molecular orbitals $\psi_1 \ldots \psi_n$ with two spin paired electrons in each orbital. These molecular orbitals are themselves linear combinations of atomic orbitals

$$\psi_i = \sum_k C_i \phi_k \tag{8.17}$$

Now only s electrons have a finite chance of being found at the nucleus; therefore in order to evaluate the 'contact' coupling term, only s atomic orbitals need be considered. The theory of Pople and Santry [23] shows that such a factorisation is possible and leads to

$$J_{AX} = h \gamma_A \gamma_X (\tfrac{2}{3}\Xi\mu)^2 \langle S_A | \delta(r_A) | S_A \rangle \langle S_X | \delta(r_X) | S_X \rangle \Pi_{S_A S_X} \tag{8.18}$$

where S is the appropriate valency shell s-wave function, the integrals $\langle S|\delta|S\rangle$ represent the s electron density at the nucleus and π is the mutual polarisibility of the orbitals given by

$$\Pi_{S_A S_X} = 4 \sum_{ij} C_{is_A} C_{is_X} C_{js_A} C_{js_X} / E_i - E_j \tag{8.19}$$

Equation (8.19) describes the mixing of excited states into the ground state; each excited state makes its own contribution to J, inversely proportional to the excitation energy.

Equation (8.18) shows that the magnitude of the coupling depends on the magnetogyric ratios of the nuclei involved; the concept of a reduced coupling constant (K) is introduced to eliminate this complication

$$K_{AX} = J_{AX}/h\gamma_A\gamma_X \qquad (8.20)$$

(b) One bond coupling

For the simple case of two nuclei joined by a σ bond eqn. (8.18) can be simplified to

$$K \propto \alpha_A^2\,\alpha_X^2\,\psi_A^2(0)\,\psi_X^2(0)\,\Delta E^{-1} \qquad (8.21)$$

where

α^2 is the s character in bonding orbital
$\psi^2(0)$ is the spin density at the nucleus of the valency s electrons
ΔE^{-1} is the excitation energy to the first excited state

and in this form predicts most of the fundamental properties of one bond couplings.

Firstly eqn. (8.21) predicts a dependence on hybridisation which is borne out in practice, e.g. the ^{13}C–H coupling

Ethane	(sp^3)	$\alpha_C^2 = 0.25$	$J =$	$125\,Hz$
Ethylene	(sp^2)	$\alpha_C^2 = 0.33$	$J =$	$156\,Hz$
Acetylene	(sp)	$\alpha_C^2 = 0.5$	$J =$	$249\,Hz$

This dependence on s character is a gross effect; much more subtle effects occur within one hybridisation state depending on the electronegativity of the substituents. For example, the value of $^1J_{CH}$ in the substituted methanes can be predicted using additive parameters dependent on the electronegativity of the group in question [24] (see section 8.6b). The larger the electronegativity of the substituent, the larger is J'_{CH}. As has been pointed out by Bent [25] electronegative substituents prefer to use orbitals of the central atom with a low s character, since the electron density is at a maximum further from the central atom in the orbitals compared with those having a larger s character. Thus the substitution of a highly electronegative group onto a carbon atom of methane results in an increase in the s character in the other three hybrids, with a corresponding increase in $^1J_{CH}$ compared with methane itself ($^1J_{CH} = 125\,Hz$ in methane and 149 in methyl fluoride).

Parallel with the change in hybridisation induced by the electronegativity of a substituent is bond polarisation. An electronegative substituent will reduce the electron density round the carbon, which can result in an increase in the effective nuclear charge of the carbon and a modification of the radial electron distribution functions; $\psi_C^2(0)$ increases and therefore so does ^1J. Which of these two effects is dominant is not clear since they both act in the same sense. Changing atomic number (Z) also affects $\psi^2(0)$ and hence ^1J. As Z increases so does $\psi(0)$ (for $Z > 10$ the increase is linear) and hence K, thus $^1K_{XH}$ (and also $^2K_{(X-C-H)}$) increases as the square of the atomic number of X [26].

(c) Two bond coupling

The basic concepts developed above also apply to coupling over more than one bond but now an extra variable, that of bond angle, has to be considered. The most common two bond coupling is the geminal H—H coupling. This, as the simple model predicts, is basically negative, at least for sp^3 carbons (in the range $-10 \sim -15\,\text{Hz}$). However for sp^2 carbons where the coupling is increased (algebraically) either sign is possible. Detailed molecular orbital calculations indicate that either sign is reasonable.

The effect of bond angle of CH_2 geminal coupling is to increase (i.e. become less negative) the value of $^2J_{(H-H)}$ as the H—C—H angle increases. The coupling can even become positive. The changes induced by substituents can however often overwhelm these geometrical effects. As expected, the effect of a substituent depends on its electronegativity; if it is bonded directly to the carbon bearing the geminal protons an electronegative substituent can greatly increase the coupling, e.g. in formaldehyde $^2J_{(H-H)} = +41\,\text{Hz}$. But if the electronegative substituent is one atom further removed its effect is frequently to decrease J. The presence of a π system adjacent to the CH_2 group also decreases J dependent on molecular geometry.

(d) Three bond coupling

Three bond (vicinal) couplings are normally positive and, in the case of proton—proton coupling, smaller than geminal coupling, as expected from the simple Dirac model. The limitations of the simple model are shown up in vicinal couplings by firstly a dependence of the coupling on interbond angles and the rule that for molecules of the type X—CH_2—CH_3 (where X is a heteronucleus) vicinal couplings are larger than geminal ones ($^2J_{(P-H)} = (-)\, 12.8\,\text{Hz}$ and $^3J_{(P-H)} = (+)$ $18.5\,\text{Hz}$ in PEt_4I).

Detailed molecular orbital and valency bond calculations, as well as experimental observation, give the following properties of proton vicinal couplings.

(a) $^3J_{(H-H)}$ increases as the C—C bond length decreases.

(b) $^3J_{(H-H)}$ increases as the H—C—C bond angle decreases.

(c) $^3J_{(H-H)}$ decreases with electronegative substitution on one of the carbons.

(d) $^3J_{(H-H)}$ depends on dihedral angle (see Fig. 8.5).

The dependence of $^3J_{(H-H)}$ on the dihedral angle φ between planes containing the two C—H bonds is given by the Karplus equation [27]

$$^3J = A + B \cos \varphi + C \cos 2\varphi \qquad (8.22)$$

The form of this equation is given in Fig. 8.5; the constants A, B and C have empirical values of 7, -1 and 5. A simple example of this dependence is given by the couplings in cyclohexane derivatives (see Table 8.1). In the case of free rotation about the C—C bond the values of vicinal couplings observed are the weighted (according to population) mean of the couplings found in the various possible

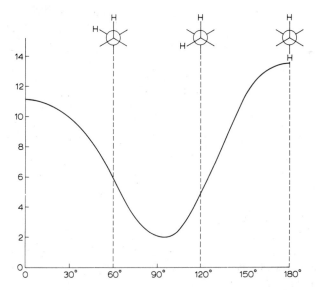

Fig. 8.5. The Karplus relationship; the effect of geometry on the $^3J(^1H-^1H)$ coupling constant.

TABLE 8.1

$^3J_{(H-H)}$ in cyclohexane derivatives

Protons	7	Typical ^3J
H_a-H_a	$180°$	14
H_a-H_e	$60°$	4
H_e-H_e	$60°$	4
$(H_{aa} \leftrightarrow H_{ee})$ average	$(180° \leftrightarrow 60°)$	9
$(H_{ae} \leftrightarrow H_{ea})$ average	$(60° \leftrightarrow 60°)$	4

conformers. The angles obtained using eqn. (8.22) should only ever be used as a relative guide, unless values of A, B and C are known from suitable model compounds since, as is to be expected, these constants are sensitive to the electronegativity of any substituents and any departure from tetrahedral bond angles.

(e) Long range coupling

Coupling through more than 3 chemical bonds is normally considered as 'long range' coupling. For protons in saturated systems long range couplings are always small ($\sim 3\,Hz$) and often negligible. Long range proton couplings are largest when the two protons are separated by carbons which are in a planar W conformation, e.g. equatorial protons in a six membered ring. If the molecule is mobile, then, as mentioned above, the weighted mean coupling will appear in the spectrum, and since in any but the above configuration ^4J will be zero, long range coupling is unlikely to be resolved. Long range coupling is greatly enhanced if one of the carbon atoms in the path is sp^2 hybridised, e.g. an allylic coupling (H—C = C—C—H) or H—C—(CO)—C—H($^4J_{H-H}$ acetone = 0.55 Hz).

(f) π bond coupling

So far only mehcanisms involving σ electrons have been considered; this is an oversimplification even for saturated systems. It is well known from hyperfine splittings found in e.s.r. spectra that correlation exists between σ and π electrons, and hence whereas the latter cannot directly contribute to the contact term they can indirectly affect electron correlation, hence spin—spin coupling. Each coupling interaction should therefore be thought of as consisting of two terms, J^σ and J^π. E.s.r. data indicate that for acetylenic and ethylenic systems σ electrons on a proton are stabilised anti-parallel to the π electron on the appropriate carbon. $^3J^\pi$ should therefore be positive, $^4J^\pi$ negative etc. J^π should show no angular dependence; it is however difficult to estimate J^π because of the uncertainty in the larger J^σ term. Since π electrons are generally delocalised J^π is especially significant in cases of long range coupling, e.g. $^9J_{\text{H-H}} = 0.4\,\text{Hz}$ in $CH_3{-}(C \equiv C)_3{-}CH_2{-}OH$. As expected from e.s.r. work, replacing a proton by a methyl in an unsaturated system changes the sign of J^π and increases its magnitude slightly. The molecule 1, 3-butadiene illustrates these nicely. $^5J_{\text{cis-cis}}$ and $^5J_{\text{trans-cis}}$ are effectively equal (0.69 and 0.60 Hz) suggesting J^π dominates, whereas for $^5J_{\text{trans-trans}}$ (where, due to the planar W arrangements, a σ contribution is possible) the 5 bond coupling is much larger (1.3 Hz).

8.4 SPECTRAL ANALYSIS [28]

So far the effects of chemical shifts and coupling constants have been considered as totally independent. This is only true if the chemical shift difference δ between the two nuclei exhibiting spin coupling is large compared with their coupling constant (J). Such spectra are termed first order spectra and obey the rules outlined previously.

A molecule having two nuclei in chemically different environments where $\delta \gg J$, would have a spectrum consisting of two doublets. This is termed an AX system. If there were two nuclei of type A (an A_2X system) then the spectrum would consist of a doublet and triplet, and so forth. Should the chemical shift between the nuclei coupling become comparable with their spin coupling constant (termed an AB spectrum or A_2B etc.) the simple rules outlined above break down and more complex or second order spectra result. First order spectra depend only on the magnitude of the couplings involved and do not show any consequences of the signs of J. When second order spectra are analysed, the signs of the coupling constants become important. As can be seen from the analysis, n.m.r. spectra do not depend on the absolute signs of the couplings: only on their relative signs. To obtain spectra which do depend on the absolute signs of the coupling constants it is necessary to place another constraint on the system, e.g. an electric field or a nematic solvent.

The analysis of complex spectra is best achieved by quantum mechanical methods. For a general system the Hamiltonian will consist of two terms the Zeeman (or chemical shift) term and the spin coupling term; thus in frequency units

$$\hat{\mathcal{H}} = - \sum \gamma_i B_0 (1 - \sigma_i) \hat{I}_{iz} + \sum J_{ij} \hat{I}_i \cdot \hat{I}_j \tag{8.23}$$

The energy levels etc., can then be obtained by solving the Schroedinger equation

$$\hat{\mathcal{H}} \psi = E \psi \tag{8.24}$$

using the appropriate wave function.

For simplicity consider two nuclei with spin $\frac{1}{2}$. The Hamiltonian is now

$$\hat{\mathcal{H}}_{AB} = - \vartheta_A \hat{I}_{Az} - \vartheta_B \hat{I}_{Bz} + J \hat{I}_A \cdot \hat{I}_B \tag{8.25}$$

As seen in section 2.3 spin $\frac{1}{2}$ nuclei have two states, described by wave functions given by the symbols α and β. For the two spin problem there are four simple base functions, $|\alpha\alpha\rangle$, $|\alpha\beta\rangle$, $|\beta\alpha\rangle$ and $|\beta\beta\rangle$ (using Dirac notation). To simplify solving the problem it is normal to remove the scalar product in eqn. (8.5) by expanding it in terms of two new operators, the raising and lowering operators $I^{\pm}$:

$$\hat{I} \pm = \hat{I}_x \pm \hat{I}_y$$

whose properties are

$$\hat{I}^+ |\alpha\rangle = 0 \qquad \hat{I}^+ |\beta\rangle = |\alpha\rangle$$
$$\hat{I}^- |\alpha\rangle = |\beta\rangle \qquad \hat{I}^- |\beta\rangle = 0$$

The Hamiltonian now becomes

$$\hat{\mathcal{H}}_{AB} = - \vartheta_A \hat{I}_{Az} - \vartheta_B \hat{I}_{Bz} + J \hat{I}_{Az} \cdot \hat{I}_{Bz} + \tfrac{1}{2} J (\hat{I}_A^+ \hat{I}_B^- + \hat{I}_A^- \hat{I}_B^+) \tag{8.26}$$

The problem is to find wave functions which are stationary states of the Hamiltonian, i.e. obey eqn. (8.24). Since I_z does convert α or β (see eqn. 2.19) inspection shows that two of the base set $\alpha\alpha$ and $\beta\beta$ are satisfactory wave functions and their energies can be found from

$$\hat{\mathcal{H}} |\alpha\alpha\rangle = [-\tfrac{1}{2}(\vartheta_A + \vartheta_B) + \tfrac{1}{4} J] |\alpha\alpha\rangle \tag{8.27}$$

and

$$\hat{\mathcal{H}} |\beta\beta\rangle = [+\tfrac{1}{2}(\vartheta_A + \vartheta_B) + \tfrac{1}{4} J] |\beta\beta\rangle \tag{8.28}$$

The other two functions however are mixed by the last term in the Hamiltonian and are therefore not satisfactory.

$$(\hat{I}^+ \hat{I}^- + \hat{I}^- \hat{I}^+)|\alpha\beta\rangle = 0 + |\beta\alpha\rangle \tag{8.29}$$

It is thus necessary to construct new functions which are linear combinations of the original functions and are not mixed by the Hamiltonian.

Equation (8.24) then becomes

$$\hat{\mathcal{H}}(C_1|\alpha\beta\rangle + C_2|\beta\alpha\rangle) = E(C_1|\alpha\beta\rangle + C_2|\beta\alpha\rangle) \tag{8.30}$$

If we define the integrals $\hat{\mathcal{H}}_{11}\ \hat{\mathcal{H}}_{12}$ as follows

$$\hat{\mathcal{H}}_{11} = \langle \alpha\beta |\hat{\mathcal{H}}| \beta\alpha \rangle \tag{8.31}$$

$$\hat{\mathcal{H}}_{12} = \langle \alpha\beta |\hat{\mathcal{H}}| \alpha\alpha \rangle \tag{8.32}$$

Equation (8.30) can be rewritten as

$$[\hat{\mathcal{H}}_{11}C_1 + \hat{\mathcal{H}}_{12}C_2]\,|\alpha\beta\rangle + [\mathcal{H}_{21}C_1 + \mathcal{H}_{22}C_2]\,|\beta\alpha\rangle - E(C_1|\alpha\beta\rangle$$
$$+ C_2|\beta\alpha\rangle) = 0 \tag{8.33}$$

or put into matrix form as

$$\begin{bmatrix} \hat{\mathcal{H}}_{11} - E & \hat{\mathcal{H}}_{12} \\ \hat{\mathcal{H}}_{21} & \hat{\mathcal{H}}_{22} - E \end{bmatrix} \begin{bmatrix} C_1 \\ C_2 \end{bmatrix} = 0 \tag{8.34}$$

The two parts of eqn. (8.34) are only consistent (two equations, one unknown C_1/C_2) if the secular determinant is zero.

$$\begin{vmatrix} \hat{\mathcal{H}}_{11} - E & \hat{\mathcal{H}}_{12} \\ \hat{\mathcal{H}}_{21} & \hat{\mathcal{H}}_{22} - E \end{vmatrix} = 0 \tag{8.35}$$

TABLE 8.2

Stationary states of AB system

$\sin 2\theta = J/[(\vartheta_A - \vartheta_B)^2 + J^2]^{\frac{1}{2}}$

State	Σm	Wave function	Energy
1	1	$\alpha\alpha$	$-\frac{1}{2}(\vartheta_A + \vartheta_B) + \frac{1}{4}J$
2	0	$\alpha\beta \cos\theta - \beta\alpha \sin\theta$	$-\frac{1}{2}[(\vartheta_A - \vartheta_B)^2 + J^2]^{\frac{1}{2}} - \frac{1}{4}J$
3	0	$\alpha\beta \sin\theta + \alpha\beta \cos\theta$	$+\frac{1}{2}[(\vartheta_A - \vartheta_B)^2 + J^2]^{\frac{1}{2}} - \frac{1}{4}J$
4	−1	$\beta\beta$	$+\frac{1}{2}(\vartheta_A + \vartheta_B) + \frac{1}{4}J$

Solution of the above equations and normalisation results in the set of wave functions given in Table 8.2.

All that remains is to calculate the transition intensities and frequencies. It can be shown that the intensity of the transition from state $|m\rangle$ to state $|n\rangle$ depends on the square of the transition probability

$$|\langle n|\gamma_A \hat{I}_A^- + \gamma_B \hat{I}_B^- |m\rangle|^2 \Delta E_{mn}^2 \tag{8.36}$$

See ref. 29, section 5.3 and ref. 30 for a fuller analysis.

If A and B are of the same nuclear species, then the relative intensities in the spectrum are proportional to

$$|\langle n|I_A^- + I_B^- |m\rangle|^2 \tag{8.37}$$

The operator has the effect of lowering the total magnetic quantum

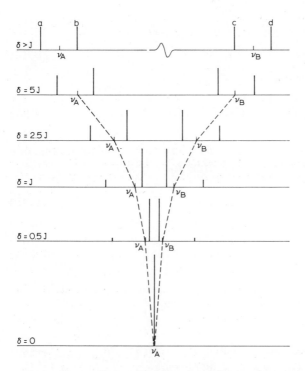

Fig. 8.6. The effect of varying the ratio of the coupling constant (J) to the chemical shift difference (δ) in a two spin system. The first spectrum is of the AX type, the last of the A_2 type. The intermediate spectra are AB.

number by 1, thus getting a selection rule of

$$\Delta m = \pm 1 \tag{8.38}$$

and the spectrum consists of four lines. The relative intensities and transition frequencies for the AB system are given in Table 8.2 and illustrated in Fig. 8.6 for various ratios of $J{:}\delta$.

The system discussed is the simplest possible; however the principles and techniques used in more complex situations are basically the same and some interesting general results can be deduced by studying the solutions arrived at. See ref. 28 for further details. Firstly consider the AX case $\delta \gg J$; this corresponds to a value of $\theta = 0$ (Table 8.2). Here the $|\alpha\beta\rangle$ and $|\beta\alpha\rangle$ states do not mix and, as shown before, the spectrum consists of 2 pairs of lines of equal intensity, each line within the pair separated by J and the pairs themselves by δ. This general conclusion can be expressed in other ways, either by saying that the last term in eqn. (8.26) is negligible or that the off diagonal elements of the matrix (eqn. 8.34) are negligible. The conclusion that when $\delta \gg J$

$$J_{AX}\hat{I}_A \cdot \hat{I}_X \longrightarrow J_{AX}\hat{I}_{Az}\hat{I}_{Xz}$$

greatly simplifies the solution of more complex spin systems and is termed the X approximation. Under these conditions spectra can be characterised simply by the spin state of X. An example of this is the ABX spectrum; in these spectra the AB region consists of 8 lines made up to two AB spectra corresponding to the two possible spin states of X and the signs of the coupling constants (see section IV—VII B of ref. 31), and can be analysed independently as two subspectra.

As the chemical shift becomes comparable with the spin coupling constant, mixing of the various spin states occurs and spectra become more complex. In the AB case the number of lines is not affected but, as can be seen from Fig. 8.6, the line intensities and positions are. The A and B regions stay as doublets separated by J but the chemical shifts of A and B are no longer in the centre of the doublets. In more complex cases many more lines result as the chemical shift becomes comparable with the coupling constant. For example, an A_3X_2 spectrum (an ethyl group) consists of two groups of lines; the A region is a 1:2:1 triplet, all the lines being separated by J, and the X region is a 1:3:3:1 quartet, again with the lines separated by J. The A_3B_2 spectrum on the other hand consists of up to 25 lines, none of which are separated by J.

TABLE 8.3

The AB system
$$D = [(\vartheta_A - \vartheta_B)^2 + J^2]^{\frac{1}{2}}; \sin 2\theta = J/D$$

Transition	Frequency	Intensity
$3 \leftrightarrow 4$	$\frac{1}{2}(\vartheta_A + \vartheta_B) + \frac{1}{2}J + \frac{1}{2}D$	$1 - \sin 2\theta$
$2 \leftrightarrow 4$	$\frac{1}{2}(\vartheta_A + \vartheta_B) + \frac{1}{2}J - \frac{1}{2}D$	$1 + \sin 2\theta$
$1 \leftrightarrow 2$	$\frac{1}{2}(\vartheta_A + \vartheta_B) + \frac{1}{2}J + \frac{1}{2}D$	$1 + \sin 2\theta$
$1 \leftrightarrow 3$	$\frac{1}{2}(\vartheta_A + \vartheta_B) - \frac{1}{2}J - \frac{1}{2}D$	$1 - \sin 2\theta$

Of some interest is the limiting case where nuclei or groups of nuclei have the same chemical shift, e.g. when $\vartheta_A = \vartheta_B$, an AB spectrum goes to an A_2 spectrum. This corresponds to a case of $\theta = 45°$ and, as can be seen from Table 8.3, the spectrum reduces to a single, double degenerate line corresponding to lines 2 and 3 of the original AB system; lines 1 and 4 have zero intensity. This result is very important since it illustrates a general rule that spin coupling within a group of 'equivalent' nuclei produces no observable coupling. The proton spectrum of methyl iodide, for example, is a single line since the 3 protons are equivalent and no coupling is observable (see section 8.12 of ref. 32 for a more detailed justification of this rule). The term 'equivalent' must be qualified. N.m.r. spectra are very sensitive as to whether nuclei are equivalent or not [33]. For example, despite free rotation the two methyl groups of valine are not

```
          H        NH
          |        |
Me _____ C _____ C _____ COOH
          |        |
          Me       H
```

equivalent; they are next to an asymmetric carbon and are therefore distinguishable in all possible conforms. As a result they have different chemical shifts and appear as two resonances. For a group or set of nuclei to be completely equivalent (i.e. show no splitting due to their mutual coupling and always perturb the spectrum as a unified group), they must not only have the same chemical shift, but they must couple equally to all other nuclei within the molecule; this is termed magnetic equivalence. If a group of nuclei have the same chemical shift but couple differently to any other nucleus in

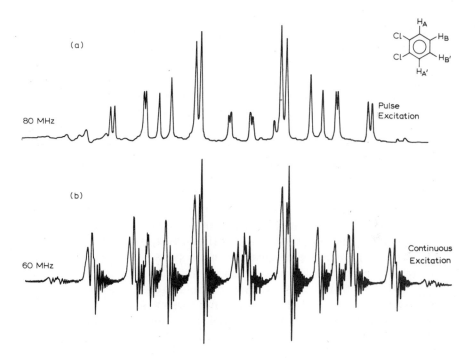

Fig. 8.7. The proton spectrum of *ortho*-dichlorobenzene in d^6 acetone. (a) At 80 MHz using the pulse excitation, (b) at 60 MHz using the sweep excitation. Note the different line positions due to the change of field and also the "ringing" which confuses the swept spectrum despite the use of a slow sweep rate (0.1 Hz/s). Both spectra have the same resolution and horizontal scale. The "lumps" to low field in spectrum (a) arise from solvent resonances folding in.

the molecule they are termed chemically equivalent. Two pairs of magnetically equivalent nuclei (an A_2X_2 system) have a spectrum consisting of two triplets. Two pairs of chemically equivalent nuclei (an AA′ XX′ system, the prime is used to identify magnetically non-equivalent nuclei within a chemically equivalent set) have a complex spectrum consisting of up to 24 lines. An example of the latter case is shown in Fig. 8.7, this being the proton spectrum of *ortho*-dichlorobenzene; the *meta* and *para* protons are chemically equivalent but are not magnetically equivalent, since they couple differently to the two *para* and *meta* protons respectively.

$J^{1,2} = 8.1$ Hz $J^{2,3} = 7.5$ Hz

$J^{1,3} = 1.5$ Hz

$J^{1,4} = 0.3$ Hz $\delta_{1,4} - \delta_{1,3} = 0.25$ ppm

Figure 8.7 also illustrates the change in the spectrum resulting from changing the observing magnetic field, and consequently altering the chemical shift difference between the A and B resonances.

In order to obtain chemical shifts and coupling constants from an observed spectrum one must first identify the type of spin system AX, A_2 etc. Information of this type is quite often known from the molecular structure that is being fitted to the spectrum or from the pattern of lines observed, especially with the aid of double resonance techniques. In the case of a simple system, e.g. a first order spectrum, chemical shifts and coupling constants can be read directly from the spectrum. In slightly more complex cases, e.g. the AB case, the observed lines can be assigned to appropriate transitions, and the required parameters calculated using analytical expressions of the type given in Table 8.3. In more complex cases a few of the more intense lines can usually be assigned and these values used to calculate the full spectrum. Should sufficient lines be impossible to assign directly then a process of iteration may be used. Probable parameters are chosen either by guesswork or experience and used to calculate a spectrum. These trial parameters are then varied until the calculated and observed spectra agree [34].

Spectrum simulation and iterative processes are obviously greatly aided by the use of a computer and most of those used in Fourier spectrometers have software available which enable them to simulate spectra resulting from up to 6 spins. The programmes normally permit iteration of some type and will print out simulated spectra on the same recorder as that used for experimental spectra. Data obtained by computer iteration must always be examined critically, since quite often there is more than one set of values which will 'fit' the experimental data. If, for example, such a procedure produces a geminal coupling with positive sign the initial premises given should be very carefully examined. Serious problems can arise when the spectrum has a large number of equivalent spins or spin groups, and in cases of high symmetry where, due to this symmetry, a relatively small number of lines appear.

Spectral analysis can often be assisted by running the sample at two fields. Chemical shifts increase with increasing field; coupling constants do not. A spectrum which is a complex ABC system at 60 MHz will probably be an AMX system at 300 MHz and consequently easy to analyse. Even for much smaller changes in field, i.e. 60 to 100 MHz spectra which are non first order at the lower field show changes which are often useful at the higher field. Comparison of simulations of the spectra at both fields can often resolve ambiguities and avoid wrong solutions of the type mentioned above. The obvious use of multiple fields in spectral analysis is to make one of the fields the largest possible and so approach as near as possible to the easy first order situation. The opposite approach can sometimes be valuable, i.e. run at a low field to induce the second order condition, and hence get spectra which are sensitive to the signs of the coupling constants involved.

8.5 CHEMICAL EXCHANGE

N.m.r. spectra are sensitive to chemical exchange. The effect of exchange on the spectrum depends on the rate of the exchange compared with the nuclear spin time scale. If the exchange is slow, then the spectra of the separate species, e.g. conformers, etc. are seen. If the exchange is fast then a spectrum corresponding to the weighted mean of the individual spectra is observed. In intermediate cases broad spectra are obtained which yield information concerning the rate constants of the processes involved.

Consider the case of two non coupled nuclei, A and X, which can exchange. If the population of both sites is equal and the rate of exchange, K, is equal in both directions, in the slow exchange limit, $K \ll (\vartheta_A - \vartheta_X)$, the spectrum consists of two lines centered at ϑ_A and ϑ_X. As the rate increases the life-time of the nucleus in each site, which is proportional to K^{-1}, is decreased and, consequently, by the Heisenberg uncertainty principle, an uncertainty in its energy of $\Delta E = K/2\pi$ Hz results. This uncertainty results in a line broadening analogous to that induced by T_2 processes (see section 2.6) with $T_2 = K^{-1}$. Each line is a Lorentzian line broadened by $(\delta\vartheta) = K/\pi$. When the lines begin to overlap significantly $K \doteq \delta$ the line shape is no longer simply the sum of two Lorentzian lines; the lines coalesce and a new maximum grows at mean frequency of A and X (the weighted mean if the populations of A and X are unequal). Finally in the fast exchange limit $K \gg (\vartheta_A - \vartheta_B)$ a single line is observed which has an exchange contribution to a line width of $\pi(\vartheta_A - \vartheta_X)^2/2K$. These cases are illustrated in Fig. 8.8.

232

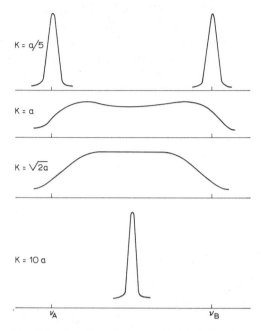

$K = a/5$

$K = a$

$K = \sqrt{2a}$

$K = 10\,a$

ν_A ν_B

Fig. 8.8. The line shapes calculated for two nuclei exchanging between two equally populated sites. K is the exchange rate and $a = \frac{1}{2}|\vartheta_A - \vartheta_B|$. The intensity of the final spectrum is half that of the others.

In the slow exchange region K may be obtained directly from the spectrum. The linewidth at half height is measured and corrected for its natural line width and any instrumental line broadening effects. Correction is normally carried out by subtracting the line width of an internal standard, and the residual line width $(\delta\vartheta)$ is equal to K/π (see section 10.7 for a discussion of the contribution of exchange to T_2). The lower limit of rate processes which can be simply studied by n.m.r. is thus in the order of a s^{-1}.

In the intermediate region K can be found if ϑ_A and ϑ_X are known by fitting the experimental curves to curves calculated from suitably modified forms of the Bloch equations for various values of K. N.m.r. can thus only be used to measure rate constants between 1 and $\Delta^2/2\pi$ s^{-1}, where Δ is the chemical shift difference (in Hz) between the exchange sites. However, it is possible to vary the temperature of the sample to bring the rate constants within this range. The range itself can be extended by increasing Δ by working at higher fields or using other nuclei as the probe, e.g. ^{13}C in organic problems.

Studies of a range of temperatures allow the determination of a number of important thermodynamic properties of the molecule undergoing the exchange process.

(A) The rate constant K can be found

    (i) Below coalescence $T < T_c$ from

$$K = \pi(\delta\vartheta) \tag{8.39}$$

where $\delta\vartheta$ is the additional broadening induced by the exchange.

    (ii) At coalescence $T = T_c$ from

$$K_c = \pi(\vartheta_A - \vartheta_X)2^{\frac{1}{2}} \tag{8.40}$$

    (iii) Above coalescence $T > T_c$ from

$$K = \pi(\vartheta_A - \vartheta_X)^2/2(\delta\vartheta)$$

(B) The free energy of activation, ΔG^*, can be obtained at the coalescence temperature from

$$K = \sigma k T \exp\left(-\frac{\Delta G^*}{RT}\right)\bigg/h \tag{8.41}$$

and if the rate is studied as a function of temperature then entropy and enthalpy can be obtained (see the work of Dalling et al. [35] on cyclohexanes as an example). The term σ in eqn. (8.41) is a transmission constant and expresses the probability that exchange will occur if the necessary energy is available, and depends on the nature of the energy barrier.

For the example discussed above, and all reactions normally studied, systems undergoing chemical exchange maintain their linearity; hence the results obtained using pulse excitation have no limitation on the pulse angle which can be used (see section 4.11).

8.6 SPECTRAL ASSIGNMENT TECHNIQUES

One of the unique features of an n.m.r. spectrum, as opposed, for example, to an infrared spectrum, is that all the lines are assignable to their transition of origin. Whether a complete assignment is a profitable exercise, especially in a complex molecule, is, of course, another question! The principal procedures available for assigning a spectrum are given below. In assigning a spectrum usually more than one method will be necessary.

(a) Correlation tables

N.m.r. has been used in chemistry for some twenty years and a wealth of data has been obtained. It is thus possible to produce tables of typical chemical shifts and correlation charts of various degress of sophistication (e.g. the simple example given in Fig. 1.1) which predict the chemical shift of a nucleus in a particular environment, or, inversely, the environment likely to yield a resonance at a particular chemical shift. Similar data presentations are possible for coupling constants. The first stage in the assignment of any n.m.r. spectrum is usually to assign the various lines or groups of lines which make up the spectrum to nuclei in a certain environment or functional group. This may take the form of assigning a hump in the 3 ppm region of a proton spectrum to CH_2 s, or a low field carbon signal to a carbonyl grouping, or a high field nitrogen resonance to a positively charged nitrogen. Gross assignments are then further refined by use of the other techniques outlined below.

Various methods of coding of chemical environments have been designed; when used with the appropriate spectra collection these enable the spectra of compounds containing nuclei in a similar environment to be easily located. An example of collections of this type is that of Johnson and Jankowski for ^{13}C [36].

For such data collections to be of much use they must be closely related to the specific application of n.m.r. in question. The reader is therefore referred to one of the books given as references for data etc. relevant to his application, or to one of the many computer-based correlation facilities now becoming available [37].

Figure 8.9 gives a simple correlation chart for proton and carbon resonances. The 'similarity' of the shift ranges for these two nuclei is useful, as most concepts learnt for one apply to the other, e.g. aromatics are down field of aliphatics. The similarity is amazing since the mechanisms giving rise to the shifts in the two nuclei are totally different, proton shifts being diamagnetic in origin and carbon shifts arising mostly from paramagnetic shielding.

With the advent of Fourier n.m.r., with its higher sensitivity, the study of weaker, and less common, nuclei is becoming more frequent. As the availability of more comprehensive data increases, so will that of correlations (see the work reported in chapter 23 of ref. 38).

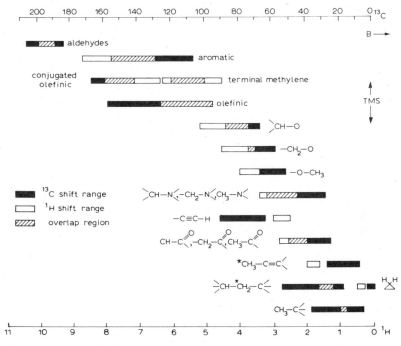

Fig. 8.9. A chemical shift correlation chart for ^{1}H and ^{13}C.

(b) Additivity rules for chemical shift and coupling constants

The various factors contributing to the chemical shift of a nucleus (see section 8.2) are to a certain extent additive. It is therefore possible to construct rules that will predict the chemical shift of a nucleus with respect to a reference compound by means of additivity parameters reflecting any changes introduced in chemical structures. The chemical shift of aromatic protons in benzene derivatives, for example, can be predicted with respect to pure benzenes in terms of correction for the number, type and position of substituent in the ring.

The best examples of additivity rules, however, are to be found in ^{13}C n.m.r., where not only are they more reproducible than in proton n.m.r., but, since the total chemical shift range is larger, they are less likely to produce ambiguous results. Consider as an example the rules of Lindeman and Adams for the ^{13}C shifts in

TABLE 8.4

Carbon-13 chemical shift parameters for paraffins [39]

Parameter	Value (ppm)	Parameter	Value (ppm)	Parameter	Value (ppm)	Parameter	Value (ppm)
B_1	6.80	B_2	15.34	B_3	23.46	B_4	27.77
A_{12}	9.56	A_{22}	9.75	A_{32}	6.60	A_{42}	2.26
A_{13}	17.83	A_{23}	16.70	A_{33}	11.14	A_{43}	3.96
A_{14}	25.48	A_{24}	21.43	A_{34}	14.70	A_{44}	7.35
γ_1	−2.99	γ_2	−2.69	γ_3	−2.07	γ_4	0.68
Δ_1	0.49	Δ_2	0.25				

alkanes [39]. Using data from some 59 paraffins they produced the following equation to calculate the chemical shift or carbon k in a simple alkane in terms of α, β, γ effects.

$$\delta_c(k) = B_s + \sum_{m=2}^{4} D_m A_{sm} + \gamma_s N_{k3} + \Delta_s N_{k4} \qquad (8.42)$$

B_s, A_{sm}, γ_s and Δ_s are constants given in Table 8.4. The subscripts s and m refer to the number of carbons bonded to carbon k and to the number of carbons bonded to those carbons respectively; N_{k3} and N_{k4} are the number of carbons 3 and 4 bonds away from carbon k. For example, the shift of carbon 5 in 3 methylheptane can be estimated thus

$$CH_3$$
$$|$$
$$CH_3 - CH_2 - CH - CH_2 - \boxed{-CH_2-} CH_2 - CH_3$$

(a) $s = 2$; the number of carbons bonded to C_5.

(b) $D_2 = 2$; the number of carbons bonded to C_5 having 2 carbons bonded to them (e.g. C_4 and C_6).

$D_3 = 0$; No carbons bonded to C_5 that are teritiary.

$D_4 = 0$; No carbons bonded to C_5 that are quaternary.

(c) $N_{k3} = 2$; There are 2 carbon 3 bonds from C_5 (C_2 and C_8).

$N_{k4} = 1$; There is 1 carbon 4 bond from C_5 (C_1).

Thus using the constants from Table 8.4 we get

$$(5) = 15.34 + 2(9.75) + 2(-2.69) + 0.25 = 29.7 \text{ ppm}$$

which in this case exactly agrees with the experimental value [40].

Similar additivity rules are available for coupling constants; Malinowski [41], for example, has derived constants where the value of $J^{13}C-H$ in a molecule of the type $CH_X YZ$ can be predicted in terms of contributions from substituents X, Y and Z thus [41]

$$J_{C-H} = E_x + E_y + E_z \tag{8.43}$$

Thus rule is very useful in ^{13}C n.m.r. where it can be used to calculate the value of $^1J(C–H)$ used in 'off resonance' assignments (see below).

(c) Integration

As was discussed in Chapter 2, the area under a peak is proportional to the number of nuclei of that type present in the sample. This property of n.m.r. spectra can be used in both an inter- and intra-molecular manner. The former permits the quantitative analysis of mixtures (see section 8.7e) and the latter enables the number of nuclei of a specific type within the molecule to be found. A proton peak due to a CH fragment is one half that of a CH_2, etc. Such information is obviously a great aid to assignment, particularly in proton spectroscopy but less so in other nuclei where the chances of more than one nucleus/molecule being in the same environment is small. If one resolved multiplet or group of resonances can be assigned (e.g. the 5 proton signal of a mono substituted benzene ring), its integral can be used to calibrate the system, and subsequently the number of nuclei giving rise to all the other multiplets can be determined.

For the proportionality between the integral and the number of nuclei to hold, the spectrum must be unsaturated and free of any nuclear Overhauser enhancement. The latter, most commonly encountered in ^{13}C n.m.r., must be removed by either gated decoupling (section 9.9) or paramagnetic reagents (section 10.9) before meaningful quantitative integrals can be obtained. To avoid 'saturation' affecting integral accuracy in pulsed excitation, *all* nuclei must have recovered from the effect of the previous pulse before a new pulse can be applied so that the signal is simply proportional to $M_0 \sin \alpha$. It takes $5T_1$ for M_0 to recover to 99% of its original value following a $90°$ pulse; therefore long delays between the pulses or small pulse angles are necessary (see section 7.5) to ensure M_z^- is equal to M_0. Integral accuracy and sensitivity therefore do not have the same requirements concerning pulse power and repetition rate. For swept n.m.r. the differences in requirement between sensitivity and integral accuracy are not so obvious since a nucleus is unlikely to be excited twice within $5T_1$; thus saturation can be avoided simply by using an excitation field such that $\gamma B_1 < (T_1 \cdot T_2)^{1/2}$.

TABLE 8.5

Filter characteristics of pre-a.d.c. filters
4 pole Butterworth

Frequency offset from pulse filter bandwidth	Signal transmitted %	Frequency offset from pulse filter bandwidth	Signal transmitted %
0	100	1.1	56.4
0.1	100	1.2	43.4
0.2	100	1.3	33.0
0.3	100	1.4	25.2
0.4	100	1.5	19.4
0.5	99.8	2.0	6.2
0.6	99.2	2.5	2.5
0.7	97.2	3.0	1.2
0.8	92.5	4.0	0.39
0.9	83.6	5.0	0.16
1.0	70.7	10	0.01

Concern is often expressed as to the accuracy of integrals obtained in Fourier n.m.r., particularly with respect to those obtainable by swept n.m.r. Provided saturation is avoided their accuracy is obviously theoretically identical, and in practice comparable. Saturation seems to be harder to avoid when using pulse excitation, mainly for psycological (sensitivity is paramount!) and not spectroscopic reasons. Fourier spectra are digital, which has an effect on integral accuracy. The latter depends simply on the accuracy of the $t = 0$ point of the f.i.d. (see chapter 3) i.e. the word-length of the a.d.c. and/or computer and the accuracy of the phase setting (see chapter 6). It does not depend, as does the *peak height*, on the number of data points used, providing of course there are sufficient to resolve the spectrum.

The filter used prior to the a.d.c. (see section 6.10) can affect integral accuracy if its bandwidth is too narrow. The filters commonly used in pulse spectrometers are of the 4 pole Butterworth type whose characteristics are shown in Table 8.5. If the filter bandwidth is set equal to the spectral width, as it often is in order to achieve maximum sensitivity, the lines at far end of the spectrum, with respect to the pulse, will have only 80% of their true intensity. This problem can be eliminated by using a larger filter band with e.g. 2Δ.

(d) The use of coupling constants

The second major parameter in n.m.r. spectra is the coupling constant, which gives information about neighbouring nuclei. The most obvious aspect here is multiplicity. If in an non-decoupled ^{13}C spectrum a carbon resonance is basically a quartet, then that carbon is bonded to 3 protons, and if it is a triplet, 2 protons etc. It must be remembered that in the absence of heteronuclear coupling, ^{13}C spectra, like those of any dilute spins (i.e. low natural abundance), are singlets. ^{13}C—^{13}C coupling does exist (e.g. $^1J(^{13}C$—$^{13}C) = 34.6\,Hz$ in ethane) but due to the low abundance is not normally visible. The abundance of a molecule with two ^{13}C's adjacent is only 1% that of molecules with only one ^{13}C in either position; thus molecules exhibiting ^{13}C—^{13}C coupling only give rise to very weak satellite lines, like the ^{13}C satellites found in proton spectra. The magnitude of the coupling will reflect the nature of the other substituents and if there is any resolvable fine structure this will give information on the number of protons on adjacent carbon atoms. The same approach can be applied in the case of homonuclear coupling; a proton resonance which is a quartet, i.e. coupling to three equivalent protons, is situated 'near' a methyl group; if the coupling is 7—8 Hz then 'near' implies on the adjacent carbon, but if 1 or 2 Hz it could imply an allylic situation, and so on.

The magnitude of a coupling constant may also depend on stereochemistry [42] (see section 8.3). To take an example, an AX or AB system in the olefinic region of the proton spectrum of a compound (i.e. 5—7 ppm) is indicative of a disubstituted olefinic system; consideration of the chemical shifts will yield information concerning the nature of the substituents and thus lead to the assignment of the proton multiplets. Their coupling constant will reflect the stereochemistry; if J is $10 \sim 14\,Hz$ then the two protons in the AX system are *trans*, and if $7 \sim 9\,Hz$ then they are *cis*. A further example of the stereochemical dependence of couplings is that of bond angle on vicinal H—H coupling discussed in section 8.2d.

A third use of coupling constants is in sequencing. Suppose from chemical shift considerations it was known that a molecule had $CHCl$, CHO and CH_2 groupings (among others). If the aldehyde is a triplet with a splitting in the 7 Hz region then it must be adjacent to the CH_2 group. Now if the CH_2 group is a four-line pattern, i.e. two doublets, then it is between the CHO and the CHCl; the sequence CHO, CH_2, CHCl has now been established as a fragment. If the CHCl proton shows further splitting then the process can be extended.

To sum up, coupling constant data can give information concerning the number of nuclei either directly bonded to the observed nucleus or 'close by', and often their stereochemical relationship. Coupling constant data can also place a nucleus 'in context' by indicating the position with respect to other nuclei within the molecule.

(e) Double resonance

The topic of double resonance is fully discussed in chapter 9. It must, however, be briefly mentioned at this stage as decoupling experiments are a very valuable assignment aid [43]. It was previously assumed that the origin of splittings within a multiplet was known; unfortunately this is not always the case. Two equal proton lines 8 Hz apart may originate from two chemically shifted resonances or may be the A part of an $A_n X$ system. Running at two fields will resolve this problem; if the separation changes with field then it is a chemical shift effect; if not it is a coupling effect. However if the splitting is shown to be a coupling it will not indicate where within the spectrum is the X part; double resonance will. Irradiation of the doublet will result (if it has its origin in spin—spin coupling) in the removal of $J(AX)$ from the spectrum, hence permitting the identification of X. The use of decoupling thus greatly facilitates the use of coupling constants for functional group sequencing as described previously.

Double resonance may also be used as an aid to spectral assignment by simplifying the spectrum. If a multiplet is complex due to spin coupling to many nuclei, the selective removal of some of the couplings will greatly assist in deciphering the origin of the remaining splitting. As a simple example, consider the AB part of an ABX system where $J(AB)$ is required for stereochemical reasons. Analysis and hence assignment, is possible (see section 8.4) but could be quite complex. However if X is decoupled, the AB part reduces to a simple four line system from which $J(AB)$ can easily be found by inspection.

Heteronuclear decoupling in addition to the uses described above can perform another very useful function, that of inter-relating the spectra coming from two or more nuclei with different magnetogyric ratios within the same molecule. The classical example, though by no means the only one, is the case of ^1H and ^{13}C spectra. Both spectra provide information peculiar to themselves, e.g. on OH groups in the former, and quaternary carbons in the latter case.

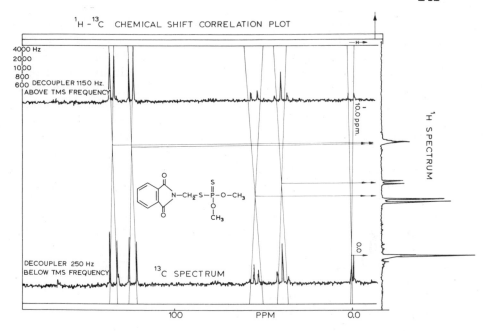

Fig. 8.10. Graphical method for cross assignment of ^{1}H and ^{13}C spectra. The horizontal spectra are off-resonance ^{13}C spectra, the vertical spectrum is the proton spectrum. The vertical position of the ^{13}C spectra corresponds to the location of the proton decoupler.

Taken as a pair they are even more powerful than they are singularly. Some regions of the carbon spectra are easy to assign uniquely, as are some, usually different, regions of the proton spectrum. Fortunately, by double resonance the two assignments can be inter-related. If a proton resonance is selectively decoupled, the effect on the ^{13}C spectrum is dramatic and simple. The large coupling between those protons and the carbon to which they are bonded is removed and that carbon appears as a tall, single (neglecting long range couplings) line. If either the proton or carbon is assigned then they both are.

The disadvantage of attempting a selective decoupling experiment of the type outlined above is that it requires one experiment for each proton/carbon pair. A more general method [44] is illustrated in Fig. 8.10. If the decoupler is off-resonance for a proton then, as is discussed in section 9.2, the observed splitting (J_r) will be reduced from the true splitting J, as given by eqn. (9.9) (or more precisely by eqn. (9.10)). This experiment is described as 'off resonance decoupling'. Within the limits defined by (9.8) J_r is linearly proportional to the offset of the decoupler frequency (ϑ_2) from the true proton frequency. A plot of line position (i.e. ϑ_1) against ϑ_2, for a doublet consists therefore of two lines of arithmetically equal slope (proportional to γB_2 and J) but of opposite sign, the intersection point corresponding to the chemical shifts of the two nuclei involved. Higher order multiplets have correspondingly more lines, but the result is the same. By such a graphical method cross-assignment is possible without having to perform a precise decoupling experiment. Presented therefore with an unknown molecule, cross-assignment between the proton and carbon spectra can be achieved by means of a series of arbitrary off-resonance experiments and a graphical analysis. For complex molecules the drawback of this method is finding the correct slope to use when drawing lines in the matrix of data points. A knowledge of the decoupler field used is very helpful since along with J it will give the slope of the lines. When there are a large number of carbons in the molecule the graph produced has many 'imaginary crossings'; these are easily distinguished from the 'real crossings' as they do not correspond to a carbon line in the noise decoupled ^{13}C spectrum.

(f) Temperature variation

Running a spectrum at more than one temperature can show changes in the spectrum which will assist in spectral assignment. If any grouping within the molecule is undergoing exchange, the resonances from this group will change with temperature; the nature of the change will depend on whether the spectra correspond to fast or slow exchange (see section 8.5). Changes in the spectrum, as a function of temperature, will also occur if any part of the molecule is undergoing a reorientation process.

(g) Solvent effects

The different effects of various solvents can often be useful in spectral assignment problems, especially when resonances overlap. If two protons in chloroform solution have overlapping resonances making assignment of the spectrum difficult, running the compound in a second solvent (e.g. benzene) may induce a differential chemical shift due to the reaction fields associated with formation of a weak solvent/solute complex. The two proton resonances may then be resolved and thus assignments made easier.

A second type of solvent effect can be used for the detection of exchangeable protons. If the proton spectrum of a molecule with an exchangeable proton, e.g. an alcohol, is run in $CDCl_3$, then resonance due to OH will normally be seen. Its line width and position, however, will vary depending on many factors such as lability, temperature, etc. Signals from such protons can be confirmed and quite equally importantly removed, by shaking the sample with a little D_2O, usually still in the n.m.r. tube. The OH will then be extensively replaced by OD via exchange with the D_2O. The proton spectrum run after shaking with D_2O will thus show effectively no resonance from the nucleus bond to the oxygen. Functions having only slow exchange will obviously be less affected on a short time scale.

A further assignment aid which may be loosely classified as a solvent effect is that of pH change. This is especially useful in biochemistry; changes can be detected in the n.m.r. spectrum of the molecule being investigated at or near the site of protonation (see ref. 45). The changes can be used in two ways, one to help assign that region of the spectrum which arises from the part of the molecule involved in the protonation or inversely to determine local pK_a values.

(h) Shift reagents [46—48]

In section 8.2e it was pointed out that shifts can be induced in n.m.r. spectra by paramagnetic reagents. The function of the so called 'shift reagents' is, as their name suggests, to induce shifts. Such reagents are usually co-ordinately unsaturated complexes of europium (inducing high field shifts) or praseodymium (low field shifts) which have a short electron relaxation time and an anisotropic susceptibility tensor. Typical ligands used are dipivalomethanate and 1,1,1,2,2,3,3-heptafluoro-7,7-dimethyl-4,6-octanedione which give complexes whose names are normally shortened to $Eu(dpm)_3$ and $Eu(fod)_3$ respectively. The latter compound has a higher solubility and is a stronger Lewis acid. These reagents form loose complexes with any polar group present in the molecule. They must therefore be used in dry non-polar solvents or their efficiency is reduced by competition between the solvent and solute for the vacant lanthanide co-ordination sites. In aqueous solution the free lanthanide ions themselves may be used; the sense of their shift is now inverted, europium shifts being to low field.

The use of shift reagents in spectral assignment is to 'spread out' the spectrum so that it approaches more to the first order situation, hence simplifying the assignment. If the co-ordination site is known and the shift is assumed to be purely dipolar in origin, then structural information can be obtained for the induced shift via eqn. (8.15). The reader is referred to ref. 46 for a comprehensive account of the uses of shift reagents. Since these reagents are paramagnetic they will to a certain extent shorten the spin—lattice relaxation time of the nucleus under observation, a fact which can be used when setting up a pulsed experiment to increase the sensitivity by the use of larger pulses.

Shift reagents can also be used in what may be described as a 'second order' manner in ^{13}C n.m.r. The addition of a shift reagent will induce shifts of the same order in both proton and carbon spectra; the shifts are, however, of less significance in the latter case due to the larger chemical shift range of carbon. As discussed previously, considerable information can be obtained from off resonance decoupling. The extra separation induced by a shift reagent permits selective decoupling effects to be observed which would not be possible, due to overlap, without the reagent. The assignment of the ^{13}C spectrum of ribose-5-phosphate by Birdsall et al. [49] illustrates this technique.

(i) Relaxation times

The spin—lattice relaxation times can be used as an assignment in some cases. For T_1 values to be meaningful, at least in any simple way, they have to be of nuclei which do not exhibit spin coupling. This condition is met for ^{13}C under noise decoupling conditions, decoupling removing the effect of the protons, and isotopic dilution removing any significant homonuclear coupling. In ^{13}C n.m.r. relaxation times can aid assignment in three ways. Firstly for protonated carbons in medium sized molecules which are dominated by dipolar relaxation, $T_1/$(number of directly bonded protons) is roughly constant for the bulk of the molecule (e.g. the skeleton of a steroid), but longer for any freely moving side chains (see section 10.4). CH_2 resonances from these two parts of the molecule can often be distinguished this way. Secondly, for quaternary carbons, where the relaxation is often still dipolar, relaxation measurements can give information on the number of adjacent protons (see the work of Wehrli [50] on the assignment of the ^{13}C spectrum of alkaloids). Finally, in small molecules where spin rotation relaxation is significant, the relationship of a carbon to the principal axis of molecular rotation can often be deduced, e.g. the assignment of the carbons in *ortho*-dichlorobenzene discussed in section 10.11.

8.7 THE APPLICATIONS OF FOURIER TRANSFORM N.M.R.

The applications of n.m.r. are numerous, varied and constantly increasing. Since, with a very few exceptions (see section 4.11) the spectra produced by Fourier techniques are equivalent to those produced by the slow sweep method, all the assignment techniques, data and applications dealt with in previous books [51, 52] on n.m.r. apply equally to Fourier n.m.r. (see the comprehensive work of Emsley, Feeney and Sutcliffe [53]. No attempt is therefore made to be comprehensive or detailed in this area; the gross outlines are given, along with reference to the appropriate text where further details may be found. The applications of Fourier Transform n.m.r. are rapidly expanding; for up to date information attention is drawn to the many review series on n.m.r. which are now available, and particularly to the annual series of specialists' periodical reports on n.m.r. (Ed. R.K. Harris) published by the Chemical Society [54].

The *new* applications opened up by Fourier transform n.m.r. arise from the ability of this technique to obtain a complete high resolution spectrum in a short period of time, in the order of T_2^*. These applications fall into two main groups, the study of relaxation times (and hence molecular motion) and secondly the study of rapid reactions. Relaxation studies are extensively discussed in chapter 10.

(a) Increased sensitivity

Increasing the sensitivity of n.m.r. is strictly speaking not an application; however since n.m.r. is by spectroscopic standards an insensitive technique, increasing its sensitivity can permit new applications and extend existing ones. As is shown in chapters 4 and 5 the gain is sensitivity per unit time depends on the typical spectral width of the nucleus being studied. The increase is about 10—20 for ^1H and 40—60 for ^{13}C, or, put the other way around, Fourier spectrometers take only 1/300th of the time taken by a swept spectrometer to get a proton spectrum (1/2000th for ^{13}C) of comparable quality. If time was not a factor, both, theoretically, could achieve the same sensitivity by time averaging, but time, of

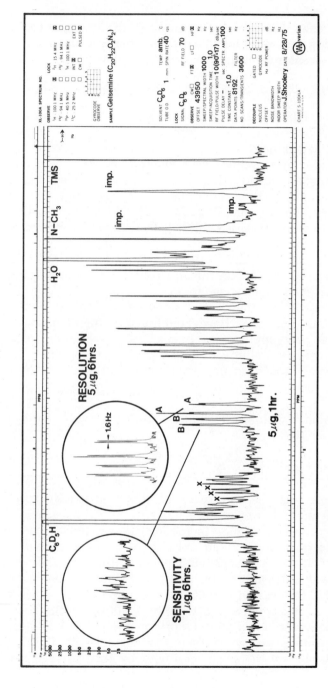

Fig. 8.11. The ^{1}H spectrum of gelsemine (5 μg). The insets show the effect of further time averaging (6 h) and lowering the concentration (1 μg) on the olefinic region (Varian Associates).

course, is always a factor and a saving of this magnitude is of immense value. The increased sensitivity of Fourier spectrometers can be utilised in two basic ways; firstly to extend the range of concentrations accessible to 'existing nuclei' e.g. ^{1}H, ^{31}P etc., and secondly to permit access, at reasonable concentrations, to the weaker nuclei e.g. ^{13}C, ^{15}N etc.

Proton n.m.r. developed rapidly in the 60's with instruments which required a hundred or so milligrams in order to obtain a reasonable spectrum. With modern Fourier spectrometers ^{13}C spectra are readily available from samples of this size and a rapid growth of ^{13}C is, as to be expected, taking place, while proton spectra can now be obtained even from thin layer chromatography samples. It is obviously difficult, if not impossible, to give lower detection or sensitivity limits for n.m.r.; so much depends on the nature of sample, the spectrometer used, and the time one is willing to allocate to the sample. For protons a typical lower limit is in the tens of micrograms range (Fig. 8.11 shows the proton spectrum obtained from only 5μg of gelsemine) and for ^{13}C the lower limit is probably in the milligram range.

Increased sensitivity not only has advantages, but it also generates some problems. The power of n.m.r. lies in the fact that every magnetic nucleus gives an assignable line; this not only applies to the compound of interest but also to the solvent and any impurities present therein. As the sensitivity of the technique improves, more care must be taken in order to obtain pure solvents. This is mainly a problem in proton n.m.r. because the proton is the most sensitive nucleus and has a small chemical shift range, hence an annoying overlap between the compound and impurity is more likely. Also, nearly all the impurities found in the solvents used in n.m.r. contain protons. The most common impurity is water, which, being at high dilution, often has a broad line or unpredictable shift which makes it quite a nuisance. Solvent impurities and artefacts, like spinning side bands and ^{13}C satellites, are obviously more of a problem at high dilution, hence the use of small sample tubes wherever possible. These, for a fixed amount of sample, use higher concentration solutions and are consequently less susceptible to solvent impurities, which are correspondingly lower. Solvent impurity problems are less severe for other nuclei (except perhaps ^{13}C, but here the basic sensitivity is low, and high concentrations are essential). Other forms of spectroscopy do not have as severe a purity problem, as the selective removal of 'active' impurities is simpler than that for the omnipotent proton.

One area where Fourier transform n.m.r. is expanding, is into the gas phase [55]. Here concentrations are invariably low (for proton work pressures of about 0.5 Atm) and sensitivity is of paramount importance. Pulse excitation is also favourable in the gas phase, as due to the rapid molecular motion, spin rotation is normally an effective relaxation process (see section 10.5) and T_1's are correspondingly shorter than in the liquid state. Some gases, however, have very long T_1's, e.g. in monatomic gases where spin rotation cannot act as a relaxation process. Study of gas phase relaxation gives information on molecular motion in the gas phase, and, in cases where loose surface adsorption occurs, on the exchange between the two phases. Ideally one would expect molecules in the gaseous state to behave as isolated molecules; they do, but only at low pressures. For example, the ^{129}Xe spectrum of XeF_6 has been studied as a function of concentration (pressure) at concentrations sufficiently low for three body effects to be ignored [55].

(b) Organic chemistry

It is in organic chemistry that n.m.r. via proton and carbon spectra finds perhaps its most significant applications, as can be ssen by the large number of books on the subject (see refs. 40, 57—61). N.m.r. has in some ways a revolutionised structure determination. At the very basic level via the integrated area under the proton and carbon spectra, the number of non-equivalent types of proton and carbons can be determined. These can then in turn be analysed in terms of likely functional groups, and finally, usually via coupling constant data, sequenced to build up structural formulae.

The significant contribution of Fourier transform techniques in this area has been to make available ^{13}C n.m.r. as a routine technique. However, two other nuclei which are also becoming more available to the organic chemist are ^{15}N and ^{3}H. Nitrogen, a nucleus commonly found in organic chemistry, has two isotopes; ^{14}N (99%) which has $I = 1$ and ^{15}N with $I = \frac{1}{2}$ [62—65]. Due to its quadrupole moment ^{14}N is limited in use, as its lines are quite often very broad. ^{15}N on the other hand has the advantage of being a spin $\frac{1}{2}$ nucleus, but has two major disadvantages, low natural sensitivity and abundance (see Table 8.5), and a negative magnetogyric ratio [65]. Both limitations are eased by Fourier n.m.r. The increased sensitivity counters the first limitation; spectra from compounds containing natural abundance ^{15}N are obtainable. The second limitation is the negative n.o.e. resulting from the negative magnetogyric ratio (see section 9.7). The nuclear Overhauser enhancement in the presence of proton decoupling can vary from its maximum value (-3 in this case) to zero depending on competitive relaxation processes, and hence can take the value -1. If the n.o.e. does equal -1 then no signal is observable when proton decoupling is used. Decoupling is normally used to simplify the spectrum and gain sensitivity due to multiplet collapse. These advantages can be maintained, while the n.o.e. is removed in a pulse experiment by using gated decoupling (section 9.9) thus opening up ^{15}N n.m.r.

Tritium is a radioactive isotope of hydrogen with spin $\frac{1}{2}$ and a higher sensitivity to n.m.r. detection than even the proton. As an isotope of hydrogen with a high sensitivity it has obvious applications as a tracer for mechanistic studies etc. The drawback is its radioactivity, both in terms of radiation hazard and radiation induced chemical decomposition. Both these problems are less severe at the lower concentrations with which it is possible to work using Fourier spectrometers, and its use must be expected to increase [66].

Most of the numerous applications of n.m.r. in organic chemistry follow directly from the concepts of spectral assignment dealt with previously, combined with the sensitivity of the chemical shift and coupling constant to molecular properties such as electronegativity bond order, stereochemistry [67], etc., and are to be found in detail in the many books and articles on the subject.

(c) Inorganic chemistry

The impact of Fourier n.m.r. will be greater in inorganic and organometallic chemistry than in organic chemistry, since it permits access, on a reasonable time scale, to many nuclei of chemical interest. In order to decide whether a nucleus is suitable for high resolution study certain factors must be considered:

(i) The natural abundance, or availability by enrichment if this is easy.

(ii) The basic sensitivity of the nucleus at a fixed field.

(iii) The spin quantum number and quadrupole coupling constant of $I > \frac{1}{2}$.

(iv) The relaxation properties.

Some of these properties are summarised in Table 8.6 for the most likely nuclei for Fourier transform n.m.r. study. The first two properties mentioned above dictate the ease of detection of a nucleus and are combined in the table to give a relative sensitivity which is normalised to ^{13}C. ^{13}C is chosen as a normalisation standard, as opposed to the proton, because its sensitivity is in many ways more characteristic than that of the proton. As can be seen, there are many nuclei which have an acceptable sensitivity when using pulse spectrometers. It should be remembered that the gain in sensitivity achievable by Fourier techniques depends on the chemical shift range (in Hz not ppm) of the nucleus being studied; the larger it is the greater the gain in sensitivity. For the heavier nuclei the chemical shift range is normally much larger than for the lighter nuclei, particularly hydrogen.

The second pair of properties from the list given above is concerned basically with the line width of the spectrum, and only indirectly with the sensitivity. Nuclei with a spin quantum number greater than $\frac{1}{2}$ possess a quadrupole moment; the nuclear charge distribution is not symmetrical and they can relax via interaction with electric fields (see section 10.8). If the quadrupole moment is large then the interaction will be very significant and broad lines will result, rendering the nucleus unsuitable for study. If, however, the moment is small, very acceptable spectra can be obtained from most molecules, e.g. ^2H, ^{17}O. Note it is the size of the quadrupole moment which is significant in determining line widths, not the quantum number itself.

The nuclei studied in inorganic chemistry can be divided into two groups, metallic and non-metallic. The non metallic nuclei, including nuclei such as ^{11}B, ^{31}P, ^{17}O etc., have fields of n.m.r. application including structure determination, such as in the boron hydrides [68], ligand conformations in phosphine complexes, such as cis/trans isomerisation in platinum complexes [69], and solvation studies, such as hydration [70]. In these cases, not only are chemical shifts of interest but also coupling data [71].

^{11}B spectra provide probably the best examples of a practical use of resolution enhancement functions of the type discussed in section 7.9. ^{11}B spectra are normally fairly broad, i.e. they will exceed the magnet inhomogeneities, but contain $^{11}B-^1H$ coupling information. Since magnetic field inhomogeneities do not contribute significantly

TABLE 8.6

N.m.r. properties of some common nuclei

Isotope	Nuclear spin	Frequency at 2.3 T	Electric quadrupole moments	Natural abundance	Relative sensitivity[a]
^{1}H	1/2	100.000	—	99.98%	5676
^{2}H	1	15.351	2.73×10^{-3}	0.02	8.53×10^{-3}
^{3}H	1/2	106.022	—	—	6071
^{7}Li	3/2	38.862	-3×10^{-2}	92.58	1537
^{9}Be	−3/2	14.054	5.2×10^{-2}	100	78.5
^{10}B	3	10.746	7.4×10^{-2}	19.85	21.28
^{11}B	3/2	32.084	3.55×10^{-2}	80.42	753.0
^{13}C	1/2	25.144	—	1.11	1
^{14}N	1	7.224	7.1×10^{-2}	99.63	5.71
^{15}N	−1/2	10.133	—	0.37	2.18×10^{-2}
^{17}O	−5/2	13.560	-2.6×10^{-2}	0.004	6.05×10^{-2}
^{19}F	1/2	94.077	—	100	4729
^{23}Na	3/2	26.452	0.14	100	523
^{27}Al	5/2	26.057	0.149	100	1176
^{29}Si	−1/2	19.865	—	4.7	2.1
^{31}P	1/2	40.481	—	100	379.0
^{33}S	3/2	7.669	-6.4×10^{2}	0.76	9.5×10^{-2}
^{35}Cl	3/2	9.798	-7.89×10^{-3}	75.53	20.17
^{37}Cl	3/2	8.155	-6.21×10^{-3}	24.47	3.88
^{39}K	3/2	4.39	—	93.08	0.26
^{45}Sc	1/2	24.503	—	100	1705
^{51}Co	7/2	23.614	0.40	100	1570
^{67}Zn	5/2	6.189	—	4.12	0.67
^{75}As	3/2	17.127	0.3	100	143
^{77}Se	1/2	19.070	—	19.07	2.97
^{79}Br	3/2	25.054	0.33	50.54	223.7
^{81}Br	3/2	27.007	0.25	49.46	276.3
^{107}Ag	1/2	4.162	—	51.53	0.19
^{109}Ag	1/2	4.251	—	48.65	0.28
^{111}Cd	1/2	21.204	—	12.86	7.20
^{113}Cd	1/2	22.181	—	12.34	7.64
^{117}Sn	−1/2	35.626	—	7.61	19.74
^{119}Sn	−1/2	37.272	—	8.58	25.55
^{133}Cs	7/2	13.113	-3×10^{-3}	100	269.3
^{199}Hg	1/2	21.50	—	33.8	18.98
^{119}Hg	1/2	17.827	—	16.84	5.47
^{203}Tl	1/2	57.150	—	29.50	13.6
^{205}Tl	1/2	57.709	—	70.50	768.7
^{207}Pb	1/2	20.900	—	21.11	11.8

[a] Normalized to ^{13}C and including corrections for natural abundance.

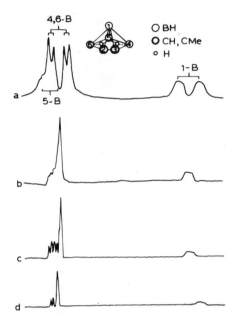

Fig. 8.12. The ^{11}B n.m.r. spectrum of $C_2MeB_4H_7$. (a) Undecoupled, (b) ^{1}H broadband decoupled, (c) line-narrowed and decoupled, and (d) $C_2B_4H_8$, line-narrowed and decoupled.

to the observed line width, the lines are Lorentzian in shape and their f.i.d.'s are experimental; they therefore respond well to being multiplied by a positive exponential in order to reduce line widths. As was seen earlier this process degrades the signal to noise ratio of the spectrum considerably; it is therefore necessary to obtain a good signal to noise ratio prior to using resolution enhancement functions. Good signal to noise ratios are easy to obtain for ^{11}B which is a relatively sensitive nucleus, especially when using pulse excitation, since T_1 and T_2 are short, and hence rapid pulse repetition rate can be used and many spectra accumulated in a short time. Figure 8.12 illustrates a nice example [72].

The applications of the metallic nuclei are as varied as the chemistry of the elements themselves. All the group I metals are accessible for study by n.m.r. and their solution chemistry is intensively studied. Aluminium has an interesting solution chemistry amenable to study by n.m.r. The ^{27}Al spectrum, for example, of a simple solution of alum (potassium aluminium sulphate) gives two lines, not just one due to Al^{3+} ($6\,H_2O$), as might be expected and as is found for other salts. The second line, which is broader, is due to a sulphate species [73, 74]. There are many other examples to be found in the literature. N.m.r. work on the heavy metal nuclei, e.g. [207]Pb, [199]Hg, [205]Tl, etc. is in its infancy. For these nuclei chemical shifts are very large, 1000's of ppm, and so is their sensitivity to solvent effects etc., e.g. $PbMe_4$ changes by 6 ppm between CCl_4 and CS_2 solutions [37]. Most of the work published so far was performed using indirect means and proton n.m.r., e.g. by INDOR. Pulse spectrometers make direct observation possible and an expansion of interest must surely follow [37, 68].

Further details of the many applications of n.m.r. in inorganic chemistry can be found in the appropriate chapter of ref. 76.

(d) Biochemistry [77]

N.m.r. is in many ways ideally suited to the study of biological systems, allowing work in aqueous solutions in the correct pH and temperature ranges. The major draw back to n.m.r. is its low sensitivity; consequently the advent of Fourier transform spectrometers has led to increased activity in this area. Unfortunately pulse spectrometers working in aqueous solution encounter problems due to dynamic range (see section 7.11) especially at the low concentrations often necessary in biochemical samples. One of the techniques previously discussed to ease this problem must be used. Solvent elimination by saturation can cause problems in biochemical systems due to the transfer of saturation effects resulting from exchange of NH and OH groups with the solvent. These effects can result in these lines disappearing from the spectrum or showing unusual Overhauser effects. The latter can be useful occasionally; for example, Dadok et al. [78] have identified non-exchangeable NH protons on the periphery of an enzyme in aqueous solution by measuring n.o.e. effects while saturating the solvent. These protons show a positive n.o.e. due to significant contribution to their relaxation from dipolar coupling between the NH and the solvent protons. Exchangeable NH's decrease in intensity under these conditions due to saturation transfer, while non-exchangeable protons buried in the molecule are unaffected. Because of its unique performance in avoid-

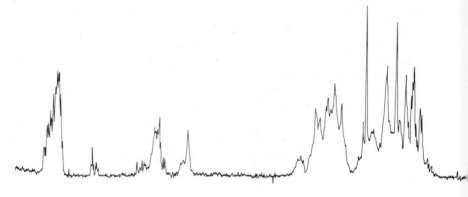

Fig. 8.13. The carbon 13 spectrum of ribonuclease.

ing the problems associated with large solvent signals in a spectrometer using digital technique, correlation spectroscopy finds its
major applications in the field of biological n.m.r.

Fourier spectrometers have encouraged the study of relaxation
times as a probe to molecular motion. When studying macromolecules,
where there is always extensive overlap of lines due to the many
nuclei in similar chemical environemts, high frequency spectrometers
using superconducting magnets are especially valuable. However,
when using pulsed high frequency spectrometers with macromolecules
one does not always get the increase in sensitivity compared with
lower frequency spectrometers that one might expect, particularly
for ^{13}C. The reason for this is that these large molecules have relatively
slow motion in solution, and with the high spectrometer frequency
the 'extreme narrowing limit' approximations (see Chapter 10) do
not apply, which results in an increased T_1 and a decreased nuclear
Overhauser effect, both of which have a detrimental effect on
sensitivity.

N.m.r. is used in macromolecular work in many ways (see refs. 77,
79—84). With the aid of paramagnetic probes (shift reagents) details
of solution conformation can be obtained by analysing the induced
shifts and assuming them to be dipolar in origin [85]. In ^{13}C n.m.r. it
is now possible to obtain spectra which show the resonances from
single discrete carbon atoms, as is seen in Fig. 8.13, which shows the
spectrum of ribonuclease. Using these resonances it is possible to
study folding and unfolding processes of these enzymes. Relaxation

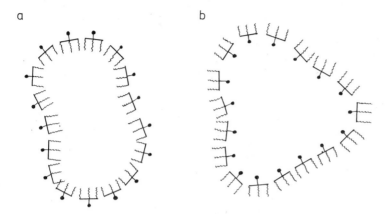

Fig. 8.14. Two phospholipid micelle configurations, (a) in aqueous and (b) in non aqueous solvents.

time data can be used to study inhibitor/substrate interactions in a very sensitive manner. By looking at the T_1 of the inhibitor, which can be present in a high concentration, and hence is easy to detect, the binding constant and site can be investigated. The T_1 of the free inhibitor which can be measured in a pure sample, is long (seconds) while that of the bound form will be fast, essentially that of a substrate (milliseconds) due to the slower motion of the large substrate. The measured T_1 is the weighted mean and hence it is possible to calculate binding constants. Such studies can be carried out using chemical shifts, but these are much less precise since the chemical shift differences involved between the free and bound inhibitor are only a few ppm, whereas the difference in relaxation times can easily be two orders of magnitude. Similarity studies of the T_1 of water in the protein solutions can give information on the protein-metal-ion-substrate complex. T_1 values can also be applied to the study of phospholipid bilayers. In a simply spherical micelle, for example, in aqueous solution (see Fig. 8.14a) molecular motion decreases away from the polar head group and hence T_1 increases along the hydrocarbon chain with increasing distance from the head group. In organic solutions the inverse is true. By studying the relaxation behaviour of a natural bilayer it can be established whether the head groups are on the inside or the outside of the structure [86].

Deuterium and phosphorus are nuclei which also have a very significant role in the biological applications of n.m.r. Phosphorous,

256

occurring naturally in many significant molecules, such as phospholipids, ATP etc., can be used to study these molecules directly. Deuterium on the other hand is incorporated into systems, e.g. biogenetically via microorganisms on deuterated amino acids. The very simple deuterium spectra are then used to study the behaviour of the complex molecule. The work of Smith et al. [87] using deuterium labelled lauric acid CD_3—$(CH_2)_9$—CD_2—COOD to study the effect of egg lecithin and cholesterol on fatty acid solutions shows dramatic changes in the deuterium line widths which can be correlated with molecular motion.

Carbon-13 has considerable application in biosynthetic studies as a stable isotope trace [88]. Various sites in a precursor are enriched

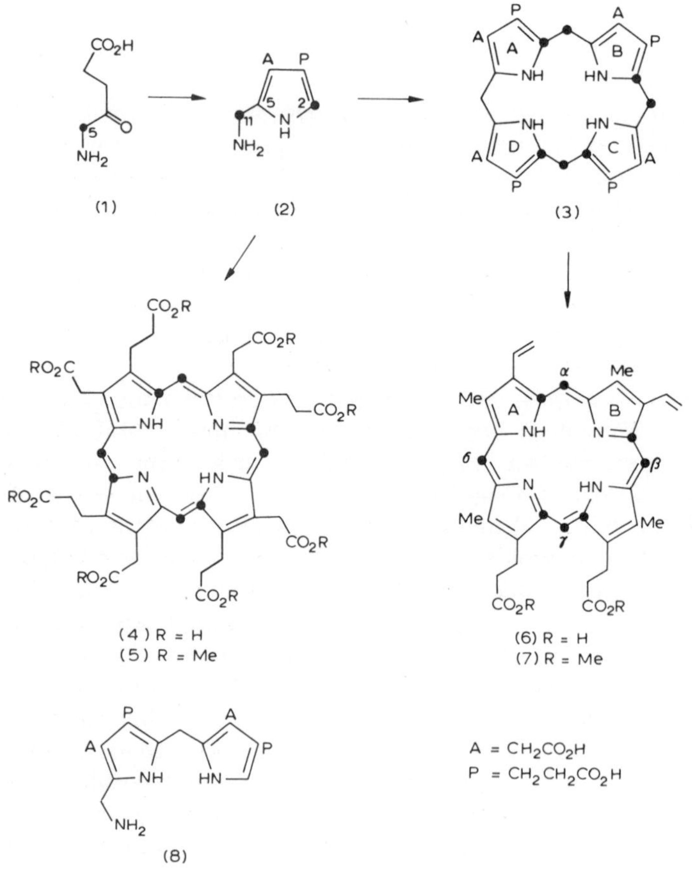

(1)   (2)   (3)

(4) R = H
(5) R = Me

(6) R = H
(7) R = Me

(8)

A = CH_2CO_2H
P = $CH_2CH_2CO_2H$

with ^{13}C, whose path is followed by n.m.r. Here Fourier n.m.r. has greatly assisted by increasing the possible detection levels etc. Even Fourier transform, ^{13}C n.m.r. does not of course approach ^{14}C work in sensitivity, but detection of the position of the label within a molecule is far simpler. A nice example of work of this type is the work by Battersby et al. [89] on the biosynthesis of Protoporphyrin IV. They started with 5-aminoleavulinic acid enriched to 90% ^{13}C in the five position (1). This was then converted to the prophobilinogen (PBG) (2) and then synthetically to symmetrical uroporphyrin 1 (4) whose ^{13}C spectrum gives 4 equivalent meso carbons with c.a. 90% as a double doublet (72 and 5 Hz) centred on a board singlet (90%) at 97.5 ppm. The small coupling is the long range $^4J(^{13}C$—C—N—$^{13}C)$. The ^{13}C enriched PBG (2) was then diluted with 4 parts unlabelled PBG and converted to (3), and then, without scrambling, converted enzymatically to the protoporphyrin IX and isolated as the dimethyl ester. Due to the dilution, the majority of molecules contained only 2 ^{13}C labels. The meso carbon region of protoporphyrin IX showed the α, β and γ carbons as 5 Hz doublets whereas the y meso carbon is a 72 Hz doublet with no 5 Hz coupling. This pattern can only be explained if PBG units AB and C are incorporated intact with PGB unit D undergoing an intramolecular rearrangement during the biosynthesis of the type III macrocycle.

(e) Quantitative analysis [90]

N.m.r. is a technique of great selectivity where the area under a peak in the absence of a decoupling r.f. field and below saturation, is proportional to the number of that type of nucleus present in the sample. It is therefore potentially a powerful method of quantitative analysis. Due to the variability of instrumental performance, etc., the absolute accuracy of an n.m.r. integral is not high. However, with the use of suitable internal standards, accuracies in the order of 1% are achievable. In the field of quantitative analysis n.m.r. must compete with other spectroscopic techniques of much higher sensitivity and lower cost. It is therefore only used for problems where its unique selectivity is necessary. Good examples of this type are found in the pharmaceutical industry [91, 92]. One extreme example of selectivity is the determination of the percentage of isotopic enrichment, e.g. ^{2}H or ^{13}C, in a sample: the resonances of the unaffected nuclei in the sample act as an internal standard. For reasons of sensitivity and freedom from complications arising from possible n.o.e. effects, proton n.m.r. is normally used in quantitative work. ^{13}C can however be successfully used in a quantitative role provided suitable precautions are taken to remove any n.o.e. effects and allow for relaxation effects. Typical applications are in determining the composition of oils and fats, particularly the degree of unsaturation and the aromatic to aliphatic carbon rates [93].

The accuracy achievable by Fourier and swept techniques is comparable provided the precautions discussed in 8.6c are observed; the former is of course more sensitive.

(f) Dynamic effects

N.m.r. can be used to study dynamic effects in chemistry by many methods [94—97]. The consequences of chemical exchange have been dealt with previously. A new area made possible by the short time scale of Fourier n.m.r. i.e. a complete spectrum in the order of $3T_2^*$, is the study of fast reactions. Reactions with half lives as short as a second (e.g. the α-chymotrypsin catalysed hydrolysis of L-phenylalamine tert-butyl ester) have been studied by Grimaldi and Sykes [98] using difference spectroscopy (see section 4.12). The reactants in aqueous solution are mixed in the probe using standard stop flow techniques, and the following sequence used

$$(\text{Push} - T_E \ ((\text{pulse} - T_a \cdot T_d)_{NT} - T_k)_{NK})_{NP}$$

T_E is a delay during which equilibrium is established and the reaction proceeds; a block of NT pulses is then taken followed by a delay of T_k, after which another block of NT pulses is taken, and so on NK times. The whole sequence can be repeated NP times with further mixes in order to improve the sensitivity by time averaging techniques. The choice of NK, NT and NP etc. depends on the concentration of the reactants and the rate of the reaction. Using difference spectroscopy has the simplifying advantage that no field/frequency lock need be set up in order to have sufficient stability for time averaging.

For slower reactions more conventional means can be used, and several spectra are recorded at a series of time intervals during the reaction. From the spectra the concentration of the various species present in the reaction can be studied as a function of time. With a Fourier spectrometer such a procedure would typically be completed

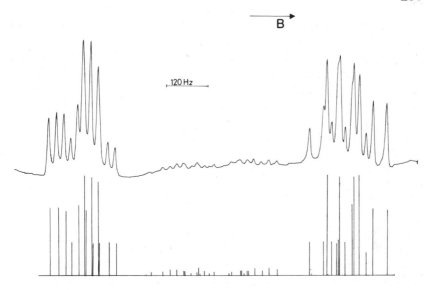

Fig. 8.15. The proton spectrum of 4-($^2H_1$) pyridine in *p*-ethoxybenzylidene-*p*-butylaniline and the calculated spectrum [99].

automatically, all the spectra being stored by the spectrometer in some backing store (e.g. on disc or magnetic tape) to be recalled later on a time scale suitable to the spectroscopist, rather that necessary for the reaction under investigation. The spectrometer's computer can frequently be used to analyse the data and calculate rate constants.

(g) Liquid crystals [99]

The spectra of compounds dissolved in nematic solvent differ greatly from those in normal solutions. In the spectrometer the solvent molecules 'line-up' parallel to the magnetic field; the solute therefore does not undergo free translational motion, but tends to have certain preferred motions i.e. anisotropic motion. The net result is that direct dipolar couplings, normally averaged to zero for isotropic motion, appear in the spectrum. These direct dipolar splittings are large (kHz) and the spectra for compounds dissolved in nematic solvents are thus wide and complex, as can be seen in Fig. 8.15. Detailed analysis of these spectra can give useful information concerning molecular geometry.

Fourier spectrometers can help in liquid crystal work, mainly by increasing the sensitivity. An unusual example of this is given in Fig. 8.15 which shows the proton spectrum of 4-($^2H_1$)-pyridine in *p*-ethoxybenzylidene-*p*-butylaniline. Without detecting the weak central lines the spectrum is 'deceptively simple' and cannot be fully analysed; with the extra lines, the spectrum is fully determined and all the parameters can be calculated.

The distinctive features of liquid crystal spectra come from their local order; if this is lost the spectra are identical to those obtained in isotropic media. One way the order can be lost is by rotating the sample at right angles to the field above a certain critical rate [100]. In a conventional magnet where the field is in the z axis, and thus at right angles to the axis of the sample tube, this simply means spinning the sample. For solenoid (superconducting) magnets, where the axis of spinning is parallel to the field, liquid crystal samples may be spun in the normal way to improve the homogeneity. If a sample in a conventional magnet is rapidly rotated by 90°, the dipolar splitting in the solute initially vanishes, and it reappears slowly as reordering takes place. With Fourier n.m.r. it is possible to obtain several complete high resolution proton spectra (via a single pulse) within the time taken for the sample to re-establish its orientation and the process studied [101].

The spectra of solutes in liquid crystal solvents are normally superimposed on the broad resonance of the solvent itself. In the time domain the signal from the solvent decays rapidly. If the start of data acquisition is delayed by a few tens of microseconds, the resulting frequency spectrum has the resonance of the solvent selectively attenuated and thus has its appearance considerably improved.

(h) Polymer chemistry

N.m.r. spectroscopy has been extensively used in polymer chemistry, initially via protons, and more recently via ^{13}C [102—104]. Each subunit and conformer in a polymer chain gives its own spectrum, the observed spectrum of the polymer consisting of the sum of all these individual parts. Polypropylene (atactic) exhibits three basic carbon resonances, one for each type of carbon,

$$\begin{bmatrix} CH_3 \\ | \\ CH_2{-}CH_2 \end{bmatrix}_N$$

Further examination of each of these resonances reveals stereo-chemical fine structure; each line of this fine structure corresponds to a subtly different substructure within the chain. The fine structure in the methyl resonances can, via model systems, be assigned to isotactic, heterotactic and syndiotactic triads. In order to assign all the methylene lines it is necessary to think in terms of tetrad and even pentad configurations.

In some polymers, e.g. natural rubber and soyabeans, it is possible to get a ^{13}C spectrum showing sharp lines from a 'solid' sample. Such spectra arise because these compounds are above their glass transition temperature in the spectrometer. Polymers above this temperature have sufficient internal motion to reduce greatly the dipolar broadening effects, producing relatively long relaxation times, ca. 100 ms, and hence fairly narrow lines. In such samples rapid pulsing can be safely used and high sensitivity achieved, thus permitting the study of very small features in the spectrum, e.g. the carbons bonded to the cross linking sulphur atoms. A further ex-ample of extra information obtainable using the higher sensitivity of Fourier n.m.r. is the study of the ^{13}C resonances from the carbons which form the small number of short chain branches occurring in highly crystalline linear polymers, such as polyethylenes. Estimates of the degree of branching from n.m.r. agree well with those obtained by other means.

Wideline relaxation measurements have long been used in polymer chemistry. Possibly the most significant contribution to polymer chemistry from Fourier techniques lies in the measurement of relax-ation time for the individual features of a high resolution spectrum. These measurements are a much more subtle probe to molecular motion, particularly when not only is the T_1 measured for each carbon, but so is the T_2 and nuclear Overhauser enhancement. Work of this type has been done by Schaeffer on systems like rubber and polybutadiene. He finds that the results (the n.o.e. of the methylene carbons of polybutadiene is below the theoretical maximum despite a T_1 of 0.5 s) cannot be explained using a single molecular correlation time as, perhaps surprisingly, seems to be possible with biopolymers. The results can be explained using distribution of correlation times to allow for the segmented motion in the polymer [105].

(i) Chemically induced dynamic nuclear polarisation (c.i.d.n.p.)
[107—111]

As has been discussed previously, at thermal equilibrium the population difference between the ground and excited states in a typical n.m.r. field is small and relaxation inefficient. In 1967 it was observed that population disturbances and even inversions leading to emission lines (see Fig. 4.14) commonly occur if the spectrum of the product of a free radical reaction is observed while the reaction is proceeding. Reactions may be induced by heating the sample, or irradiating it with light in the n.m.r. probe. Since such reactions are unlikely to produce radicals with electron spin state in equilibrium with their surrounds, the effect appeared to originate in an Overhauser process and was called chemically induced dynamic nuclear polarisation. It was soon shown that such an explanation was incorrected but the name c.i.d.n.p. remains. The nuclear spin polarisation occurs during the reaction process of two free radicals of a radical pair. The major applications of the technique are mechanistic; using the rules given below, it is possible to decipher a complex reaction pathway. The mere observation of a polarisation effect is sufficient to demonstrate a radical pathway.

The mechanistic information in c.i.d.n.p. spectra lies in the phases of polarised signals. Simple qualitative rules have been given by Kaptein [112] who defines two quantities Γ_n and Γ_m to describe the net polarisation (i.e. the total effect on a multiplet) and the multiplet polarisation (i.e. the relative effects within a multiplet).

$$\Gamma_{ni} = \mu \epsilon A_i \Delta_g \tag{8.44}$$

$$\Gamma_{mij} = \mu \epsilon A_i A_j J_{ij} \sigma_{ij} \tag{8.45}$$

μ is $+$ ve if the radical pair is formed either from a triplet precursor or by the encounter of free diffusing radicals, and is $-$ ve if the pair results from a singlet precursor.

ϵ is $+$ ve for the products of cage reactions and $-$ ve for the products of scavenging reactions.

σ_{ij} is $+$ ve if i and j belong to the same radical, $-$ ve if they do not.

Δ_g is $+$ ve for radical with the higher g value, $-$ ve for the lower.

A_i, A_j and J_{ij} have their normal meaning and have their appropriate signs (these can be found if all the other parameters are known).

Using these definitions one calculates the sign of the Γ's given in eqns. (8.44) and (8.45). If Γ_{ni} is $+$ ve then the resonance of nucleus i will show enhanced absorption (A), but if $-$ ve it will be an emission signal (E). If Γ_{mij} is $+$ ve the multiplet will be EA, and if $-$ ve, AE. In practice the values of Γ is determined experimentally and then used to evaluate the reaction mechanism via μ, ϵ and σ.

Consider the aldehyde proton of benzaldehyde, which shows enhanced absorption on irradiation of a mercury lamp [107]. Under the action of the light, an intersystem cross occurs to form a triplet state which abstracts hydrogen from a ground state molceular to form a radical pair (μ + ve).

$$PhCHO \text{ (triplet)} + PhCHO \rightarrow Ph\dot{C}HOH + Ph\dot{C}O$$

The radicals are PhCO ($g = 2.0006$) and PhCHOH ($g = 2.0033$, $A_\alpha = -1.59\,mT$). The spin polarisation could occur either by a polarised ketyl radical transferring a hydrogen to the benzoly radical within the cage (ϵ + ve), or to a ground-state benzaldehyde molecule in a scavenging reaction (ϵ − ve). Equation (8.44) gives + = + ϵ − + and hence ϵ is − ve and the reaction must be

$$Ph\overset{*}{\dot{C}}HOH + PhCHO \rightarrow Ph\overset{*}{\dot{C}}HO + Ph\dot{C}HOH$$

There are also emission peaks due to the formation of benzoin; here ϵ is − ve showing that benzoin is the result of a cage recombination. Similar arguments can be applied to the multiplet effect. For example the CH and CH$_3$ resonances of photolysed 2-iodomethane both show an EA multiplet effect. This arises from transfer of iodine from molecules outside the cage to spin polarised isopropyl radicals escaping (ϵ − ve) from $Pr^i \cdot Pr^i$ pairs formed by random diffusive encounters (μ + ve). Since A is negative for α(CH) protons and + ve for β(CH$_3$) protons eqn. (8.45) gives $\Gamma_m = + - + - + + = +$, as observed.

C.i.d.n.p. spectra do not arise from systems in thermodynamic equilibrium and therefore the spectra produced by swept and pulsed excitation will differ unless small pulse angles are used [113]. This problem has been discussed in section 4.11. The rules given above are for the spin system transfer function, i.e. that obtained by slow passage condition, and should only be applied to spectra where the pulse angle did not exceed 35°. For more powerful pulses the net effect is unchanged, but the multiplet effect which arises when two radicals of the sample type result is averaged if the coupling is homonuclear in origin and may result in the disappearance of the resonance (see Fig. 4.14 and eqns. 4.33 and 4.34). Having to use small pulse angles reduces the sensitivity advantage enjoyed by pulse excitation over swept excitation to about 50% of its maximum value; this form is, however, still the most efficient method.

264

REFERENCES

1 J.J. Musher, Adv. Magn. Res., 2, (1966) 177.
2 R.B. Mallion in R.K. Harris (Ed.), Nuclear Magnetic Resonance, Vol. 3, The Chemical Society, London, 1974, p. 1 and references therein.
3 W.G. Lamb, Phys. Rev., 60, (1941) 817.
4 W.C. Dickinson, Phys. Rev., 80, (1951) 563.
5 N.F. Ramsey, Phys. Rev., 78, (1950) 699.
6 T.D. Gierki and W.H. Flygare, J. Amer. Chem. Soc. 94, (1972) 7277 and references therein.
7 R. Freeman, G.R. Murray and R.E. Richards, Proc. Roy. Soc., A 242, (1957) 455.
8 H.M. McConnell, J. Chem. Phys., 27, (1957) 226.
9 J.S. Waugh and R.W. Fessenden, J. Amer. Chem. Soc., 80, (1958) 6697.
10 E.D. Becker, R.B. Bradley and C.J. Watson, J. Amer. Chem. Soc., 83, (1961) 3743.
11 A.D. Buckingham, Can. J. Chem., 38, (1960) 300.
12 M. Karplus and T.P. Das, J. Chem. Phys., 34, (1962) 1683.
13 W.D. Knight, Phys. Rev., 76, (1949) 1259.
14 H.M. McConnell and D.B. Chesnut, J. Chem. Phys., 28, (1958) 107.
15 D.R. Eaton and W.D. Philips, Adv. Magn. Res., 4, (1965) 103.
16 C.C. Hinckley, J. Amer. Chem. Soc., 91, (1969) 5160.
17 J.W. Fuller and G.N. La Mar, Tetrahedron Lett., (1974) 699.
18 J. Homer, Appl. Spect. Rev., 9, (1975) 1.
19 M. Barfield and D.M. Grant, Adv. Magn. Res., 1, (1965) 149.
20 W.G. Proctor and F.C. Yu, Phys. Rev., 81, (1951) 20.
21 H.S. Gutowsky and D.W. McCall, Phys. Rev., 82, (1957) 768.
22 N.F. Ramsey, Phys., Rev., 91, (1953) 303.
23 J.P. Pople and D.P. Santry, Mol. Phys., 8, (1964) 1.
24 N. Muller and D.E. Pritchard, J. Chem. Phys., 31, (1959) 1471.
25 H.A. Bent, Chem. Rev., 61, (1961) 275.
26 E.J. Wells and R.W. Reeves, J. Chem. Phys., 60, (1964) 2036.
27 M. Karplus, J. Amer. Chem. Soc., 85, (1963) 2870.
28 R.J. Abraham, The Analysis of High Resolution NMR Spectra, Elsevier, Amersterdam, 1971, p. 324.
29 P.L. Corio, Structure of High Resolution NMR Spectra, Academic Press, New York, 1967.
30 R.M. Lynden-Bell and R.K. Harris, Nuclear Magnetic Resonance Spectroscopy, Nelson, London, 1971.
31 F.A. Bovey, Nuclear Magnetic Resonance Spectroscopy, Academic Press, New York and London, 1969.
32 A. Carrington and A.D. McLachlan, Introduction to Magnetic Resonance, Harper and Row, New York and London, 1967.
33 T.H. Siddall and W.E. Stewart, Progr. NMR Spectrosc., 5, (1969) 33.
34 C.W. Haigh, A Simple Guide to the Use of Iterative Computer Programs in the Analysis of NMR Spectra, Ann. Repts. NMR Spectrosc., 4, (1971) 311.
35 D.K. Dalling, D.M. Grant and L.F. Johnson, J. Amer. Chem. Soc., 93, (1971) 3676.

36 L.F. Johnson and W.C. Jankowski, ^{13}C NMR Spectra: a Collection of Assigned, Coded and Indexed Spectra, Wiley-Interscience, New York, 1972, p. 680.

37 W.W. Simons and M. Zanger, The Sadtler Guide to NMR Spectra, Heyden, 1972, p. 600.

38 T. Axenrod and G.A. Webb, Nuclear Magnetic Resonance of Nuclei other than Protons, Wiley, New York, 1974.

39 L.P. Lindeman and J.Q. Adams, Anal. Chem., 43, (1971) 1245.

40 G.C. Levy and G.L. Nelson, ^{13}C NMR for Organic Chemists, Wiley-Interscience, New York, 1972, p. 222.

41 E.R. Malinowski, J. Amer. Chem. Soc., 83, (1961) 1479.

42 V.F. Bystrov, Russ. Chem. Rev., 41, (1972) 281.

43 W. Von Philipsborn, Angew. Chem. Intern. Edn., 10, (1971) 472.

44 B. Birdsall, D.J.M. Birdsall and J. Feeney, J.Chem. Soc. Chem. Commun., (1972) 316.

45 J. Batchelar and J. Feeney,

46 J. Reuben, Progr. NMR Spectrosc., 9, (1973) 1.

47 R.E. Sierers, Nuclear Magnetic Shift Reagents, Academic Press, New York, 1973.

48 B.C. Mayo, Che,. Soc. Rev., 2, (1973) 49.

49 B. Birdsall, J. Feeney, J.A. Glasel, R.J.P. Williams and A.V. Xavier, Chem. Commun., (1971) 1473.

50 F.W. Wehrli and T. Wirthlin, Interpretation of [13]C N.M.R. Spectra, Hayden, London, 1976.

51 W. McFarlane and R.F.M. White, Techniques of High Resolution NMR Spectroscopy, Butterworths, London, 1972, p. 132.

52 J.W. Akitt, NMR and Chemistry, An Introduction to NMR Spectroscopy, Chapman and Hall, London, 1973, p. 182.

53 J.W. Emsley, J. Feeney and L.H. Sutcliffe, High Resolution NMR Spectroscopy, Vols. I, II, Pergamon, London, 1966.

54 R.K. Harris (Ed.), Nuclear Magnetic Resonance (Specialist Periodical Report), Chem. Soc., London Annual.

55 G. Govil, Appl. Spectrosc. Rev., 7, (1973) 47.

56 C.J. Jameson, K.A. Jameson and M.S. Cohen, J. Chem. Phys., 62, (1975) 4224.

57 L.M. Jackman and S. Sternhell, Applications of NMR Spectroscopy in Organic Chemistry, Pergamon, London, 1969.

58 J.B. Stothers, Appl. Spectrosc., 26, (1972) 1.

59 J.B. Stothers, ^{13}C NMR Spectroscopy, Academic Press, London and New York, 1972, p. 559.

60 W. Von Philipsborn, Pure, Appl. Chem., 40, (1974) 159.

61 K.W. Bentley and G.W. Kirby (Eds.), Elucidation of Organic Structures by Physical and Chemical Methods, in Techniques of Chemistry, Vol. 4, Wiley, New York, 1972.

62 M. Witanowski and G.A. Webb, Ann. Repts. NMR Spectrosc., 5A, (1972) 353.

63 J.P. Kintzenger and J.M. Lehn, Helv. Chim. Acta, 5i, (1975) 905.

64 E.W. Randall and D.G. Gillies, Progr. NMR Spectrosc., 6, (1971) 119.

65 R.L. Lichter in (Eds.), ^{15}N NMR in Determination of Organic Structures by Physical Methods, F.C. Nachod and J.J. Zuckerman, Vol. 4, 1971, Ch. 4, pp. 195—232.

66 J.M.A. Al-Rawi and J.A. Elvidge, J. Chem. Soc., Perkin Trans., 2, (1975) 449, and references therein.
67 F.A. Anet and R. Anet, in F.C, Nachod and J.J. Zuckerman (Eds.), Determination of Organic Structures by Physical Methods, Vol. 3, 1971, Ch. 7, pp. 344—420.
68 A.O. Clouse, D.C. Moody, R.R. Rietz, T. Roseberry and R. Schaeffer, J. Amer. Chem. Soc., 95, (1973) 2496.
69 F.H. Allen and S.N. Sze, J. Chem. Soc. (A), (1971) 2054.
70 J.P. Hunt, Water Exchange Kinetics in Labile Aquo and Substituted Aquo Transition Metal ions by Means of Oxygen-17 NMR Studies, Vol. 7, 1971, pp. 1—10.
71 J.C. Tebby, in S. Trippett (Ed.), Organophosphorus Chemistry (Specialist Periodical Reports), Vol. 6, The Chemical Society, London, 1975, pp. 221—238.
72 J.W. Akitt and C.G. Savory, J. Magn. Res., 17, (1975) 122.
73 J.W. Akitt, Annual Reports on NMR Spectroscopy, 5A, (1972) 446.
74 J.W. Akitt and R.H. Duncan, J. Magn. Res., 15, (1974) 162.
75 R.H. Cox, Magn. Res. Rev., 3, (1974) 207.
76 B.E. Mann in N.N. Greenwood (Ed.), Spectroscopic Properties of Inorganic and Organometallic Compounds (Specialist Periodical Reports), Vol. 7, The Chemical Society, London, 1974, pp. 1—166.
77 R.A. Dwek, Nuclear Magnetic Resonance in Biochemistry, Oxford University Press, London, 1974.
78 T.P. Pinter, J.D. Glickson, J. Dadok and G.R. Marshall, Nature, 250, (1975) 582.
79 I.D. Robb and G.J.T. Tiddy R.K. Harris (Ed.), Nuclear Magnetic Resonance, The Chemical Society, London, 1973, p. 79 and references therein.
80 G.A. Grey, Crit. Rev. Biochem., 1, (1973) 247.
81 A.F. Casey, PMR Spectroscopy in Medical and Biological Chemistry, Academic Press, London and New York, 1971, p. 426.
82 J.C. Metcalfe, N.J.M. Birdsall and A.G. Lee in A.T. Bull, J.R. Lagnado, J.O. Thomas and K.F. Tipton, (Eds.), Companion to Biochemistry, Longmans, London, 1974, Ch. 3, pp. 139—162.
83 D.P. Hollis, Meth. Pharmacol., 2, (1972) 191.
84 D.R. Kearn and R.G. Shalman, Accounts Chem. Res., 7, (1974) 33.
85 S.J. Ferguson, Techniques and Topics in Bioinorganic Chemistry, (1975) 305.
86 A.G. Lee, N.J.M. Birdsall and J.C. Metcalfe, Chem. Brit. 9, (1973) 116.
87 H.H. Mentsch, H. Saito, L.C. Leitch and I.C.P. Smith, J. Amer. Chem. Soc., 19, (1974) 256.
88 M. Tanabe in T.A. Gussman (Ed.), Vol. 3, Biosynthesis, Chemical Society, London, p. 247.
89 A.R. Battersby, E. Hunt and E. McDonald, J. Chem. Soc. Chem. Commun. (1973) 642.
90 R.A. Lalancette, NMR Spectral Interpretation. A Workbook, Hulley, Merion, Pennsylvania, 1972.
91 D.M. Rackham, Proc. Soc. Anal. Chem., 9, (1972) 20.
92 D.M. Rackham, Proc. Soc. Anal. Chem., 11, (1974) 335.

93 J. Shoolery, Varian Applications Notes.

94 B.D. Sykes and M.D. Scott, Ann. Rev. Biophys. Bioeng., 1, (1972) 27.

95 K. Vrieze and P.W.N.M. Van Leeuwen, Progr. Inorg. Chem., 14, (1971) 1.

96 L.M. Jackman and F.A. Cotton (Eds.), Dynamic NMR Spectroscopy, Academic Press, New York, 1975, p. 660.

97 F.W. Dallquist, K.J. Longmuir and R.B. Dullernet, J. Magn. Res., 18, (1975) 406.

98 J.J. Grimaldi and B.D. Sykes, J. Amer. Chem. Soc., 97, (1975) 273.

99 J. Emsley and J.C. Lindon, N.M.R. Using Liquid Crystal Solvents, Pergamon, 1975.

100 J.W. Emsley, J.C. Lindon, G.R. Luckhurst and D. Shaw, Chem. Phys. Lettets, 19, (1973) 345.

101 B.M. Fung, J. Magn. Res., 15, (1974) 171.

102 F.A. Bovey, Progr. Polym. Sci., 3, (1971) 1.

103 F.A. Bovey, High Resolution NMR of Macromolecules, Academic Press, London and New York, 1971.

104 V.D. Mochel, J. Macromol. Sci. Rev., Macromol. Chem., C8, (1972) 289.

105 O. Jardetzky and N.G. Wade-Jardetzky, Ann. Rev. Biochem., 40, (1971) 605.

106 W.W. Simons and M. Zanger, The Sadtler Guide to the NMR Spectra of Polymers, Sadtler Research Labs., Philadelphia, 1973, p. 298.

107 P.G. Frith and K.A. McLauchlan, Nucl. Magn. Res., 3, (1974) 378.

108 H.R. Ward, Accounts Chem. Res., 5, (1972) 18.

109 R.G. Lawler, Accounts Chem. Res., 5, (1972) 25.

110 G.L. Closs, Adv. Mag. Res., (1974) 157.

111 R.G. Lawler, Progress NMR Spectrosc., (9, (1973) 145.

112 R. Kaptein, Chem. Commun., (1971) 732.

113 S. Schaublin, A. Hohner and R.R. Ernst, J. Magn. Res., 13, (1974) 196.

CHAPTER 9

MULTIPLE RESONANCE

9.1 INTRODUCTION AND HISTORY

So far only experiments where the sample has been subjected to a single exciting field have been considered. (A single field is not completely accurate in that while performing a field/frequency lock we are subjecting the sample to two exciting fields, but these are arranged to stimulate the nuclei of two different molecules within the sample). In considering multiple or, as is normally the case, merely double resonance we consider the situation where two (or more) nuclei within the same molecule are excited simultaneously, the signal from one being detected.

Multiple resonance experiments can be classified initially into two types depending on whether all the nuclei are of the same species (where the term used is homonuclear) or of different species (where the term is obviously heteronuclear). These two basic groups are further subdivided depending on the power of the second radio frequency field used. Table 9.1 gives such a classification. The power of the exciting field is expressed in terms of the coupling between the two spins under investigation. By far the most common double resonance experiment is the 'spin-decoupling' experiment where high powers are used and spin coupling effects removed. It was realised very early (by Bloch in 1954) that the application of a strong r.f. field could be used to collapse spin multiplets in weakly coupled spin systems [1]. The first realisation of this concept was achieved a few months later by Virginia Royden in an experiment to determine the magnetogyric ratio of ^{13}C with respect to that of the proton by finding the frequency which gave the optimum collapse of the $^{13}C-H$ multiplet [2]. It is ironical that the first double resonance experiment should involve ^{13}C and ^1H considering the importance now of ^{13}C n.m.r. with proton noise decoupling. The original experiment was of course the inverse of the present experiments: the proton was observed and the carbon decoupled. The

TABLE 9.1

Classification of double resonance experiments

Amplitude of B_2 (Hz)	Effects observed	Terminology
$\gamma B_2 \gg \Sigma\, n \cdot J$ (n depends on the multiplicity of the line)	Spectral simplification	Spin decoupling
$\gamma B_2 \sim J$	Selective removal of spin coupling	Selective decoupling
$\gamma B_2 \sim (T_2^*)^{-1}$	Additional splitting	Spin tickling
$\gamma^2 B_2^2 (T_1 T_2) \approx 1$	Changes in relative intensities	Generalised Overhauser effect

first homonuclear double resonance experiments, ^{1}H-{^{1}H}, were reported two years later by Anderson [3]. For instrumental reasons most of the early double resonance work was of the heteronuclear type, proton being the detected nucleus. With the perfection of homodecoupling techniques on commercial spectrometers proton–proton decoupling has become a very important experiment, particularly when applied to the problem of spectral assignments. Finally the introduction by Ernst of noise modulation techniques was one of the major factors permitting the recent explosive growth of ^{13}C n.m.r. [4].

In the great majority of multiple resonance experiments the result is independent of the type of excitation used to detect the observing nucleus. In other words all the theory, experience, etc. developed with continuous wave spectrometers can be applied when using pulse spectrometers. In this chapter we will consequently only give a brief account of the basic theory and applications of double resonance and refer readers to some of the many excellent review articles already published on this topic, e.g. that of Baldeschweiler and Randall [5] (early work, particularly heteronuclear), Hoffman and Forsen [6] (basic double resonance theory) Von Philipsborn [7] (applications) and the appropriate chapters of the annual review series for more 'in depth' background material. We will concentrate more on the areas of double resonance interrelated with pulse excitation, e.g. nuclear Overhauser studies [8].

9.2 DOUBLE RESONANCE THEORY

New terminology must be introduced at this stage. In a multiple resonance experiment the extra radio frequency fields are referred to as ϑ_2, ϑ_3 or B_2, B_3 etc. depending on whether one is describing their frequency or amplitude; they are applied in the usual manner i.e. as linearly oscillating fields along the x axis. When describing a double resonance experiment it is conventional to place the nucleus being irradiated in brackets, thus ^{1}H-$\{^{13}$C$\}$ means observing protons while decoupling carbon (the original decoupling experiment) while ^{13}C-$\{^1$H$\}$ represents the more conventional proton decoupling experiment.

The effect of ϑ_2 on an n.m.r. spectrum is very dependent on its amplitude with respect to the spin coupling between the two nuclei under investigation. When B_2 measured in frequency units (i.e. γB_2) is below the appropriate coupling constant, subtle effects occur, e.g. 'tickling effects', which will be discussed later; when the field becomes comparable with the coupling constant then spin decoupling occurs. The phenomenon of spin decoupling may be understood in terms of different models. In one simple model the field B_2 is considered as causing rapid oscillations of the irradiated spins between their energy levels. When this oscillation occurs at a frequency (γB_2) which is higher than J, the manifestations of spin coupling are removed from the spectrum. It should be noted that power required for spin decoupling ($\gamma B_2 > J$) is greater than that required to cause saturation ($\gamma B_1 \simeq (T_1 T_2)^{-1/2}$). For this reason spin decoupling effects are sometimes explained in terms of saturation, an effect with which they have nothing in common.

To analyse the effect of a second r.f. field, the simple case of an AX system will be considered. As shown for a spin system, the general Hamiltonian is

$$\hat{\mathcal{H}}^0 = \hbar \left[\sum_R \vartheta_R \hat{I}_z(k) - \sum_{R < 1} J_{k1} \hat{I}(k)\hat{I}(1) \right] \tag{9.1}$$

for the AX case this becomes

$$\hat{\mathcal{H}}^0 = -\hbar [\vartheta_A \hat{I}_z + \vartheta_X \hat{I}_z - J(\hat{I}_A \cdot \hat{I}_x)] \tag{9.2}$$

Under normal experimental conditions the appropriate Schoedinger equation has a solution with transitions at

$$\vartheta_A + mJ \quad \text{and} \quad \vartheta_X + mJ$$

where m is the magnetic quantum number I, $(I-1)$, etc; if we further simplify the situation to the case where both nuclei are of spin $\frac{1}{2}$ then the spectrum is simply two doublets centered at ϑ_A and ϑ_X respectively, with a splitting of J, i.e. transitions at

$$\vartheta_A \pm \tfrac{1}{2}J \quad \text{and} \quad \vartheta_X \pm \tfrac{1}{2}J$$

If we now introduce two radio frequencies we add two further time dependent terms to the Hamiltonian describing the system which now becomes

$$\mathcal{H} = \mathcal{H}^0 + \mathcal{H}_1(t) + \mathcal{H}_2(t) \tag{9.3}$$

the added terms are given by

$$\mathcal{H}_1(t) = -B_1 \sum_i \gamma_i [\hat{I}_x(i) \cos 2\pi\vartheta_1 t - \hat{I}_y(i) \sin 2\pi\vartheta_1 t] \tag{9.4}$$

and

$$\mathcal{H}_2(t) = -B_2 \sum_i \gamma_i [\hat{I}_x(i) \cos 2\pi\vartheta_2 t - \hat{I}_y(i) \sin 2\pi\vartheta_2 t] \tag{9.5}$$

The time dependence of one but not both of these parts of the Hamiltonina can be removed, as shown previously, by transforming into a rotating frame of reference. Since the B_2 field is the strong perturbation, double resonance experiments are normally solved by transforming into a frame rotating at ϑ_2. The observed r.f. field B_1 now represents a weak time-dependent perturbation in the rotating frame; transition probabilities due to this field can be calculated using first order perturbation theory. The problem of calculating a double resonance spectrum thus consists of two parts; firstly the diagonalisation of the double resonance Hamiltonian to find the energy level and eigen functions. This is equivalent to the procedure used in chapter 8, except that we are now working totally in a rotating frame (with consequent modifications to the spin Hamiltonian). The second part involves calculating the populations of these levels. If the system is more complex than an AX system a full density matrix solution is necessary [9].

We will not pursue a detailed solution of double resonance spectra any further; an excellent discussion is given in the article by Hoffman and Forsen [6], but a more pictorial, though less generally rigorous approach, will now be considered based on the work of Bloom and Schoolery [10]. Consider the spin system in a frame rotating at

ϑ_2; the static magnetic field now becomes $B_0 - \vartheta_2/\gamma$ (see section 2.8). For a resonance which manifests spin decoupling, the field along the z axis will depend on the spin state of the nucleus to which it is spin coupled and the resonance will have a band width (separation between the outer lines of the X multiplet) of B_0. Even when ϑ_2 is centered at ϑ_X the effective field at X will have a z component with a magnitude of up to $\frac{1}{2}\Delta B_0$. However, if B_2 is made large compared with ΔB_0 then the effective field in the rotating frame will be along B_2, which is perpendicular to B_0. The ϑ_2 will, however, not be close enough to the Larmor frequency of any other group to make B_2 significant compared with $(B_0 - \vartheta_2/\gamma)$ for any other nuclei in the spectrum. Consequently all other nuclei will see fields effectively along the z axis and will be quantised along that axis. A and X are hence quantised along orthogonal axes and since nuclear spin coupling depends on the scalar products, $J_{AX}\hat{I}_A \cdot \hat{I}_X$, the interactions vanish.

The model is illustrated in Fig. 9.1 for the AX case. For the nucleus X the effective magnetic field will depend on the spin state of the A nucleus ($m_A = \pm \frac{1}{2}$); the two cases are given in the a and b parts of the figure. The energy of nucleus X, however, depends on whether its spin is parallel or antiparallel with the resultant field ($\gamma_X B_{eff}$ or $\gamma_X B'_{eff}$); likewise the energy of nucleus A depends on whether or not its spin is parallel or antiparallel with B_0. In order to specify the total magnetic energy of the system one therefore has to define (i) whether diagram a or b is applicable and (ii) whether

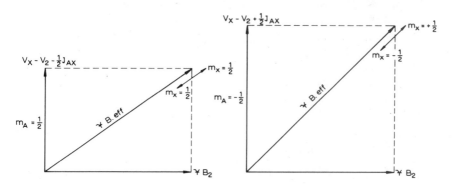

Fig. 9.1. Diagrams representing the effective magnetic field acting on the irradiated nucleus X in an AX system in the rotating frame. The left-hand diagram (a) pertains to molecules in which the A spin is opposed to $B_0(m_A = -\frac{1}{2})$. The right-hand diagram (b) is for $m_A = +\frac{1}{2}$.

$m_X = \pm \frac{1}{2}$ in that diagram. An A transition corresponds to an interchange between the a and b diagram or vice versa, but in addition may also involve a change in the value of m_X since the effective fields B_{eff} and B'_{eff} acting on X in the two diagrams are not themselves parallel. As a consequence, whether X is parallel with B_{eff} in one diagram does not indicate whether X is parallel or antiparallel with B'_{eff} in the other diagram, i.e. after an A transition. Since the frequency of a A transition is determined by the energy difference between the initial and final states, four possible transitions are possible. These transitions are centered on ϑ_A and are separated by plus and minus the sum or difference in the lengths of the two diagonal vectors in the two diagrams i.e.

$$\vartheta_A \pm \gamma_X (B_{eff} \pm B'_{eff}) \tag{9.6}$$

The phenomena of spin decoupling can now easily be explained. If ϑ_2 is centered on ϑ_X then the z components in Fig. 9.1 a and b will be equal but of opposite sign and two of the four lines predicted by eqn. (9.6) will coalesce at ϑ_A. Further if B_2 is strong compared with J_{AX} then B_{eff} and B'_{eff} will become equal and parallel; any transition involving a change in the spin of X thus becomes forbidden. The spectrum of A has now changed from a doublet in the unperturbed state to a single line. In the intermediate case the four possible line frequencies are simply given by eqn. (9.6). The line intensities (transition probabilities) are not so easily dealt with, and are given by the product of two complex factors, firstly a term equal to the probability of a radiation induced A transition, and secondly a term involving the square of the appropriate co-efficient in the expansion of the X spin functions along B_{eff} in terms of those along B'_{eff}. We will return to this situation when considering 'spin tickling'.

If the decoupling frequency is strong but not precisely set at ϑ_X, i.e. it is 'off resonance', then the spectrum at A will have the same multiplicity as in the non-decoupled spectrum (a doublet in the AX case) but with a reduced splitting. This reduced or residual splitting is given by the symbol J_r and can be deduced from Fig. 9.1 to be

$$J_r = [(\delta\vartheta - \tfrac{1}{2}J)^2 + (\gamma B_2)^2]^{\frac{1}{2}} - [(\delta\vartheta + \tfrac{1}{2}J)^2 + (\gamma B_2)^2]^{\frac{1}{2}} \tag{9.7}$$

where $\delta\vartheta$ is the offset of the decoupling field from ϑ_X. The degree of decoupling decreases rapidly as $\delta\vartheta$ increases. If the B_2 is sufficiently strong that

$$|\gamma B_2| \gg |\delta\vartheta|, \quad |J| \tag{9.8}$$

then eqn. (9.7) reduces to

$$J_r = J\delta\vartheta/\gamma B_2 \qquad (9.9)$$

Although we have only deduced eqns. (9.7) and (9.9) for the AX case the effect is perfectly general [4] and has considerable analytical significance in the assignment of ^{13}C spectra (see section 8.6e). Equation (9.9) holds only over a limited range because of the limiting conditions given by inequality (9.8). A linear relationship between J_r and $\delta\vartheta$ which is valid over the much wider condition

$$\gamma B_2 \gg \tfrac{1}{2}|J - J_r|$$

can be obtained using the equation [11]

$$\delta\vartheta = J_r \left[\frac{(\gamma B_2)^2 + \tfrac{1}{4}(J - J_r)^2}{(J^2 - J_r^2)} \right]^{\frac{1}{2}} \qquad (9.10)$$

9.3 BLOCH-SIEGERT EFFECTS

It has been implicitly assumed that the presence of the second radio frequency field has no effects on the spectrum other than those resulting from the perturbation of the irradiated group. This assumption is only true when $\gamma B_2 < |\vartheta_A - \vartheta_X|$, otherwise two further effects are observed which are generally grouped together under the term 'Bloch-Siegert effects' [12] and result in the shifting of resonance frequencies from their 'true' frequency. The shifts depend on the power of the second field and are only generally significant when performing spin decoupling experiments.

The first Bloch-Siegert effect of a second r.f. field at the X part of an AX system results in a shift of the frequency of the A resonance to a new value

$$\vartheta_A + \frac{B_2^2}{2(\vartheta_A - \vartheta_2)} \simeq \vartheta_A + \frac{B_2^2}{2(\vartheta_A - \vartheta_x)} \qquad (9.11)$$

away from the irradiated nucleus. When considering the effective field in the rotating frame acting on nucleus A, one must also take into account the extra field within that frame resulting from B_2; eqn. (9.11) does so and quantifies the resultant shift. A second shift arises from the requirement that for optimum spin decoupling the effective field observed by A and X must be perpendicular. Allowing for second order effect results in the frequency of optimum decoupling being shifted from ϑ_X as given by

$$\vartheta_X + \frac{(\gamma B_2)^2}{\vartheta_A - \vartheta_X} \tag{9.12}$$

As a consequence of Bloch-Siegert effects, the line frequencies either observed or implied (i.e. from the frequency of optimum decoupling) in double resonance experiments must be regarded only as approximations to the values in the unperturbed spin system unless very low powers are used [12].

9.4 THE GENERALISED OVERHAUSER EFFECT

The last section was concerned with double resonance experiments where a powerful r.f. field was used to remove spin coupling effects; whereas this type of experiment is the most common form of multiple resonance, it is by no means the only one. As the amplitude of the second frequency is progressively increased, certain changes in the n.m.r. spectrum occur. The first observable effects are those concerned with saturation. When ϑ_2 is situated on a specific transition (line) these effects become detectable where $(\gamma B_2)^2 T_1 T_2$ becomes in the order of unity. As the line widths in high resolution n.m.r. are often limited by magnet inhomogeneities, powers below the observable line widths may induce partial saturation [6].

Figure 9.2 shows the energy levels of an AX system. Consider a weak ϑ_2 field $(\gamma B_2 < 1/\pi T_2)$ set at line A_1; population changes will

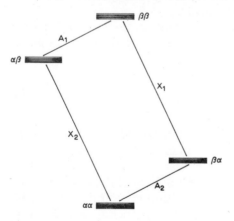

Fig. 9.2. Energy level diagram for the AX system.

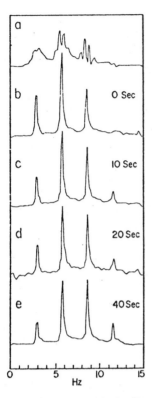

Fig. 9.3. The acetaldehyde quartet with irradiation of the high field methyl line (a). The same quartet observed at various times (b—e) after B_2 has been switched off. From R. Freeman, J. Chem. Phys., 53 (1970) 457.

occur and a new set of populations established in a time in the order of T_1. These changes result in a generalised Overhauser effect. The Overhauser effect, as originally defined, is the polarisation of nuclei which results from saturating the spins of the conduction electrons in a metal [13]. However, common usage has made the term describe the changes in population which result in internuclear polarisation based on relaxation mechanisms (dipolar coupling etc.). The intensity changes caused by low power irradiation are small and must be detected by comparison with lines unaffected by the second field. The generalised Overhauser effect is normally utilized in the INDOR technique (see section 9.6).

Using pulsed excitation it is possible to obtain a complete high resolution spectrum in less than T_1; it is thus possible to study the growth and decay of the generalised Overhauser effect with time [14]. Figure 9.3 shows the quartet of the CHO proton in acetaldehyde measured at a variable time after removal of a weak second radio frequency field from one line of the methyl doublet. The spectra show the intensities of the quartet relaxing back to their 1:3:3:1 equilibrium ratio. An exact analysis of generalised Overhauser effects is complex as it depends on the detail of the various relaxation processes involved.

9.5 SPIN-TICKLING [15]

When the strength of the second field becomes comparable with the natural line-widths within the spectrum, not only do population changes result, but mixing between the original unperturbed spin states occurs. In the AX case, if, for example, line A_2 (Fig. 9.2) is irradiated, mixing results between the original $\alpha\alpha$ and $\beta\alpha$ spin states to produce two extra spin states. Any line with an energy level in common with the irradiated line (i.e. X_1 and X_2 in this case) is split into a doublet centred on the original frequency.

Two different types of transition can be distinguished with respect to the irradiated line. In one type, the total change in magnetic quantum number for such a transition, and the irradiated transition, is zero; X_2 is the example in our case. A_2 and X_2 are said to be regressively connected. A_2 and X_1 are of the other progressive type where $\delta m = 1$. When ϑ_2 is off resonance to high field of a line, progressively connected transitions are shifted to high field of the original degenerate position, regressive lines move to low field, etc. Even when B_2 is exactly on resonance, lines which are progressively and regressively connected to the irradiated line are distinguishable. Differentiation is possible since magnetic field inhomogeneities contribute differently in the two cases; for the progressive case inhomogeneities add together and the lines of the resultant are broader than the original lines. For regressive lines inhomogeneities subtract, and a sharper doublet is produced. The mean line width of the four new transitions is, however, the same as the mean line width of the original lines. For further detail see the original work of Freeman and Anderson [15]. Figure 9.4 shows a spin tickling on the AMX system of 2, 3-dibromopropionic acid. The progressively and regressively connected doublets are clearly distinguishable. The irradiated line is saturated and hence does not appear in the spectrum.

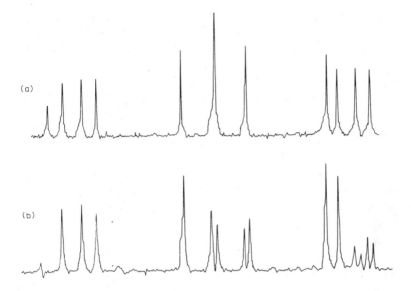

Fig. 9.4. Spin tickling on the AMX system of 2, 3-dibromopropionic acid. The line A irradiated.

Spin tickling is a very precise double resonance experiment. Also, since it only requires low powers, it is free from complications arising from Bloch-Siegert effects. Tickling can therefore be used to measure accurately the frequency of one line while observing another. The power of tickling used in this way is that the line which is irradiated may be undetectable, either 'hidden' under other resonances, or, if it is a heteronucleus, unobservable with the instrumentation available. The $\emptyset$CH = C signal in styryldiphenylphosphane oxide, which is hidden under the phenyl protons, can be detected using spin tickling [7].

Heteronuclear tickling is used in many ways, such as the determination of the chemical shifts of ^{129}Xe in the xenon fluorides [16] (see also ref. 7). Spin tickling may also be used to determine the relative signs of coupling constants [7].

9.6 INDOR

INDOR is probably the only experiment possible with swept excitation which has no direct parallel when using pulsed excitation.

INDOR is a double resonance technique originally proposed by Baker [17] where the intensity of a line is *continuously* monitored while a second low power field is swept through the spectrum of interest. Changes in intensity of the monitor line occur whenever ϑ_2 corresponds to the frequency of a line that has an energy level in common with the monitor transition at ϑ_1. The resulting intensity changes are plotted as a function of ϑ_2. INDOR signals are, with a power less than a line width (e.g. using the generalised Overhauser effect), positive for progressively connected transitions and negative for regressive ones. If powers in the order of a line width are used (i.e. utilizing tickling effects) the result is an increase in the positive lines. Two experiments have been proposed which perform the same function as INDOR yet work with pulsed excitation. The first of these which is applicable to the homo- and heteronuclear cases is double resonance difference spectroscopy (d.r.d.s.) [18]. Here the normal spectrum is recorded and then subtracted from a spectrum run under identical conditions, except that the line which would be the monitor line in INDOR is irradiated with a weak ϑ_2 field. The only lines which appear in the output of the subtraction are those

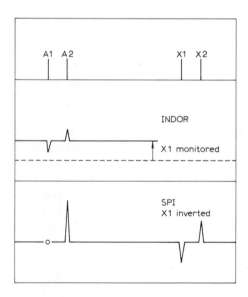

Fig. 9.5. Comparison of the INDOR and s.p.i. techniques applied to the AX system. In both cases the X_1 line is the "monitor". From A.A. Chalmers, K.G.R. Pachler and P.L. Wessels, J. Magn. Res., 15 (1974) 419.

whose intensities have been changed by generalised Overhauser effects. The frequency and sign of these transitions are identical to those obtained by INDOR since the pro- or regressive nature of a pair of transitions does not depend on which one is irradiated. D.r.d.s. does not gain the full sensitivity advantage of pulsed excitation as in general two runs are necessary (i.e. a loss of $2^{1/2}$ in sensitivity); it is nevertheless much more sensitive than normal INDOR.

The second experiment is selective population inversion (s.p.i.) [19, 20]. The principle is best illustrated by comparing s.p.i. with INDOR for the AX system (see Fig. 9.5). In INDOR with X_1 as the monitor line, as ϑ_2 passes through transitions A_1 and A_2, spins are 'pumped' from the lower to higher energy levels resulting in changes in intensity of X_1. The maximum effect occurs when the A transitions are saturated, i.e. Δn_A is zero. The changes in intensity are then $\pm \frac{1}{2}\Delta n_A$ (where Δn is the population difference resulting from Boltzman effects), the sign depending on whether the transitions are progressively or regressively connected. If, however, a selective $180°$ pulse is applied to X_1, the populations of the $\beta\beta$ and $\beta\alpha$ energy levels will be inverted. The power level of the pulse should be low enough that only X_1 is affected ($\gamma B_2 \ll J_{AX}$) and the pulse length short compared with T_1 in order to minimise secondary effects due to relaxation during the pulse. The A spectrum will now contain lines whose intensity has changed by $\pm \Delta n_X$ depending again on the type of connection between the transitions. In the homonuclear case an inverted X_1 line and the unperturbed X_2 line will also be seen, see Fig. 9.5. S.p.i. is easily applicable to either homo- or heteronuclear studies and conventional time domain spectrometers.

The power of the INDOR technique (or its pulse equivalents) like spin tickling lies in its ability to detect hidden transitions and measure the relative signs of spin coupling constants. For example if an X line of an AX system is monitored then the frequency of the A transitions can be determined whether they themselves can be detected or not. An excellent example of the use of homo-INDOR techniques is to be found in the work of Von Philipsborn on the thermal dimerisation product of 11, 13-dioxo-12-methyl-12-aza-4, 4,3-propellane. [7]. Here the technique of consecutive INDOR is used; one proton well removed from the aliphatic envelope is used as an initial monitor line, a spin coupled resonance from an adjacent proton is identified, this frequency is then itself used as a monitor

282

line, and a further proton detected, etc., thus permitting the elucidation of proton sequences.

The main attraction of heteronuclear INDOR is to use high sensitivity nuclei like ^{1}H and ^{19}F as monitors in studies of less sensitive
nuclei like ^{13}C, ^{15}N, ^{31}P, ^{29}Si, etc. Some very elegant ^{13}C work has
been performed using the ^{13}C satellite lines, especially from methyl
signals, as the monitor transition [21]. A secondary advantage of
hetero-INDOR is that little extra instrumentation is required to
study another nucleus, assuming it spin couples to a detectable
nucleus; a stable frequency source of the correct value is sufficient.
The cost of such a study is therefore low. The development of more
sensitive and flexible spectrometers makes hetero-INDOR less attractive than in the past.

9.7 THE NUCLEAR OVERHAUSER EFFECT

The 'nuclear Overhauser effect' (n.o.e.) is distinguishable from the
'generalised Overhauser effect' discussed previously in that (i) it can
be either inter- or intra-molecular in nature (i.e. scaler coupling
between the nuclei in equation is not essential) and (ii) it involves
irradiation of the *total* signal (i.e. not just a single line of a multiplet)
due to one type of nucleus while the second one is observed. The
effect was first demonstrated on the protons of chloroform while
irradiating the cyclohexane protons in a mixture of the two compounds, and can either be of a homo- or heteronuclear type. In the
former case its applications are in structural stereochemical work
where advantage is taken of its r^{-6} dependence, and in the latter case,
e.g. ^{13}C-{^{1}H}, it is also a powerful tool in the study of molecular
motion. The nuclear Overhauser effect is itself the subject of a book

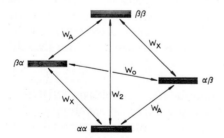

Fig. 9.6. The transition probabilities for an AX system.

by Noggle and Schirmer [8] to which the reader is referred for greater detail than is given here.

The nuclear Overhauser effect is due to the fact that in a double resonance experiment one is not dealing with an equilibrium system, and it can be understood by considering an AX system, as illustrated in Fig. 9.6. (N.B. although the spin coupling is not essential, it is quite often present). n_i is the population of the i state at equilibrium and W_{ij} is the transition probability of a transition between the i and j energy levels. The rate equations that govern the population dynamics are [22]

$$dn_1/dt = (n_2 - n_1) W_{12} + (n_3 - n_1) W_{13} + (n_4 - n_1) W_{14} \qquad (9.13)$$

$$dn_2/dt = (n_1 - n_2) W_{12} + (n_3 - n_2) W_{13} + (n_4 - n_2) W_{23} \qquad (9.14)$$

$$dn_3/dt = (n_1 - n_3) W_{13} + (n_2 - n_3) W_{23} + (n_4 - n_3) W_{24} \qquad (9.15)$$

$$dn_4/dt = (n_1 - n_4) W_{14} + (n_2 - n_4) W_{24} + (n_3 - n_4) W_{34} \qquad (9.16)$$

For a two spin system the transition probabilities have a degree of symmetry. One can therefore define new terms thus

$$W_A = W_{12} = W_{34}$$

$$W_X = W_{13} = W_{24}$$

$$W_2 = W_{14}$$

$$W_0 = W_{23}$$

To evaluate the nuclear Overhauser effect one must find the intensity of an A line $[(n_1 - n_2) = (n_3 - n_4)]$ (a) under normal Boltzman conditions (i), and (b) when the X spectrum is saturated (i'). In the latter case the populations of levels 1 and 3, and 2 and 4 are equalised i.e. $(n_1 - n_2) = (n_3 - n_2)$ etc. These intensities can be calculated from eqns. (9.13—16) to be

$$\frac{i'}{i} = 1 + \left(\frac{W_2 - W_0}{W_0 + 2W_1 + W_2} \right) \frac{\gamma_X}{\gamma_A} \qquad (9.17)$$

The bracketed terms are known as the nuclear Overhauser enhancement η.

The relative significance of these contributions to each transition probability can be shown in terms of T_1^{DD} and T_1^* which represent contributions from other mechanisms to be [22]

$$W_A = (2T_{1A}^*)^{-1} + 3 (20T_1^{DD})^{-1} \qquad (9.18)$$

$$W_X = (2T_{1X}^*)^{-1} + 3\,(20T_1^{DD})^{-1} \tag{9.19}$$

$$W_2 = 3(5T_1^{DD})^{-1} \tag{9.20}$$

$$W_0 = (10T_1^{DD})^{-1} \tag{9.21}$$

Thus only dipolar relaxation contributes to the nuclear Overhauser effect, all other mechanisms, being in the denominator, reduce it. If one assumes, as is quite often the case, especially in ^{13}C n.m.r., that dipole—dipole is the dominant relaxation mechanism, then W_0: $W_A = W_X:W_2$ is $1/6:1/4:1$ and η becomes equal to $\frac{1}{2}(\gamma_X/\gamma_A)$. Thus for the case of $^1H\text{-}\{^1H\}$ the enhancement factor is $\frac{1}{2}$ and for $^{13}C\text{-}\{^1H\}$ it is 1.988. The main interest in the n.o.e., apart from the fact that it increases a signal's intensity, is that it originates purely from dipolar relaxation. Thus, as is discussed in the next chapter, it can be used to measure the relative contribution of dipole—dipole relaxation. For the case of ^{13}C under conditions of proton decoupling the n.o.e. is given by [23, 24]

$$\text{n.o.e.} = 1 + \frac{\gamma_H}{\gamma_C \chi}\left[\frac{6\tau_C}{1 + 4\pi^2(\vartheta_H + \vartheta_C)^2\tau_C^2} - \frac{\tau_C}{1 + 4\pi^2(\vartheta_H - \vartheta_C)^2\tau_C^2}\right]$$

where $\tag{9.22}$

$$\chi = \frac{\tau_C}{1 + 4\pi^2(\vartheta_H - \vartheta_C)^2\tau_C^2} + \frac{3\tau_C}{1 + 4\pi^2(\vartheta_C^2)\tau_C^2}$$

$$+ \frac{6\tau_C}{1 + 4\pi^2(\vartheta_C^2 - \vartheta_C^2)\tau_X^2}$$

and is thus field dependent (see section 10.4 and Fig. 10.4). However, under the limiting condition $\vartheta_C\tau_C \ll 1$ (the extreme narrowing limit), eqn. (9.22) reduces to [23]

$$\text{n.o.e.} = 1 + \frac{\gamma_H}{2\gamma_C} = 2.988 \tag{9.23}$$

It is important to note that the enhancement factor depends on the sign of the magnetogyric ratio; for nuclei where the sign is negative e.g. ^{15}N and ^{29}Si when using proton decoupling γ can be negative. Since η can range from zero up to a maximum value given by eqn. (9.23), dependent on the contribution of mechanisms other than dipole—dipole to the total relaxation, it can have a value of -1. Under these conditions no signal can be observed unless the

Overhauser effect is removed by either instrumental or chemical means, the latter being a relaxation reagent e.g. $Cr(acac)_3$ (see section 10.9).

The homonuclear Overhauser effect can be used in structural chemistry, as was first realised by Anet and Bourne [25]. The property which is used here is its dependence on the distance between the two nuclei involved. Chemical bonding and scalar coupling are not required. If two nuclei are physically close within a molecular, mutual dipole—dipole relaxation will be their dominant relaxation process and saturation of one nucleus will enhance the intensity of the other by up to 50%. For example, in the case of dimethylformamide the amide proton and the *cis*-methyl exhibit mutual dipolar relaxation and hence a n.o.e. can be observed at one while saturating the other. The distance between the amide proton and the *trans*-methyl is too great for a n.o.e. to be observed. Bell and Saunders [26] have described a correlation of a large number of n.o.e. values with the internuclear distance and confirmed the inverse sixth power law (see Fig. 9.7).

Great care should be taken when applying homonuclear n.o.e. techniques. A negative result in isolation should never be used to prove anything. A positive n.o.e. will *only* be observed if dipole—dipole relaxation dominates; if, for example, there is a paramagnetic

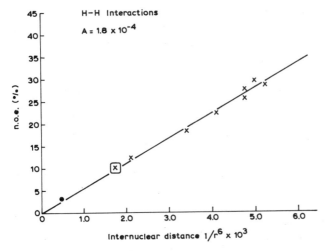

Fig. 9.7. N.o.e. enhancements versus internuclear distance for two-spin H—H interactions. From R.A. Bell and J.K. Saunders, Can. J. Chem., 48 (1970) 1114.

species such as dissolved oxygen, in the solution, this may provide the dominant relaxation process for all nuclei. No n.o.e. can then be observed, no matter how close the two protons being investigated are (within practical limits). Care must therefore be taken to exclude oxygen from solutions used for n.o.e. work. In practice, when one has a choice as to which 'half' of the system to irradiate, one should always observe the part with the fewest alternative relaxation mechanisms since this one is most likely to be dominated by dipole—dipole relaxation and hence show a n.o.e. If, for example, one has the choice between a single proton and a methyl group, one should irradiate the methyl (this having the possibility of spin rotation and internal dipole/dipole relaxation) and observe the proton.

When considering ^{13}C n.o.e. determinations the experimental problems are not as severe as in the proton case; carbon nuclei being in the centre, not on the periphery of the molecule, are less influenced by intermolecular effects. Relaxation is dominated by the shorter range directly bonded C—H dipolar effects. Also there is, in general, no homonuclear coupling. When studying ^{13}C n.o.e.'s, oxygen is only a problem with molecules where the T_1 is longer than about 10 s. As is illustrated in section 9.9 it is possible by means of 'gated decoupling' experiments to separate the n.o.e. from intensity changes caused by multiplet collapse. The major value resulting from a knowledge of the n.o.e. lies in being able to establish the contribution of dipolar mechanisms to the total relaxation, and hence being able to use the structural information available from this mechanism.

The technique of 'gated decoupling' is not easily applied to proton n.m.r. where a proton will generally be coupled to and relaxed by more than one proton; consequently it would be necessary to irradiate simultaneously with more than one decoupling frequency. Experiments of this type, however, have been achieved using tailored excitation, where any desired region of the spectrum may be differentially irradiated [27].

9.8 TECHNIQUES OF SPIN DECOUPLING

Spin decoupling is achieved by providing a second r.f. field at the Larmor frequency of the nucleus that is to be decoupled. In the previous section it was assumed that frequency was continuous and coherent and that the magnetic field was fixed. The latter will always be the case for pulsed excitation, but need not be the case for swept excitation. It is obvious from the basic Larmor equation (2.11) that in a sweep experiment either the field or the frequency may be swept keeping the other fixed. For single resonance work the result (ignoring instrumental artifacts related to modulation levels) is independent of the variable; for multiple resonance work this is not the case.

The difference between the field and frequency sweep in c.w. double resonance can be visualised easily. The frequency sweep case is simple; with a fixed field, a fixed frequency ϑ_2 will always irradiate a set multiplet. However, for a field sweep experiment, a fixed frequency ϑ_2 will only irradiate a given multiplet at a certain point with the experiment. It must be so arranged that $(\vartheta_1 - \vartheta_2)$ is equal to the chemical shift difference between the detected and decoupled nuclei. This is quite a limitation; it means that if a nucleus M is spin coupled to two other nuclei A and X then two experiments are necessary to demonstrate both couplings, one with $(\vartheta_1 - \vartheta_2)$ equal to $(\vartheta_A - \vartheta_M)$ and one with $(\vartheta_1 - \vartheta_2)$ equal to $(\vartheta_M - \vartheta_X)$. Using frequency sweep or pulsed excitation only one experiment $(\vartheta_2 = \vartheta_M)$ would be necessary. An interesting parallel occurs in mass spectrometry, when considering the using of the Nier-Johnson and inverse Nier-Johnson geometry to study metastable ions. The two double resonance experiments also differ in the detail of their off resonance spectra (see ref. 12). The field sweep method was originally used because of instrumental simplicity; it is now no longer used for obvious reasons.

Another assumption made in the preceding sections was that the irradiating frequency was coherent. Suppose we wish to decouple simultaneously two chemically shifted nuclei: we could either use a high power from a single source or two frequencies from different sources. Now take the argument one stage further. Suppose we wish to decouple many groups of chemically shifted nuclei, e.g. all the protons while observing ^{13}C. For this we need a decoupling field with a wide flat power distribution. In 1966, long before the now familiar concepts of pulse and stochastic excitation, Ernst pointed out that noise modulation of the decoupling frequency could achieve this goal [4]. This technique is now normally referred to as 'noise decoupling'; the output from the decoupler is modulated (either by phase shifting or gating) with the output of a pseudo random shift register. The decoupler then has a power spectrum given by [4]

$$B_2(\vartheta) \propto \left[\frac{\sin (\pi\vartheta_2/\vartheta_M)}{(\pi\vartheta_2/\vartheta_M)}\right] \tag{9.24}$$

$$J_r = J\delta\vartheta/\gamma B_2 \tag{9.25}$$

In order to quantify the efficiency of noise decoupling, define a residual broadening b_r Hz as the increment in the line width at half height due to incomplete decoupling

$$b_r = J^2/(\gamma b_2)^2 \tag{9.26}$$

where b_2 is the power spectral density. For random noise modulation of the decoupler frequency, b_2^{max} is given in terms of the amplitude of ϑ_2 by

$$b_2^{max} = B_2(\pi\vartheta_M)^{-\frac{1}{2}} \tag{9.27}$$

The power spectral density also determines the life time of a nucleus subjected to noise decoupling. This life time (τ) is important, since, as is discussed in section 10.7, in the case of ^{13}C spectra recorded with proton noise decoupling, the achievable line width may be

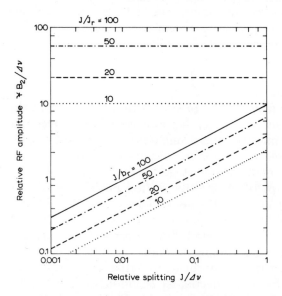

Fig. 9.8 Comparison of the residual coupling (J_r) and broadening (Jb_r) which occur in coherent and noise decoupling as a function of the decoupling power used and the relative offset. From R.R. Ernst, J. Chem. Phys., 45 (1966) 3845.

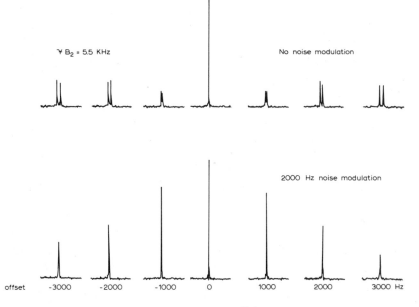

Fig. 9.9. The effect of decoupler offset on ^{13}C spectrum of benzene with and without noise modulation.

limited to τ^{-1}.

$$\tau = 2[\pi(\gamma b_2)^2]^{-1} \tag{9.28}$$

The efficiency of the two decoupling techniques can be compared by considering the power necessary to reduce the residual broadening or splitting below a certain value (b_{min} and J_{min}) over a width of $\Delta\vartheta$. For coherent decoupling from eqn. (9.25) this is

$$\gamma B_2 = (J/J_m)\Delta\vartheta \tag{9.29}$$

and for noise decoupling from eqns. (9.26) and (9.27) it is [4]

$$\gamma B_2 = (2\Delta\vartheta/b_m)^{\frac{1}{2}} J \tag{9.30}$$

Equation (9.30) is only an approximate solution since eqn. (9.27) only gives the maximum field and does not allow for the fall in power as a function of offset. For a wide noise bandwidth ($\vartheta_2 \geqslant 2\Delta\vartheta$) this is a reasonable approximation. Equations (9.29) and (9.30), which are illustrated graphically in Fig. 9.8, show that when the range of chemical shifts get larger than the coupling constants involved, noise decoupling is the most efficient. The latter is the case when decoupling all proton couplings from a heteronucleus e.g. ^{13}C. Figure 9.9 illustrates spectra produced as a function of the offset from the position of optimum decoupling for the ^{13}C resonance of benzene using the same decoupler power, with and without noise modulation. Without noise modulation, even slight offsets result in residual splitting; this is not the case with noise modulation.

Homonuclear spin decoupling has been achieved using tailored excitation by synthesizing a power spectrum which has a 'spike' in it at the frequency of the line that is to be decoupled; an example is shown in Fig. 9.11.

An alternative method of generating a broad band of frequencies is to modulate the phase of the carrier with a regular sequence of square waves [27A] instead of the pseudo random sequence suggested by Ernst. If the modulation (ϑ_m Hz) has a 50% duty cycle, i.e. the phase shifted by $180°$ every $(2\vartheta)^{-1}$s and spends equal times in each state then a series of sidebands are generated. These sidebands have frequencies of $\pm(2n + 1)_m$ Hz and intensities of $2\gamma B_2(2n + 1)$ where $n = 0, 1, 2$. If these sidebands are sufficiently close (ϑ_m typically $50\sim100$ Hz) compared with their power then decoupling can be effected over a wide band width. Square wave modulation makes more efficient use of the available power than does noise decoupling when the range of nuclei to be decoupled is comparable with, or less than the decoupler power available. A further advantage of square wave modulation is that it does not introduce a stochastic component into the relaxation of the non irradiated nucleus. Thus in the case of broadband decoupling of protons while observing ^{13}C, narrower lines are found than when using noise modulation.

9.9 GATED DECOUPLING EXPERIMENTS [28—30]

The nuclear Overhauser effect is established on the time scale of the appropriate spin—lattice relaxation time whereas the associated decoupling effect is established effectively instantaneously (in the order of ϑ_0^{-1}). Particularly in ^{13}C n.m.r. it is useful to be able to use this difference in time scale to separate the two effects, firstly to eliminate the unpredictable n.o.e. enhancement and permit quantitative work, secondly to measure the contribution of the dipole—dipole mechanism to the total relaxation rate, and thirdly to have

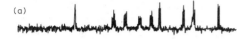

(a)

(b)

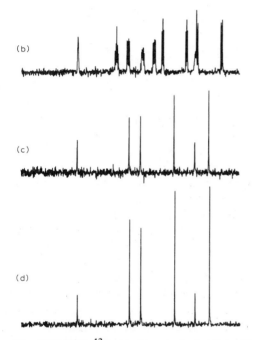

(c)

(d)

Fig. 9.10. The ^{13}C spectrum of nicotinamide. (a) With no proton decoupling; (b) gated decoupling to permit n.o.e. while maintaining spin coupling; (c) gated decoupling to eliminate n.o.e. while removing spin coupling; (d) normal noise decoupling.

the advantage of the n.o.e. enhancement while studying non-decoupled spectra [28]. These goals can be achieved by using the so-called 'gated decoupling' experiments as shown in Fig. 9.10 where the decoupler is switched on and off at the appropriate parts of the pulse cycle.

Figure 9.11 shows the timing diagram for the two possible experiments. In both experiments a delay is introduced between the end of the data acquisition and the initiation of the next pulse. During

Function	No spin decoupling and maximum NOE	Minimum NOE while spin decoupling
B_1		
Receiver		
My' i.e. signal		
B_2		
Overhauser enhancement		

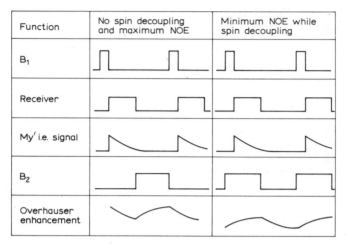

Fig. 9.11. Timing diagram for gated decoupling experiments.

the delay period the spin decoupler is either switched on or off in apposition to its state during the period of data acquisition. If the n.o.e. enhancement is not required, during the delay period (T_p) the decoupler is off and the n.o.e. relaxes (ignoring cross relaxation) as

$$\eta(t) = \eta_0 \exp\left(-T_p/T_1\right) \qquad (9.31)$$

where η_0 is the value of the n.o.e. at the start of the delay. If T_p is made sufficiently long then η becomes negligible at the time of the exciting pulse, and the resulting spectrum contains effectively no n.o.e. During the period of data acquisition the n.o.e. builds up. The growth of the n.o.e., however, has no effect on the intensity of the lines, since, as shown in section 3.2, the area under the peak in the frequency domain only depends on the value of $m(t)$ at zero time. If the n.o.e. is not required, the inverse sequence is applied. After three or four passes through the cycle a steady state is reached. The fraction of the total available n.o.e. present in the steady state for the experiment where the decoupler is off during data acquisition is given approximately by

$$\eta = \frac{(1 - E_d)}{(1 - E_a E_d)} \quad \text{where} \quad \begin{aligned} E_a &= \exp -(T_a/T_1) \\ E_d &= \exp -(T_d/T_1) \end{aligned} \qquad (9.32)$$

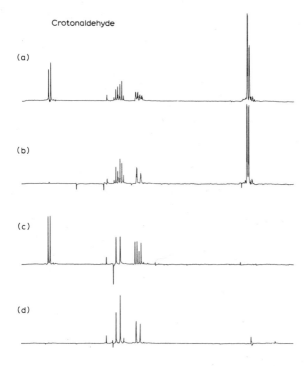

Fig. 9.12. Homonuclear decoupling using tailored excitation for crotonaldehyde. (a) Normal spectrum; (b) irradiation of low field doublet; (c) irradiation of high field multiplet; (d) irradiation of both high and low field multiplets. From H.D.W. Hill [31].

For the inverse experiment

$$\eta = \frac{E_d(1 - E_a)}{(1 - E_a E_d)} \tag{9.33}$$

These functions are illustrated in Fig. 7.6.

9.10 APPLICATIONS OF DOUBLE RESONANCE TECHNIQUES

There are numerous examples of the application of double resonance techniques to every facet of n.m.r. spectroscopy; the main areas are summarised below.

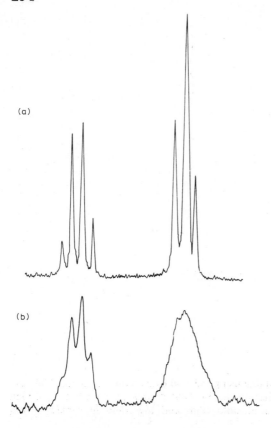

Fig. 9.13. Proton spectrum of As(Et)$_4$ Br in D$_2$O. (a) With double irradiation at the ^{75}As frequency; (b) without such irradiation.

(a) Spin decoupling

This is the removal of spin coupling from a spectrum. (i) To establish its origin. (ii) To simplify a complex spectrum. Figure 9.12 illustrates this application in the homonuclear case for croton aldehyde. Both the methyl and aldehyde protons can be decoupled leaving the olefinic protons as a simple AB system. Figure 9.13 describes the heteronuclear decoupling where the broadening from a quadrupolar nucleus (^{75}As in this case) is removed by decoupling. (iii) To enhance the signal to noise ratio by spectral simplification and generating a nuclear Overhauser effect, e.g. ^{1}H noise decoupling in ^{13}C n.m.r., as shown for undecane in Fig. 9.14.

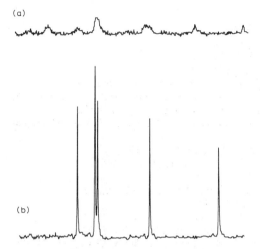

Fig. 9.14. ^{13}C spectrum of undecane without (a) and with (b) proton noise decoupling. Both spectra were recorded in the same spectrometer time.

(b) Spin tickling

Spin tickling is used mainly for the accurate location of a hidden peak. Tickling effects are very sensitive to the precise frequency of ϑ_2 with respect to the resonance of the line being irradiated. Being a low power experiment it does not suffer from Bloch-Siegert effects. Tickling is also used in the determination of the relative signs of coupling constants.

(c) INDOR (s.p.i. or d.r.d.s.)

This has similar applications to the above except that a prior knowledge of 'the place to irradiate' is unnecessary.

(d) N.o.e.

Determination of the nuclear Overhauser enhancement yields information on the degree of mutual dipolar relaxation between the two nuclei involved. This information can be used in the ^{1}H-$\{^1$H$\}$ case to estimate ^{1}H—^{1}H internuclear distances etc. (see section 9.7) or in the ^{13}C-$\{^1$H$\}$ case to study molecular motion, etc. (see section 10.10).

REFERENCES

1 F. Bloch, Phys. Rev., 94, (1954) 496.
2 V. Royden, Phys. Rev., 96, (1954) 543.
3 W.A. Anderson, Phys. Rev., 102, (1956) 151.
4 R.R. Ernst, J. Chem. Phys., 45, (1966) 3845.
5 J.D. Baldeschwieler and E.W. Randall, Chem. Rev., 63, (1963) 81.
6 R.A. Hoffman and S. Forsen, Progr. NMR Spectroscopy, 1, (1966) 15.
7 W. Von Philipsborn, Angew. Chem., Intern. Edn., 10, (1971) 472.
8 J.H. Noggle and R.E. Schirmer, The Nuclear Overhauser Effect: Chemical Application, Academic Press, New York, 1971.
9 R.K. Wangness and F. Bloch, Phys. Rev., 48, (1953) 134.
10 A.L. Bloom and J.N. Shoolery, Phys. Rev., 151, (1965) 102.
11 K.G.R. Pachler, J. Magn. Res., 7, (1972) 442.
12 W.A. Anderson and R. Freeman, J. Chem., Phys., 37, (1962) 85.
13 A. Overhauser, Phys. Rev., 89, (1953) 689.
14 R. Freeman, J. Chem. Phys., 53, (1970) 457.
15 R. Freeman and W.A. Anderson, J. Chem. Phys., 39, (1963) 1518.
16 T.H. Brown, E.B. Whipple and D.M. Verdier, J. Chem. Phys., 38, (1963) 3029.
17 E.B. Baker, J. Chem. Phys., 37, (1962) 911.
18 J. Feeney and P. Partington, J. Chem. Soc. Chem. Commun., (1973) 611.
19 A.A. Chalmers, K.G.R. Pachler and P.L. Wessels, J. Magn. Res., 15, (1974) 419.
20 S. Sorensen, R.S. Hansen and H.J. Jacobsen, J. Magn. Res., 14, (1974) 243.
21 R. Freeman, S. Wittekoek and R.R. Ernst, J. Chem. Phys., 52, (1970) 1529.
22 I.D. Campbell and R. Freeman, J. Magn. Res., 11, (1973) 143.
23 K.F. Kuhlmann, D.M. Grant and R.K. Harris, J. Chem. Phys., 52, (1970) 3439.
24 D.M. Doddrell, V. Glushke and A. Allerhand, J. Chem. Phys., 56, (1972) 3683.
25 F.A.L. Anet and A.J.R. Bourne, J. Amer. Chem. Soc., 87, (1965) 5250.
26 R.A. Bell and J.K. Saunders, Can. J. Chem., 48, (1970) 1114.
27 R. Freeman, H.D.W. Hill, B.L. Tomlinson and L.D. Hall, J. Chem. Phys. 11 (1974) 4466.
27A J.B. Grutzner and R.E. Santini, J. Magn., Res., 19, (1975) 173.
28 J. Feeney, D. Shaw and P.J.S. Pauwells, Chem. Commun., (1970) 554.
29 R. Freeman and H.D.W. Hill, J. Magn. Res., 5, (1971) 278.
30 R. Freeman and H.D.W. Hill and R. Kaptien, J. Magn., Res., 7, (1972) 372.
31 H.D.W. Hill, private communication.

CHAPTER 10

RELAXATION

In order to describe completely and understand an n.m.r. spectrum it has always been necessary to include consideration of the relevant relaxation times. This, however, is very rarely done. Along with high resolution pulsed spectrometers, introduced primarily for their sensitivity, has developed, almost as a by-product, an increasing awareness of relaxation times and their analytical and chemical value. This awareness has occurred for two basic instrumental reasons. On the one hand, to optimise a pulsed spectrometer one needs to be aware of the spin-lattice relaxation times of the nuclei within the sample, and on the other, pulsed spectrometers can easily measure T_1, and with less ease, T_2. Parallel with (resulting from?) these instrumental changes in technique has come the renaissance of ^{13}C n.m.r. which has itself also led to an upsurge in interest in relaxation times. ^{13}C relaxation data is simpler to interpret than proton data as it is mainly intramolecular in origin and uncomplicated by homonuclear spin coupling, and will therefore form the bulk of this discussion. Bearing all these points in mind it is not surprising that more space will be devoted to the study of relaxation times in this work than in those based on n.m.r. from a swept view point.

10.1 DEFINITION OF T_1 AND T_2

It is expedient to recapitulate the meaning of T_1 and T_2 at this point (see section 2.5) before progressing to their detailed study. Both relaxation times are time constants used to characterise what are assumed to be first order rate processes. Indeed, when considering relaxation phenomena we sometimes consider relaxation rates instead of relaxation times ($R_{1,2} = (T_{1,2})^{-1}$) as they are additive. The spin lattice relaxation time T_1 quantifies the rate of transfer of energy from the nuclear spin system to its surroundings (the lattice). It describes the rate of return of the M_z component of the total magnetisation to equilibrium after a perturbation.

$$\frac{dM_z}{dt} = -\frac{1}{T_1}(M_0 - M_z) \tag{10.1}$$

T_1 can also be defined as the sum of the reciprocal of all the transition probabilities (W) available to the spin system

$$(T_1)^{-1} = \sum_{m=0}^{m} n\, W_{\Delta m} = R_1 \tag{10.2}$$

where Δm is the change in magnetic quantum number in the transition and n is the possible number of such transitions.

The spin—spin relaxation time (T_2) on the other hand involves the phase coherence of the individual nuclear moments in the $x'y'$ plane after a perturbation and is defined by the following differential equation.

$$\frac{dM_{x'y'}}{dt} = -\frac{1}{T_2}M_{x'y'} \tag{10.3}$$

As was pointed out previously, T_1 is an energy effect, and T_2 is an entropy effect.

10.2 THEORY OF MOLECULAR MOTION

The two relaxation processes defined above occur by the interaction of the nuclear spin system with fluctuating local magnetic fields. These fields are generated by other molecules in the sample and their fluctuation is governed by the motion of these molecules. If the locally induced magnetic fields, which act like microscopic radio frequency fields, have components at the appropriate Larmor frequency, they can interact and cause spin relaxation. The larger such a component the quicker relaxation can occur. To explore relaxation effects we need therefore to study the motional behaviour of molecules in solution, paying particular respect to the frequency components of their motion. The range of relevant frequencies is in the MHz region; fast motions, e.g. electronic motions and molecular vibrations, are thus going to be inefficient and of little importance. Brownian motion (rotational and diffusional) is important here, as are certain molecular torsionial and rotational motions.

Consideration is now due to what methods are available to solve this problem. In solution molecular motion is effectively a random process and as such any property generated from it e.g. $b(t)$ will have zero average, i.e.

$$\overline{b(t)} = 0 \tag{10.4}$$

its mean square average however will not be zero.

$$\overline{b_{\text{loc}}^*(t) \cdot b_{\text{loc}}(t)} \neq 0 \tag{10.5}$$

This property of random motion can easily be verified by taking a set of random numbers with a zero average e.g. $(-1, -3, 4)$ and finding their square average (8.66). A square can never be negative; a finite average must therefore result. Thus this average will be a useful property in describing molecular motion. If we Fourier transform any random function $b(t)$ we obtain $B(\vartheta)$ which is itself random with a zero average but a non-zero square average. For example, if $b(t)$ were the amplitude of a wave form then $|B(\vartheta)|^2$ would represent the total energy available at frequency ϑ, and would be infinite, unless one limited the time under consideration (T). One can usefully employ $B(\vartheta)$ by defining a power spectrum thus (see section 3.4).

$$J(\vartheta) = \lim_{T \to \alpha} \frac{\pi}{T} B_{\text{loc}}^*(\vartheta) \cdot B_{\text{loc}}(\vartheta) \tag{10.6}$$

In this context the power spectrum is normally called the spectral density function and given the symbol J. It is this function for molecular motion we wish to evaluate as it tells us the power (energy /unit time) available in the molecular motion as a function of ϑ. If we know $J(\vartheta)$, then we can find the magnitude of the fluctuating fields at the Larmor frequency, which is what we require to study relaxation processes. The problem is finding $b(t)$. All we know of it is that it has to be random and that we could calculate a value for its mean square average $b^*(t) b(t)$.

A useful property in describing random molecular motion is that of a correlation time (τ_c) which can be defined as the 'average' time between molecular collisions for translational motion, or the average time for a molecule to rotate by one radian for reorientational motion. If a molecule is in one state of motion for τ_c s, one would expect there to be frequency components in the motion spread around τ_c^{-1}.

The correlation time of a molecule in solution will depend on many factors, such as molecular size, symmetry and solution viscosity. If the molecule is small (mol. wt. < 100) then in solutions of normal viscosity τ_c is about $10^{-12} \sim 10^{-13}$ s; for larger molecules (mol. wt. $= 100 - 300$) τ_c may increase to as much as 10^{-10} s. Molecular symmetry will obviously have an effect on the correlation time; a symmetrical molecule, causing less disordering of the solvent as it rotates, will move faster than an asymmetric one. Viscosity describes the ease with which reordering can be achieved in the solution and depends on both the solute and solvent. The correlation time for molecular reorientation used in this section is equal to one third of the τ_D of the Debye theory of liquids [1] and can thus be expressed, using that theory, in terms of the solution viscosity (η) and the molecular radius (a). The latter can in turn be calculated in terms of molecular volume if required.

$$\tau_c = 1/3 \, \tau_D = 4\pi\eta \, a^3 / 3kT \tag{10.7}$$

Chapter 3 examined the auto-correlation function which measures the persistence of the consequences of an effect. Using the concept of a correlation time, one can set up a reasonable model for the auto-correlation function by assuming that the effect of a molecular collision decays exponentially with a time constant τ_c. From eqn. (3.32) it follows that for this simple model the auto-correlation function will have the following form.

$$\rho(\tau) = \int_{-\infty}^{\infty} b^*(t) \cdot b(t) \exp - (|\tau|/\tau_c) dt \tag{10.8}$$

The first part of eqn. (10.8) is simply the square average of the local magnetic field; we could thus rewrite this equation as

$$\rho(\tau) = \overline{b_{loc}^2} \exp - (|\tau|/\tau_c) \tag{10.9}$$

In this form it has interesting similarities with the equation described in the f.i.d. (eqn. 4.7), τ_c having the same function as T_2. Such a similarity is not surprising; T_2 represents the phase memory of the nuclear spins after a pulse, while τ_c is defined as a time which characterises the decay of the consequences of a molecular collision.

To investigate the frequency dependence of the auto-correlation function one must Fourier transform it. As seen in chapter 3, the transform of an auto-correlation function is none other than its power spectrum or spectral density function.

$$\rho(\tau) \xrightarrow{\ \mathcal{F}^-\ } J(\vartheta) \tag{10.10}$$

which, as seen previously, is a useful function in the study of relaxation processes. Thus from eqn. (10.9) we obtain

$$J(\vartheta) = \overline{b_{\text{loc}}^2} \, \frac{2\tau_c}{1 + (2\pi\vartheta)^2 \tau_c^2} \tag{10.11}$$

This is a very basic equation in the understanding of nuclear relaxation. The spectral density function is mapped in Fig. 10.1 for three* limiting cases: (i) fast molecular motion where $2\pi\vartheta_0\tau_c \ll 1$; (ii) the intermediate case where $2\pi\vartheta_0\tau_c \approx 1$ and (iii) slow motion where $2\pi\vartheta_0\tau_c \gg 1$.

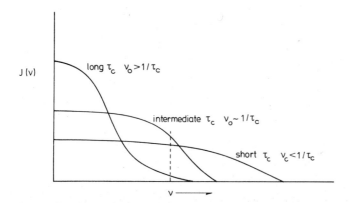

Fig. 10.1 The spectral density function plotted against frequency for fast, intermediate and slow molecular motion.

Figure 10.1 uses a semi-log plot, hence the curves are not immediately recognisable as Lorentzian. In all these cases the area under the curve is constant, being equal to $\overline{b_{\text{loc}}^2}$ (the value of $\rho(\tau)$ at $\tau = 0$, see section 3.2), which corresponds to the total energy available in the motion. Using Fig. 10.1 we can see how molecular motion, represented by τ_c, affects $J(\vartheta_0)$, the value of the spectral density at the Larmor frequency (which itself is directly related to the relaxation rate).

If τ_c is short, molecular motions are distributed over a wide frequency range ($0 \sim \tau_c^{-1}$ Hz) and all motional frequencies within that range have equally probable chance of occurring. If τ_c is long, then low motional frequencies have a higher chance of happening, while high frequency has an almost zero probability. Since the area under the curve in Fig. 10.1 is constant, $J(\vartheta_0)$ must have a maximum value as a function of τ_c; this maximum occurs when $2\pi\vartheta_0\tau_c = 1$. The theory therefore predicts that relaxation should be at its most efficient and T_1 a minimum when $\tau_c = (2\pi\vartheta_0)^{-1}$. This prediction was verified experimentally by Bloembergen, Purcell and Pound [2] who studied the spin—lattice relaxation time of the protons in glycerine as a function of the sample viscosity, which in turn is related to the molecular correlation time ($\tau_c \propto \eta/T^0$). They showed that T_1 did indeed have a minimum as the temperature was changed; the position of this minimum depended on the experimental field used. The dependence of T_1 on the molecular correlation time is shown in Fig. 10.2.

The above discussion is only applicable to random re-orientational motion, but it can be shown that any relaxation process can be represented by an equation which has the following general form [3].

$$R = \overline{b}_{\mathrm{loc}}^2 \, f(\tau_c) \tag{10.12}$$

The value and origin of b_{loc} and τ_c will depend on the mechanism under consideration.

10.3 RELAXATION MECHANISMS

Local magnetic fields in solution can be generated in many ways. The principal source is the dipolar field of the nucleus. The modification of the basic magnetic field by any chemical shift anisotropy, effects transmitted by scalar coupling, fields generated by molecular rotation etc., are secondary sources of local magnetic fields. Table 10.1 summarises the mathematical form of these various magnetic fields and the appropriate molecular correlation time. Later sections will deal with each of these mechanisms in turn.

The fluctuation of the magnetic field generated by the effects described above has different effects with respect to T_1 and T_2 processes. One must consider the directional components of the local fields in order to understand this difference. First any field is resolved into its three components, thus

$$b = i\,b_x + j\,b_y + k\,b_z \tag{10.13}$$

TABLE 10.1

Summary of relaxation mechanisms

Mechanism	$\overline{b_{loc}^2}$	Correlation time	Example	Comments
Dipole/dipole				
(a) nuclear — nuclear	$\dfrac{(\gamma_1 \gamma_S \hbar)^2 I(I+1)}{r_{IS}^{-6}}$	reorientational	$^1\text{H} - {}^{13}\text{C}$	Dominant mechanism for protonated carbons. Relaxation reagents.
(b) electron — nuclear		translational	O_2	
Spin rotation	$IkT(2C_h + C_{\parallel})^2/h^2$	angular momentum	$^{13}\text{CS}_2$; ClO_3F	Important in small molecules, increases with temperature
Chemical shift Anisotropy	$\gamma^2 B_0^2 (\sigma_{\parallel} - \sigma_h)^2$	reorientational	$\text{CH}_3\,{}^{13}\text{CO OH}$	Usually insignificant, increases with field
Scalar coupling	$2/3 S(S+1)\pi^2 J^2$	T_2^S	$^{13}\text{CH Br}_3$	Depends on the properties of the "other nucleus"
Quadrupolar	$(eq\, Q/\hbar)^2$	reorientational	$^{14}\text{NH}_3$	Usually dominant for nuclei with $T > \tfrac{1}{2}$

Such a resolution can be carried out in either the laboratory or the rotating frame. b_z is static in both frames; however $b_{x'}$ and $b_{y'}$ are static only in the rotating frame, when they have a motion at the Larmor frequency. It follows therefore that a field b can only relax a nucleus when it has a static component in a frame rotating at the nuclear Larmor frequency. Under these conditions the strength of the interaction is proportional to the vector product of the field and the nuclear magnetic moment, i.e.

$$R = (b \times M)' = (i\,b_{x'} + j\,b_{y'} + k\,b_{z'}) \times (i\,M_{x'} + j\,M_{y'} + k\,M_{z'})$$
(10.14)

If this equation is expanded, and use made of the property of unit vectors that $i \times i = 0$ etc. then

$$R = i(b_{y'}M_{z'} - b_{z'}M_{y'}) + j(b_{z'}M_{x'} - b_{x'}M_{z'}) + k(b_{x'}M_{y'} - b_{y'}M_{x'})$$
(10.15)

T_1 processes, by definition, depend only on the k terms in eqn. (10.15); T_2 processes are those involving the i and j components. It follows therefore that T_1 processes are not sensitive to b_z whereas T_2 processes are. This is significant in that b_z is the same in both the laboratory and rotating frames; thus a static field on the laboratory frame can be a source of spin—spin relaxation but *not* of spin—lattice relaxation. Herein lies the major interest in T_2 and the major experimental problem in measuring T_2. Since T_2 is sensitive to zero frequency local magnetic fields and T_1 is not, T_2 is quite often shorter than T_1 and is never longer. Contributions to T_1 and T_2 can also result from other frequencies which correspond not only to the basic Larmor frequency but also to all the possible transitions of the spin system including, in the case of dipolar relaxation, components at the sum of the Larmor frequencies of the nuclei involved [4].

From this discussion and the previous sections, the expected form of any equations describing nuclear relaxation resulting from random molecular motion will be

$$1/T_1 \propto \overline{b_{loc}^2} \left[\frac{C_1 \tau_c}{1 + 4\pi^2 \vartheta^2 \tau_c^2} \right]$$
(10.16)

$$1/T_2 \propto \overline{b_{loc}^2} \left[C_2 \tau_c + \frac{C_3 \tau_c}{1 + 4\pi^2 \vartheta^2 \tau_c^2} \right]$$
(10.17)

Although it is more logical when dealing with relaxation phenomena to use the term of 'relaxation rates', these being additive, it is conventional to talk in terms of 'relaxation times' for historical notational and experimental convenience.

In studying relaxation times so far the effect of any spin coupling which might be present has been neglected. Such neglect can be shown to be justified if the spin coupling is removed by double resonance techniques, as is normally the case when studying ^{13}C relaxation times and n.o.e.'s [4]. Such neglect is not justified in proton and non-decoupled ^{13}C spectra where scalar coupling is present. In the presence of spin coupling 'cross relaxation' or 'three spin effects' occur which result in non-exponential relaxation [5]. The relaxation of the nucleus under investigation now also depends on the relaxation properties of any nucleus to which it is spin coupled. Cross relaxation effects considerably complicate the interpretation of results, especially in proton spectra. The effects are significant in ^{13}C too; for example, the spin lattice relaxation time of the carbon nucleus in formic acid is 10 s under proton noise decoupling conditions, but only 7 s when the protons are not decoupled [6]. Another term has been added to eqn. (10.17). When the protons are decoupled, the values of T_1 measured by pulse and adiabatic rapid passage methods are the same; in the non-decoupled case they are not, the value from pulsed excitation being higher, e.g. for the case of benzene $T_{1c}^{pulse} = 28.0$ s; $T_{1c}^{a.r.p.} = 24.5$ s [7]. This difference arises because in pulsed techniques all components of the spin multiplet are excited simultaneously, whereas they are excited singly and sequentially in the a.r.p. method. Differences in saturation behaviour between pulse and swept spectra also result from effects which differ between simultaneous and sequential excitation.

10.4 DIPOLE-DIPOLE RELAXATION [3]

The principal source of nuclear relaxation for spin $\frac{1}{2}$ nuclei is via dipole—dipole (DD) interactions. Consider the relaxation of a nucleus I by a magnetic particle S (an unpaired electron or a nucleus). The local field generated at I by S is given by the classical equation

$$B_{loc}^{DD} = \pm \mu_S (3 \cos^2 \theta - 1) r_{IS}^{-3} \tag{10.18}$$

in terms of the magnetic moment of S (μ_S), the distance r_{IS} between I and S, and the angle (θ) between the static field and the axis joining I and S. These local fields are very large $\sim 2\,mT$ but, as a rule, no net effect is seen in high resolution spectra as molecular motion results in an averaging of θ over all values when $\cos^2\theta = 1/3$ and B_{loc} averages to zero. In solids and liquid crystals complete averaging does not occur and large dipole—dipole splitting is observed.

Equation (10.18) coupled with the master eqn. (10.16) is informative about dipole—dipole relaxation. Firstly, as the local field depends on μ_S^2, the proton, with the largest *nuclear* magnetic moment, must be the most powerful source of internuclear dipolar relaxation. If present, an unpaired electron whose magnetic moment is 860 times that of the proton is even more efficient. Secondly, because of the inverse sixth power dependence on the inter-dipole distance, dipole—dipole relaxation is very much a short range effect.

A detailed derivation of the full equations for dipolar relaxation can be found in ref. 3. The results are, assuming rotational motion, for the case of N spins of type S relaxing a spin I,

$$\frac{1}{T_1^{DD}} = \frac{1}{10}\hbar^2 \gamma_I^2 \gamma_S^2 I(I+1) \sum_N r_{IS}^{-6} K \tag{10.19}$$

where

$$K = \frac{\tau_c}{1 + 4\pi^2(\vartheta_S - \vartheta_I)^2\tau_c^2} + \frac{3\tau_c}{1 + 4\pi^2\vartheta_I^2\tau_c^2} + \frac{6\tau_c}{1 + 4\pi^2(\vartheta_I + \vartheta_S)^2\tau_c^2} \tag{10.20}$$

and

$$\frac{1}{T_2^{DD}} = \frac{1}{20}\hbar^2 \gamma_I^2 \gamma_S^2 I(I+1) \sum_n r_{IS}^{-6} \left[K + 4\tau_c + \frac{6\tau_c}{1 + 4\pi^2\vartheta_S^2\tau_c^2} \right] \tag{10.21}$$

It is informative and rewarding to compare the detailed results given above with the simple ideas developed earlier. It can be seen firstly that the terms at the frequencies corresponding to $\Delta m = 0, 1, 2$ transitions have the form predicted by eqn. (10.16) and secondly that T_2 has indeed got a term which does not involve the frequency of the experiment. The form of these equations is illustrated in Fig. 10.2.

Initially T_1 and T_2 decrease as molecular motion decreases until the limit of $2\pi\vartheta_0\tau_c = 1$ after which T_1 increases, and T_2 decreases

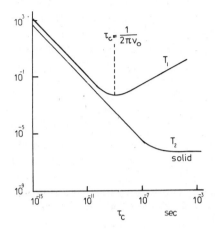

Fig. 10.2. The effect of molecular correlation time on T_1 and T_2 dipolar.

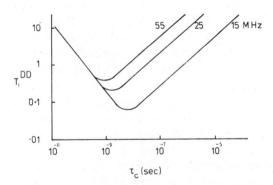

Fig. 10.3. The dependence of T_1 dipolar for $^{13}C\text{-}^1H$ relaxation at various operating frequencies [8].

further, reaches a lower limit and then remains constant, even into the solid state. For molecular motion slower than $\tau_c = (2\pi\vartheta_0)^{-1}$ T_1^{DD} increases, i.e. it becomes a less efficient process. Other mechanisms therefore become relatively more significant. Molecular correlation times are related to molecular size: the larger the molecule the longer τ_c; the minimum value of T_1 therefore depends on two factors, molecular size and (as is shown in Fig. 10.3) the experimental frequency (field). For large molecules (mol. wt. > 1500) motion is

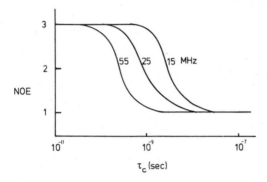

Fig. 10.4. The dependence of the ^{13}C nuclear Overhauser effect on correlation time at various operating frequencies. From D. Doddrell, V. Geushke and A. Alleshand, H. Chem. Phys., 56 (1972) 3683.

slow enough ($\tau_c \sim 10^{-8}$s) that the value of T_1^{DD} will depend on the type of spectrometer being used [8]. The higher the spectrometer's field, the smaller the molecule must be in order to ensure that it lies on the fast motion side of the T_1 minimum. This result is of practical significance from the view point of ^{13}C sensitivity. Since, for motion slower than $\tau_c = (2\pi\vartheta_0)^{-1}$, the dipolar mechanism becomes less efficient, the nuclear Overhauser effect decreases as is illustrated in Fig. 10.4. Also, any increase in the $T_1 : T_2$ ratio reduces the sensitivity of pulsed excitation (see eqn. 5.45). Both these effects will decrease the sensitivity of ^{13}C spectra measured by pulsed excitation while proton noise decoupling. For large molecules, therefore, the use of superconducting spectrometers does not necessarily lead to higher sensitivity compared with lower field iron magnets, despite the higher basic sensitivity of the former [8].

For mobile liquids where molecular re-orientation is fast, and thus τ_c is short enough that $\tau_c\vartheta_0 \ll 1$, (i.e. to the left of Fig. 10.2 and 3) and for nuclei with $I = \frac{1}{2}$, eqns. (10.19) and (10.21) greatly simplify. In this limit, the 'extreme narrowing' limit, T_2 becomes equal to T_1 [3].

$$(T_1)^{-1} = (T_2)^{-1} = \hbar^2 \gamma_I^2 \gamma_S^2 \sum_N r_{IS}^{-6} \tau_c \qquad (10.22)$$

The quantitative description of dipolar relaxation, even in the extreme narrowing limit, becomes complex when the overall molecular motion is no longer isotropic and cannot be described by a single correlation time [8, 9]. As an example, consider a freely spinning methyl group attached to a slow moving molecule. The methyl group will have a shorter correlation time (τ_m) for rotation about its axis which is at an angle θ to the principal molecular axis. The basic molecular correlation time is τ_c. An extra factor ψ must now be included in the expression for T_1.

$$(T_1^{DD})^{-1} = \hbar^2 \gamma_H^2 \gamma_C^2 \sum_N r_{CH}^{-6} \psi \tau_c \tag{10.23}$$

where

$$\psi = \tfrac{1}{4}(3\cos^2\theta - 1)^2 + 18(5 + \epsilon)^{-1} \sin^2\theta \cos^2\theta + 9/4$$
$$(1 + 2\epsilon)^{-1} \sin^4\theta$$

and

$$\epsilon = (\tau_c + \tau_m)/\tau_m$$

in the limit where methyl rotation is very fast compared with the total molecular motion, ψ reduces to

$$\psi = \tfrac{1}{4}(3\cos^2\theta - 1)^2 \tag{10.24}$$

As a result, for a methyl group where $\theta \sim 109°$ the contribution to relaxation from the dipole–dipole mechanism is greatly reduced and the T_1 of the methyl carbon can be up to 9 times longer than that of a carbon which is part of the molecular skeleton. If other types of segmental motion are available, even larger differential effects can be observed. Any reduction in the rate of dipole–dipole relaxation will of course decrease the n.o.e. at that carbon; contributions from other mechanisms then become significant.

10.5. SPIN ROTATION

After the dipole–dipole mechanism the next most important relaxation mechanism for spin $\tfrac{1}{2}$ nuclei, especially in small molecules is that of spin rotation. The fluctuating magnetic fields associated with this mechanism are generated by the motion of the *molecular* magnetic moment arising from the electronic distribution within the molecule. Consider a molecule whose moment of inertia is I; when it is in the Jth rotational state its rotational frequency V is given by

$$V = \hbar \ J/T \tag{10.25}$$

Any electron in the molecule undergoing such rotation will generate a local magnetic field at the nucleus since it behaves like a circulating electric current. Molecular collisions causing changes in both direction and rotation will modulate this field and provide a relaxation pathway. It can be shown that for molecules undergoing isotropic re-orientation the relaxation rate is given by

$$1/T_1^{SR} = (2IKT°/3\hbar^2)(2C_h^2 + C_{11}^2)\tau_J \tag{10.26}$$

where C is the appropriate spin-rotation coupling constant and τ_J is the angular momentum correlation time. For temperatures below the liquid boiling point τ_J is related to the molecular re-orientation correlation time used in the dipole—dipole mechanism by the following equation

$$\tau_c \cdot \tau_J = I/6kT° \tag{10.27}$$

From eqns. (10.26) and (10.27) it follows that τ_J is inversely proportional to solution viscosity; as the solution temperature increases, τ_J therefore also increases, unlike τ_c. From the form of equations, it is evident that T_1^{SR} is most efficient for small symmetrical molecules at high temperatures. At low temperatures R_1^{SR} is inefficient and R_1^{DD} is efficient due to the slow molecular motion; as the temperature increases, R_1^{DD} in general decreases slowly (within the extreme narrowing limit) due to the decrease in τ_c, until R_1^{SR} becomes dominant and T_1 begins to decrease with temperature. If the relaxation mechanism were pure spin rotation, T_1 would decrease linearly with temperature. Nuclei relaxed partially by spin rotation therefore show a non-linear behaviour in T_1 as a function of temperature. This enables spin rotation to be detected in the presence of other relaxation processes.

Spin rotation is a very significant mechanism when dealing with molecules in the gas phase [10]. It is also important for small molecules and for nuclei with a large chemical shift range e.g. ^{19}F, ^{13}C, ^{15}N, ^{31}P etc. In ^{13}C it is especially important, even for fairly large molecules, for quarternary carbons where proton dipolar relaxation is not as efficient as it is for proton bearing carbons, due to its inverse sixth power dependence on internuclear distance. The relationship between R_1^{SR} and chemical shift range lies in the fact that both depend on the electronic distributions within the molecule. A distribution which results in large chemical shifts also leads to large spin-rotation interactions [11]. Considering this parallel, it is not surprising that spin rotation is the dominant relaxation mechanism found in ^{19}F.

10.6. CHEMICAL SHIFT ANISOTROPY

The magnetic field experienced by a nucleus is not the primary magnetic field, but is modified by the molecule. This modification is expressed in terms of a screening constant σ, which is itself a tensor, thus

$$B_{\text{loc}} = B_0(1 - 1/3\,(\sigma_{xx} + \sigma_{yy} + \sigma_{zz})) \qquad (10.28)$$

In solution, due to Brownian motion, only the average value of these three components, the trace of the tensor, is observed in the form of a chemical shift. If, however, the three components of σ are not equal, the chemical shift is anisotropic, and molecular motion produces a fluctuating magnetic field which can act as a relaxation mechanism. For the case of axial symmetry (for the general case see Abragam, ref. 3) this interaction is given by

$$1/T_1^{CA} = 2/15\,\gamma^2 B_0^2\,(\sigma_{11} - \sigma_{h})^2 \left[\frac{2\tau_c}{1 + (2\pi\vartheta_0)^2\tau_c^2}\right] \qquad (10.29)$$

$$1/T_2^{CA} = 1/45\,\gamma^2 B_0^2\,(\sigma_{11} - \sigma_{h})^2 \left[4\tau_c + \frac{3\tau_c}{1 + (2\pi\vartheta_0)^2\tau_c^2}\right] \qquad (10.30)$$

which is of the general form predicted in section 10.3. In the extreme narrowing limit these equations, like those for the dipole—dipole mechanism, greatly simplify as the denominator becomes equal to one; T_2, however, does not equal T_1, being only 6/7th of it.

Chemical shift anisotropy is an inefficient mechanism and is very rarely found to contribute significantly to spin relaxation except perhaps in liquid crystal solvents. The mechanism can be detected by its dependence on the square of the operating field. Two examples where R_1^{CSA} is known to be significant are $CH_3{}^{13}COOH$ [12], and $Ph\text{-}C \equiv {}^{13}C - C \equiv C - Ph$ [13].

10.7. SCALAR RELAXATION

If a nucleus (1) is spin—spin coupled with a second nucleus (S) it is possible for S to provide a fluctuating magnetic field, and hence a relaxation mechanism, at the first nucleus via scalar interactions involving the bonding electrons. Since this mechanism relies on scalar spin coupling it is called scalar relaxation. Fluctuations can occur from two sources, firstly from any time dependence of their spin coupling constant resulting, for example, from chemical exchange, and secondly from the time dependence of the excited state of spin S. To be efficient for spin—lattice relaxation these fluctuations must be rapid (in the order of the Larmor frequency), and hence the mechanism is only efficient when exchange is fast (see later) or the life time of the excited state of the S nucleus is short, i.e. it has a broad line and in practice is usually a quadrupolar nucleus The first set of conditions gives rise to 'scalar relaxation of the first kind'; the other set gives 'scalar relaxation of the second kind'.

In the presence of chemical exchange the local field at nucleus I is $\pm JS/2\gamma_I$ when there is a covalent bond and zero when there is not. If the exchange rate is faster than the spin—lattice relaxation times of both I and S, as well as that of their spin coupling constant J, then no multiplet structure will be observed, only a single line. If the exchange life time is τ_e, then by arguments similar to these in section 10.2, it can be shown that the equations for scalar relaxation are [3]

$$1/T_1^{SC} = 8/3 \, \pi^2 J^2 S(S+1) \left[\frac{\tau_e}{1 + 4\pi^2 (\vartheta_I - \vartheta_S)^2 \tau_e^2} \right] \qquad (10.31)$$

$$1/T_2^{SC} = 4/3 \, \pi^2 J^2 S(S+1) \left[\tau_e + \frac{\tau_e}{1 + 4\pi^2 (\vartheta_I - \vartheta_S)^2 \tau_e^2} \right] \qquad (10.32)$$

For scalar relaxation 'of the second kind' it is the excited state life time of $S(\equiv T_{2S})$ which replaces the exchange life time in the above equations.

The form of eqns. (10.31) and (10.32) show when scalar relaxation is significant and show that this mechanism can have very different effects on T_1 and T_2 processes. In the heteronuclear case, e.g. $^{13}C-^1H$, due to the difference in Larmor frequencies appearing in the denominator, scalar relaxation does not contribute significantly to T_1. The exceptions are where the Larmor frequencies are very close, e.g. $^{13}C-^{81}Br$, and when T_{2S} is so short that it is comparable with the difference in Larmor frequencies [14]. Scalar relaxation does however contribute significantly to spin—spin relaxation since R_2^{SC} contains a second zero frequency term which does not have $\vartheta_I - \vartheta_S$ in the denominator.

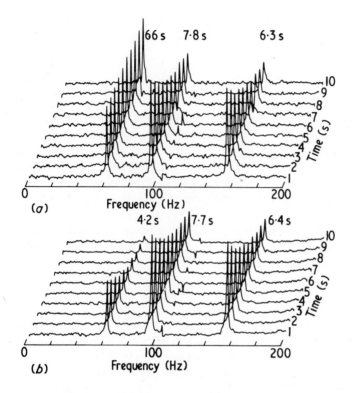

Fig. 10.5. ^{13}C spectra from *ortho*-dichlorobenzene showing the decay of the longitudinal magnetisation in (a) the laboratory frame (T_1) and (b) in the rotating frame ($T_{1\rho} = T_2$). From R. Freeman and H.D.W. Hill, J. Chem. Phys., 55 (1971) 1905.

The differential significance of scalar relaxation is given by the example in Figure 10.5. This shows the measurement of T_1 and $T_{1\rho}$ (which equals T_2 under high resolution conditions) for the carbons of *ortho*-dichlorobenzene [15]. For the proton bearing carbons T_1 equals T_2 as expected, but for the chlorine bearing carbons T_1 is long, as proton dipole relaxation is relatively inefficient due to the increased distance between the nuclei. T_2 on the other hand is much shorter due to scalar relaxation with the quadrupolar chlorine nucleus.

At first sight, it may seem that scalar relaxation of the second kind will be a significant spin—lattice mechanism in the presence of homonuclear coupling, since the frequency difference in the denominator of eqn. (10.31) is simply the chemical shift difference between the nuclei (δ). If δ is larger than the line width involved, eqn. (10.31) becomes

$$(T_1^{SC})^{-1} = (J/\delta)^2 (I/2T_{2S}) \tag{10.33}$$

which indicates that under high resolution conditions ($T_{2S} < 1$) scalar relaxation is less efficient than dipolar relaxation for weakly coupled systems. For strongly coupled systems, the simple concepts used in this chapter are no longer valid and to talk of a single T_1 is meaningless. Full density matrix theory is necessary. The situation with scalar relaxation of the first kind in the presence of chemical exchange is complex due to saturation transfer (see chapter 7 of ref. 12).

The contribution of scalar relaxation to carbon T_2's can be very significant and of great practical importance when proton noise decoupling is used. For this system T_{2c}^{SC} is given by

$$1/T_{2c}^{SC} = 4/3 \pi^2 J_{CH}^2 T_{2H} \tag{10.34}$$

consequently, if as is normally the case, $(T_{2H})^{-1} \ll J_{CH}^2$ then T_{2c} must always be shorter than T_{2H} [16]. No carbon line, therefore, can be narrower than any proton line to which it is spin coupled. In the extreme narrowing limit dipolar relaxation contributes equally to both T_{1c} and T_{2c}, whereas scalar relaxation only affects T_{2c} which is therefore shorter than T_{1c}. For example, in $^{13}CH_3I$ T_{1c} is 13.4 s, but since $T_{1H} = 10.9$ s T_{2c} is only 3.9 s [12]. In the presence of a noise decoupling field the proton excited state life time (τ_H), which equals T_{2H} in the absence of B_2, becomes a function of the decoupling field (see eqn. 9.28). At low powers, under conditions of incomplete decoupling where the proton lines are broad and $\tau_H^{-1} > J$, the carbon resonances are considerably broadened. As B_2 further increases, τ_H increases as the protons become more efficiently decoupled (see eqn. 9.28). $\tau_H^{-1} \ll J$ the scalar mechanism again becomes inefficient compared with other relaxation processes and $T_{1c} = T_{2c}$. To achieve the latter condition requires high decoupling powers [16].

Scalar relaxation also plays a significant role in proton line widths especially when the proton is bonded to a quadrupolar nucleus such as ^{14}N or ^{11}B where T_2 is in the correct range ($10 \sim 100$ ms) with respect to the spin coupling constant for R_2^{SC} to be significant. Effects are, however, not seen from nuclei like ^{35}Cl where T_2 is too short compared with the proton coupling constant for scalar relaxation to be efficient.

On the face of it, it seems surprising that scalar relaxation is ever significant compared with dipolar relaxation, as scalar coupling constants are usually smaller than dipolar couplings, but in fact the correlation times involved can be in the range $(\vartheta_I - \vartheta_S)\tau \approx 1$, even for small molecules, and the mechanism then becomes very efficient. For dipolar relaxation on the other hand, usually $\vartheta_I \tau_c \ll 1$ and the mechanism is not at its most efficient. From the above discussion it can be seen that the significance of scalar relaxation can depend on the field at which the experiment is performed, and the size and mobility of the molecule under investigation.

10.8 QUADRUPOLAR RELAXATION

So far in this chapter attention has been directed to nuclei whose spin quantum number is $\frac{1}{2}$ and to the various relaxation mechanisms available to them. If a nucleus with a quantum number greater than $\frac{1}{2}$ is considered there is an additional relaxation process available which occurs via interactions with a fluctuating *electric* field, as opposed to magnetic field. A spin $\frac{1}{2}$ nucleus has a spherical nuclear charge distribution; for nuclei with a higher quantum number the charge distribution is non-spherical, resulting in their having a quadrupole moment Q. Quadrupolar nuclei do not have an electric dipolar moment, hence their energy is independent of orientation in a uniform electric field. However, in the presence of an electric field gradient, they precess about the net electric field, and in doing so provide a relaxation mechanism. The electric field gradients are, as usual, modulated by molecular motion and under the extreme narrowing limit $2\pi \vartheta_0 \tau_c \ll 1$

$$1/T_1^Q = 1/T_2^Q = \frac{3}{40}\left(\frac{2I+3}{I^2(2I-1)}\right)(1+\eta^2/3)\left(\frac{e^2qQ}{\hbar}\right)^2 \tau_c \quad (10.34)$$

where η is the quadrupole's asymmetry parameter and q is the electric field gradient at the nucleus [3]. The term $(e^2qQ/\hbar)$ is normally referred to as the quadrupole coupling constant, and, since it depends on the electric field gradient present in the compound, can show quite a range of values even for a specific nucleus. In the highly symmetrical $^{14}NH_4^+$ ion, for example, it is zero [17], $T_1 \simeq 50\,s$ and relaxation being $^{14}N-H$ dipolar, whereas in $CH_3C^{14}N$ it is 4 MHz with $T_1 = 22\,ms$ [18]. Quadrupole couplings can be as large as 100 MHz, resulting in relaxation times in the order of microseconds.

For nuclei with $I > \frac{1}{2}$ quadrupole relaxation dominates unless the molecule possesses high electrical symmetry e.g. BH_4^-, NH_4^+ but also, surprisingly in CH_3NC where $(e^2qQ/\hbar)$ is zero. Quadrupole nuclei depend almost entirely on intramolecular effects; thus if the quadrupole coupling constant is known (e.g. from nuclear quadrupole resonance) T_2 gives a simple measure of τ_c.

An interesting example of the sensitivity of quadrupolar nuclei to electric field gradients is given by the $AsMe_4^+$ ion [19]. In aqueous solution the proton line displays broadening resulting from efficient scalar relaxation of the protons by the ^{75}As nucleus, which in turn implies a fairly long T_{2As}. However, with chloroform as the solvent, where the species in solution is an ion pair, i.e. $AsMe_4^+X^-$, the proton line is sharp, implying a short T_{2As} arising from rapid quadrupole relaxation. The rapid relaxation results from interaction of the arsenic nucleus with the field gradient introduced by the ion pair. Indeed the sensitivity of the ^{75}As nucleus to the asymmetry of electric field is such that the $AsMe_3Et^+$ has sharp proton resonances even in aqueous solution, the nuclear quadrupole detecting and reacting even to such a small departure from tetrahedral symmetry.

10.9 ELECTRON—NUCLEAR RELAXATION

The electron is a magnetic dipole; an unpaired electron will therefore generate a local magnetic field, which in general will have a non-zero average value, i.e. cause a Knight shift (see section 8.2e). This local field will, as usual, be randomly modulated by molecular motion, and consequently provide a relaxation mechanism. The interaction is dipolar in nature and its efficiency therefore depends on the square of both the magnetic moment of this electron and that of the nucleus. The electron's *magnetic* moment is about 10^3 times larger than that of the proton, which itself has the largest nuclear moment. Electron—nuclear dipolar relaxation is thus a much more efficient relaxation process by a factor of 10^6 than the nuclear—nuclear dipole relaxation discussed so far. Against this, the increased efficiency is only partially offset by the larger characteristic mean inter-dipole distance. The efficiency of this form of relaxation means that it has detectable effects even at very low concentrations of the paramagnetic species. The most common example of electron—nuclear relaxation is that resulting from oxygen dissolved in the solution. Dissolved oxygen can lead to line broadening, particularly in proton spectra where line widths below a few tenths of a Hertz normally cannot be achieved without degassing the sample [20]. Even in ^{13}C where the mean distance of oxygen approach is much larger than that found for protons and consequently the effects much weaker, oxygen still can contribute significantly to the relaxation of quaternary carbons.

Electron—nuclear relaxation is normally utilised in n.m.r. by the addition of small quantities of paramagnetic ions to the solution. If the ions' paramagnetism comes purely from spin angular momentum its magnetic moment is given by

$$\mu = \gamma_e \hbar \, [S(S + 1)]^{1/2} \qquad (10.36)$$

If there are other contributions, an effective magnetic moment μ has to be used. As a rule the relaxation rates R_1 and R_2 are directly proportional to N_s, the concentration of paramagnetic species, and roughly proportional to μ_{eff}^2. A detailed analysis of the system is complex (see section 13.4 of ref. 21) but basically it follows the approach used in section 10.2 and results, in the familiar extreme narrowing limit in [22]

$$1/T_1^e = (4S(S + 1) \, \gamma_e^2 \gamma_I^2 \, /3\hbar^2 r^6)\tau_c \qquad (10.37)$$

$$1/T_2^e = 1/T_1^e + (S(S + 1)a^2/\hbar^2)\tau_e \qquad (10.38)$$

In the second (scalar) term of this expression for spin—spin relaxation, a is the hyperfine electronuclear spin coupling constant. The correlation time τ_e is related to the electron relaxation time and also to the exchange life time τ_h of the molecular complex between the ion and the molecule being relaxed [21]:

$$\tau_e^{-1} = T_{2S}^{-1} + \tau_h^{-1} \qquad (10.39)$$

The effect of paramagnetic ions on nuclear relaxation has important applications in chemistry and biology, where they can be applied to the measurement of exchange reactions. It is possible, for example, to study the properties of the binding of metal ions to DNA enzymes etc. by measuring the relaxation times of the solvent water molecules [23]. The 'relaxation reagents' used in ^{13}C n.m.r. provide another example of electron—nuclear relaxation. These reagents are usually transition metal complexes, the trisacetonylacetonate of chromium and iron (Fe^{3+}) being the most common, and their function is to dominate ^{13}C relaxation in the sample. In doing so they achieve two 'desirable' effects; firstly they shorten all T_1's, thus speeding up a pulsed experiment and secondly they eliminate the n.o.e. They are especially useful in quantitative ^{13}C work and in studying ^{15}N where they can eliminate the negative n.o.e.

The details of the interaction complex between the relaxation reagents and the organic molecule is far from clear. In some cases a loose complex between the reagent and solute appears to form based on an interaction with any polar group. Under these circumstances the specific electron—nuclear relaxation rate depends as one would expect, on the inverse sixth power of the electron—nuclear distance. Behaviour of this kind is typified by the interaction of Borneol with Fe(acac)$_3$ [24]. Relaxation is more efficient for the carbons closest to the OH group as expressed by a large specific nuclear relaxation rate (R_1^e) and least effective at the remote 5-C and 9-CH$_3$ positions. (R_1^e) is defined as follows

$$(R_1^e) = (1/T_1)/\eta N_s \tag{10.40}$$

solution viscosity is used (see eqn. 10.7) to replace τ_c and give a parameter which reflects the efficiency of the relaxation agent. For Cr(acac)$_3$ $(R_1^e) \simeq 30$ for Fe(acac)$_3$ $(R_1^e) \simeq 50$.

10.10 THE RELATIVE SIGNIFICANCE OF RELAXATION MECHANISMS

The observed relaxation time or rate is the sum of the relaxation rates of all the various mechanisms outlined in the previous six sections.

$$R_{1,2} = R_{1,2}^{DD} + R_{1,2}^{SC} + R_{1,2}^{CSA} + R_{1,2}^{SR} + R_{1,2}^{EQ} + .. \tag{10.41}$$

Each mechanism can give different chemical information; consequently, in order to gain the full value from a study of relaxation data, the contribution of each mechanism must be resolved.

As a simple example of the resolution of relaxation mechanisms consider the chloroform molecule: the various relaxation times observed are [12]

$T_{1H} = 90\,\text{s}$ $\qquad\qquad$ $T_{2H} = 10\,\text{s}$

$T_{1C} = 33\,\text{s}$ $\qquad\qquad$ $T_{2C} = 0.35\,\text{s}$

$T_{1Cl} = T_{2Cl} = 34\,\mu\text{s}$

For the proton there are two probable relaxation mechanisms, proton—proton dipolar and scalar with the ^{35}Cl (which itself will be relaxed by the quadrupole mechanism). We can therefore state

$$1/T_{1H}^{obs} = 1/T_{1H}^{DD} + 1/T_{1H}^{SC}$$

$$1/T_{2H}^{obs} = 1/T_{2H}^{DD} + 1/T_{2H}^{SC}$$

$$1/T_2^{obs} = 1/T_2^Q$$

Now, as was pointed out in section 4, for a molecule of the size of chloroform T_1^{DD} will equal T_2^{DD}; also, since $\vartheta_H - \vartheta_1$ is about 40×10^6 Hz and T_2 is 34×10^{-6} s, from eqn. (10.31) it follows that T_{1H}^{SC} will be effectively infinite as is usually the case; scalar relaxation is inefficient for spin—lattice relaxation. We can therefore say that

$$1/T_{1H} = 1/T_1^{DD}$$

and

$$1/T_{2H} - 1/T_{1H} = 1/T_{2H}^{SC} = \tfrac{1}{3}[2\pi J_H]^2 S(S+1)T_2^Q \qquad (10.42)$$

thus deducing $^2J(H-H)$ is ~ 6.9 Hz. Similarly the ^{13}C data gives $^1J(C-Cl)$ as 23 Hz. By these arguments the relative contributions of three mechanisms have been assigned and two coupling constants have been deduced which are not directly observable [12].

Attention must now be directed to ^{13}C spin—lattice relaxation times. For proton bearing carbons in medium and large molecules, if they are within the 'extreme narrowing limit', their relaxation is predominantly dipolar in nature. The significance of the dipolar contribution can be measured by determining the nuclear Overhauser enhancement factor η, which is the ratio of the line intensities under continuous decoupling conditions to those when gated decoupling or paramagnetic ions are used to suppress the Overhauser effect.

$$T_1^{DD} = 2T_1^{obs}/\eta \qquad (10.43)$$

The remaining relaxation rate arises from contributions to the total relaxation from the other mechanisms discussed, and in general it will be small. For small molecules and quaternary carbons, however, these contributions may be very significant. Consider pure benzene; $\eta = 1.6$ and $T_1^{obs} = 29.3$ s, therefore using eqn. (10.43), $T_1^{DD} = 37$ s leaving a $T_1^{other} = 146$ s; in this case spin rotation is presumably the other mechanism [25]. In general spin rotation is the next most efficient mechanism, $T_1^{SR} > 20$ s, followed by chemical shift anisotropy. The significance of these processes may be evaluated by measuring T_1 as a function of temperature and field respectively. Scalar relaxation need only be considered in very special cases, e.g. in the case of a bromine bearing carbon.

If any of the nuclei in the sample exhibit T_1 of longer than about 10 s, oxygen must be eliminated since it can act as a relaxation agent $T_1^{e(O_2)} \simeq 100$ s (see section 10.9) and mask more informative data. As an example of these effects again consider benzene [25]

Benzene	T_1^{obs}	T_2	T_1^{DD}	T_1^{SR}	$T_1^{e(O_2)}$	s
degassed	29.3	1.60	37	146	—	s
undegassed	23.0	1.30	35	—	106	s

10.11 APPLICATIONS OF CARBON T_1 DATA

Carbon spin—lattice relaxation data is becoming increasingly used in chemistry. Detailed accounts of all the many applications are dealt with elsewhere [26]. This section will merely sketch the areas of application and point out a few pit-falls in interpreting such data. Almost invariably applications are based on the use of T_1^{DD} obtained from T_1^{obs} in proton noise decoupled spectra as outlined above (eqn. 10.43) and assuming the extreme narrowing limit (eqn. 10.22).

Relaxation data can be used as an assignment aid (see ref. 26, section 4B), especially for quaternary carbons where directly bonded proton coupling data is not available or difficult to measure. Subtle differences in the relaxation times, for example, have been used in the assignment of the quaternary carbons in the alkaloids codine, brucine and reserpine; these differences arising from the number of 'close' protons [26]. The dependence of T_1^{DD} on the

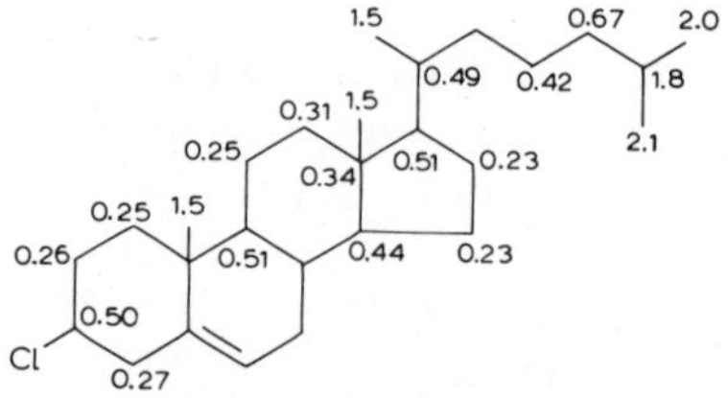

Fig. 10.6. [13]C spine—lattice relaxation times, in seconds, for cholesteryl chloride, $1 M$ in CCl_4, measured at 15.1 MHz and 42°C. From A.A. Allerhand, D. Doddrell and R. Komoroski, J. Chem. Phys., 55 (1971) 189.

number of directly bonded protons can be used for assignment. Figure 10.6, for example, illustrates the $^{13}C\,T_1$ data for cholesteryl chloride, here $NT_1 = 0.52\,s$ (where N is the number of directly bonded protons) for the carbons in the ring system indicating a correlation time of $9 \times 10^{-11}\,s$ [27]. These values are typical of molecules of this size. In an incompletely assigned case, the appropriate value of NT_1 must be measured for an assigned carbon; this is then used to find the value of N for the unassigned carbons. The values of N can also be obtained from proton off-resonance decoupling experiments. Such an experiment is far from easy to interpret, however, due to the enormous number of overlapping lines which are present in such spectra.

Relaxation times can only be used to estimate the number of directly bonded protons when, as is the case for the steroid skeleton in the above example, one molecular correlation time is adequate to describe the molecular motion. The carbons in the C-17 side chain of cholesteryl chloride do not have a value of NT_1 equal to $0.52\,s$, nor do the angular methyls. This fact can be used, with care, as an assignment aid. The NT_1 values of those carbons are longer, due to the side chain having faster motion, hence less efficient dipolar relaxation. Their relaxation is still, however, dipole dominated. It is in studies of segmental and isotropic motion that relaxation times are most powerful.

As an example of segmental motion, consider n-decanol, where the T_1 values [28]

$$CH_3{-}CH_2{-}CH_2{-\!\!-}CH_2{-}CH_2{-\!\!-}CH_2{-}CH_2{-}CH_2{-}CH_2{-\!\!-}CH_2OH$$

3.1	2.2	1.6	1.1	0.84	0.84	0.84	0.77	0.77	0.65s

show a steady increase with distance from the OH carbon. This behaviour results from the OH end of the molecule being 'anchored' by hydrogen bonding; segmented motion increasing down the hydrocarbon chain τ_c being 36 ps for C-1 decreasing to 5 ps for C-10. It is not essential to have hydrogen bonding in order to see segmental motion; a large molecular fragment e.g. the steroid skeleton in cholesteryl chloride can be a sufficient 'anchor'. The relaxation times of the α, β, γ and δ carbons of N_1N_1-di-n-butylformanide (3.1, 2.3, 1.6, 1.1 s) show a parallel behaviour to those of decanol, as do those of a butyl side chain attached to the massive dipalmitoyllecithin molecule, indicating a similar degree of segmental motion in all cases [29].

Care must be taken interpreting angular methyl rotation in terms of effective correlation times where 'gearing' effects are possible. In 1,2,3-trimethylbenzene the sterically hindered 2-CH$_3$ is essentially a free rotor while the 1- and 3-CH$_3$'s are significantly hindered in their rotation as can be seen from their T_1 values [30] (see Table 10.2).

TABLE 10.2

Contributions to relaxation of aromatic methyl groups [30]

	T_1^{OBS}	T_1^{DD}	T_1^{SR}
CH$_3$ (benzene ring)	25 26	270 32	28 147
CH$_3$ / CH$_3$ ... CH$_3$ (benzene ring)	12 12	28 14	20 98

Compared with C-1 and 3 dipolar relaxation is less efficient for C-2 while spin rotation is more efficient, as is consistant with a faster rotation of the 2-CH$_3$ group.

It is known that for small molecules rotational motion can be anisotropic due to preferential rotation about a specific axis. A well studied case is the mono-substituted benzenes where preferential rotation about the C$_2$ axis leads to unequal relaxation times for the *ortho/meta* carbons on one hand and the *para* carbon on the other. Dipolar relaxation in these cases is described by eqn. (10.23), where θ is 60° and 120°. Rotation about the C$_2$ axis has no modulation effect on the *para* carbon and consequently no effect on its relaxation, but the rotation reduces the efficiency of dipolar relaxation for the *ortho* and *meta* carbons by ψ, which, in the limit of fast rotation, for 60° or 120° is 1/64; thus $T_{1(o,m)}^{\text{DD}} = 64\,T_{1(p)}^{\text{DD}}$. Such an extreme ratio has not yet been observed, but the principle holds as can be seen by diphenyldiacetylene [31]

The difference in T_1 values for the *ortho* and *meta* carbons of *ortho*-dichlorobenzene (Figure 10.5) can be explained in terms of a preferential axis of rotation between the two chlorines. The *ortho* carbons are subject to the more significant motion about this axis than the *meta* ones, and hence have the longer T_1 (7.8 s).

A great deal of effort is being made in the investigation of polymers [32] and biopolymers [23]. Before a detailed motional analysis can be undertaken on these large systems by relaxation studies, their T_1, T_2 and n.o.e. values must be fully characterised, allowing for the possibility that in such molecules the 'extreme narrowing' approximation will probably not be valid. Recent work indicates that a single τ_c or even several τ_c's may not be adequate to describe these systems and that τ_c must be treated as a statistical distribution of values with significant contributions from very slow motions [33].

In all the above examples, it has been implicitly assumed that the C-H bond distance is constant within the molecule. This may not be the case and, since this distance appears in the expression for dipolar relaxation to the sixth power, even small changes can be significant. In pyridine the γ carbon has a significantly shorter C-H bond length than the *ortho* carbons, but it is enough to a account for a T_1 of 17.0 s while the α and β carbons have a T_1 of 18.2 and 19.1 s [34].

10.12 THE MEASUREMENT OF SPIN LATTICE RELAXATION TIMES

(a) Techniques

It has always been possible to measure spin—lattice relaxation times under high resolution conditions. The classical method is the adiabatic rapid passage technique; i.e. sweeping through the resonance with a high power sufficiently rapidly that the spin populations were inverted. The recovery curve for the magnetisation is then monitored by sweeping through the signal at a normal power and sweep rate τ s later. Also spin—lattice relaxation times have been measured for a long time under 'wide line' conditions, using pulsed techniques, where an 'average' T_1 is measured. With the advent of Fourier techniques the simpler pulse techniques for measuring T_1 could be applied to high resolution spectra. This advance, originally suggested by Vold et al. in 1968 [35], along with wide spread acceptance of ^{13}C n.m.r. where relaxation data is more easily applied to chemical problems, has led to an upsurge in interest in relaxation time measurements. This section is written mainly with ^{13}C in mind, but unless

324

attention is drawn to the contrary, the *experimental* techniques will apply equally to other nuclei.

The basic principle of the pulse Fourier transform approach to relaxation time measurement is to use the speed of the technique to take 'snap shots' of the sample as it relaxes, i.e. it is possible to record all the data required to produce a high resolution spectrum within the time scale of relaxation time being measured. The three commonly used techniques for measuring T_1 will be dealt with individually, and they will then be compared with respect to their efficiency and accuracy.

(i) Inversion recovery

The basis of the inversion recovery experiment is the $180°$—τ—$90°$ pulse sequence. The behaviour of the nuclear magnetisation during such a sequence is shown in Fig. 10.7 along with the mathematical expression describing it. The $180°$ pulse which is assumed to be sufficiently brief that no significant relaxation occurs during it, inverts the spin population and rotates the equilibrium magnetisation on to the $-z$ axis. Following this pulse the magnetisation relaxes back towards its equilibrium value by an exponential process with a rate constant T_1. After some time τ the sample is subjected to a $90°$ pulse which rotates the residual longitudinal magnetisation on to the y' axis; a free induction decay results which may be transformed to produce a spectrum. This spectrum is in effect a 'snap shot' of the state of the spin system τs after the spin population has been inverted. For a small value of τ the spectrum contains inverted lines;

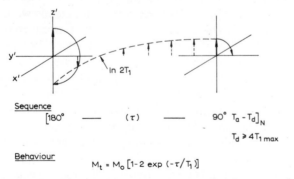

Fig. 10.7. The inversion recovery sequence for measuring T_1.

$a - M_z$ value gives a $- M_{y'}$ component and hence a negative signal. When τ is equal to $\ln 2 T_1$ no signal will appear, as at this time M_z is zero. For longer values of τ larger and larger signals are observed until, when τ is 4 or 5 times T_1, a signal of 'normal' intensity is observed, (i.e. equal to that which would follow a single 90° pulse). The inversion recovery sequence is illustrated in Fig. 10.15 for a mixture of benzene and dioxane.

A modified form of the inversion recovery sequence, originally proposed by Freeman and Hill [36], is sometimes used. In this experiment the f.i.d. resulting from the 180°—τ—90° sequence is subtracted from the f.i.d. following a single 90° pulse. The resulting spectrum shows lines which decay from $2M_0$ to zero as the interval between the pulses increases. The value of this sequence is that it is less sensitive to changes in spectrometer performance over a long period of time, as each sequence determines $(M_0 - M_\tau)$, a key parameter with respect to accuracy. Figure 10.8 shows a set of ^{13}C spectra obtained from 3.5-dimethylcyclohex-3-en-5-one using the modified inversion recovery sequence [36]. Note the large difference in relaxation times between the protonated and non-protonated carbons. Figure 10.9 shows the data from this experiment plotted in semi-log form, $\ln(M_0 - M_\tau)$ against τ; the slope of this graph gives T_1.

Needless to say the spectra shown in Fig. 10.8 are not the result of a single pulse sequence, but the time average of many. Since at the start of each sequence equilibrium is assumed, $M_z = M_0$, a delay of about $4T_1$ must be allowed after the 90° pulse before the next pulse sequence, $(T_a + T_d) > 4T_1$. The necessity for such long delay between pulses (possibly many minutes if quaternary carbons are being investigated) makes T_1 measurements long experiments.

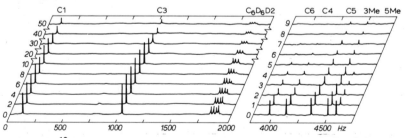

Fig. 10.8. ^{13}C spin—lattice relaxation spectra from 3,5-dimethylcyclohex-2-en-1-one using the modified inversion recovery sequence (180°—τ—90°—T_d—90°—T_d). From R. Freeman and H.D.W. Hill, J. Chem. Phys., 54 (1971) 3 367.

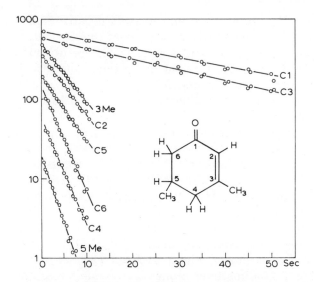

Fig. 10.9. Semi log plot of the inversion recovery (squares) and progressive saturation data for 3,5-dimethylcyclohex-2-en-1-one. From R. Freeman and H.D.W. Hill, J. Chem. Phys., 54 (1971) 3367.

(ii) Progressive saturation

A second method of measuring T_1 is based on the principle of saturation. As we have seen (Fig. 7.4), if the sample is subjected to a series of 90° pulses more frequently spaced than $3\,T_1$, the line intensities are detectably reduced by saturation. By studying line intensities as a function of the interval between the 90° pulses, T_1 can be measured. The main limitation of this technique is that the smallest

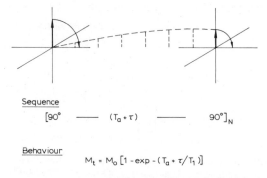

Sequence

$$[90° \quad\underline{\hspace{2cm}}\quad (T_a + \tau) \quad\underline{\hspace{2cm}}\quad 90°]_N$$

Behaviour

$$M_t = M_0\,[1 - \exp - (T_a + \tau/T_1)]$$

Fig. 10.10. The progressive saturation sequence for measuring T_1.

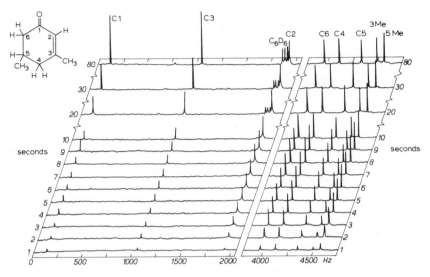

Fig. 10.11. ^{13}C spin—lattice relaxation spectra from 3,5-dimethylcyclohex-2-en-1-one using the progressive saturation technique. From R. Freeman and H.D.W. Hill, J. Chem. Phys., 54 (1971) 3367.

interval between pulses is limited by the data acquisition time. Figure 10.10 shows the behaviour of the magnetisation and Fig. 10.11 shows a family of spectra, again obtained for 3,5-dimethylcyclohex-2-en-5-one [36]. A similar semi-log plot to that used for the inversion recovery data yields T_1. Until a steady state has been established, no data should be recorded, as the results will be atypical.

(iii) Saturation recovery

A spin system will also recover from saturation via an exponential process characterised by T_1; following the recovery will therefore permit a measurement of T_1. The sample may be saturated in two ways, either by a series (about 5 ~ 10) of closely spaced 90° pulses [37], or by a single 90° pulse followed by a homogeneity spoiling pulse (h.s.p.) [38]. At this point in the experiment there is no net magnetisation along any axis. The system then returns towards equilibrium; at time τ after the first 90° pulse a second 90° pulse is initiated and the resulting f.i.d. recorded, transformed, etc. Any residual transverse magnetisation is destroyed by a second homo spoiling pulse and the sequence repeated. The spectrum produced

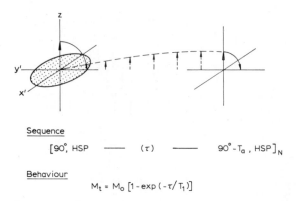

Sequence

$$[90°, \text{HSP} \quad \underline{\qquad} \quad (\tau) \quad \underline{\qquad} \quad 90° - T_a, \text{HSP}]_N$$

Behaviour

$$M_t = M_0 [1 - \exp(-\tau/T_1)]$$

Fig. 10.12. The saturation recovery sequence for measuring T_1.

grows as $(1 - \exp - \tau/T_1)$ as is shown in Fig. 10.12 and again T_1 is measured from a semi-log plot.

(iv) N.o.e. growth

A fourth approach to the measurement of T_1, only really applicable in ^{13}C or a nucleus coupling to ^1H, is by use of the property that the nuclear Overhauser enhancement grows exponentially with time constant T_1 (not just T_1^{DD}) [39]. The sequence of events is to switch on the proton decoupler, and τ s later to sample the magnetisation with a 90° pulse. As τ increases η increases up to η_{max}, thus

$$\eta_t = \eta_{max} (1 - \exp - \tau/T_1) \tag{10.44}$$

which may of course not be the theoretical maximum. A delay of $4 T_1$ is again necessary before the sequence can be repeated. The attraction of this sequence is that the η_{max} is measured at the same time; thus T_1^{OBS} and T_1^{DD} are obtained from a single experiment [39].

(b) The accuracy

Many effects influence the accuracy of a T_1 measurement; the major ones are listed in Table 10.3. The first group are chemical and their significance should by now be apparent. This section is concerned with the instrumental factors which limit the precession with which T_1 values can be measured in a given sample.

The limitations associated with probe geometry are related to r.f.

TABLE 10.3

Factors which influence the accuracy of T_1 measurements

(a) Chemical	
	Concentration, mainly via viscosity effects
	Paramagnetics, principally dissolved O_2 (only for $T_1 > \sim 60$ s)
	Molecular diffusion, important if B_1 is inhomogeneous and/or short samples are used
(b) Instrumental	
	Coil geometry, B_1 inhomogeneity effects
	Errors in pulse angles, varies with pulse sequence
	Temperature stability, again mainly via viscosity changes
	Resolution/sensitivity stability, particularly over long periods
	Digitisation, intensity errors due to the finite number of data points per line
	Choice of pulse sequence

inhomogeneity and molecular diffusion [40]. Greatly simplified, the problem lies in the fact that during the delay time τ, into the active volume of the receiver coil may diffuse molecules which were not subjected to the same pulse power as those in the active volume at the time of the first pulse. Such molecules will give an anomalous contribution to the sampled f.i.d., and hence affect the measurement value of T_1. Effects of this type are almost exclusively confined to single coil probes whose B_1 field is relatively inhomogeneous when compared with that of a crossed coil probe. Diffusion effects are not very important if short samples (~ 30 mm) are used or if T_1 values are less than 50 s. For single coil probes T_1^{OBS} may be a function of sample length if longer T_1's are being considered [41]. A similar problem, that of gas phase relaxation, can arise if solute molecules are volatile and if the temperature is near the solution boiling point. Under these circumstances liquid/vapour molecular exchange occurs, resulting in anomalously short T_1 values and low n.o.e. values. T_1 is reduced by a contribution from the spin rotational relaxation which is very efficient in the gas phase. Should this problem be serious it can be overcome by constructing the surface of the sample either by means of a vortex plug or the use of spherical microcells.

A detailed analysis of the accuracy of the various T_1 sequences has been undertaken by Freeman et al. [42] and Jones [43] and Levy et al. [41] among others. A summary of their conclusions is given in Table 10.5. The pulse sequences differ with regard to the effect of finite pulse power and timing accuracy; the IR sequence is more demanding on pulse power as it requires a $180°$ pulse and is quite sensitive to any drooping of power across the spectrum due to the finite length of the pulse; it is, nevertheless, insensitive to the accuracy of the pulse. Progressive saturation on the other hand is less sensitive to offset effects, but is very sensitive to the accuracy of the $90°$ pulse. Saturation recovery, unlike progressive saturation, is insensitive to the precise value of the pulse angle but, like inversion recovery, sensitive to offset effects.

The above remarks apply even assuming that the pulses used are homogeneous and reproducible. The effects resulting from finite pulse power and r.f. inhomogeneity are often hard to separate. If a significant contribution from r.f. inhomogeneity is suspected, it may be checked for by comparing signals from a $90°$ and $270°$ pulse. They should have the same magnitude, and be $180°$ out of phase; if they are not, then the B_1 field is significantly inhomogeneous. When doing such a test the line must be close to the carrier to avoid offset effects. The most significant problem caused by inhomogeneity effects, and also by the finite nature of the pulse (i.e. B_{eff} is not constant over the total spectrum), is the resultant transverse magnetisation left after the $180°$ pulse. The presence of resultant magnetisation gives rise to a spectrum showing phase errors which cannot be corrected for using the normal phasing routines (Fig. 10.13a). Fortunately, these effects can be overcome either by using a homogeneity spoiling pulse after the $180°$ pulse to destroy the residual magnetisation (as was done in Fig. 10.14b) or by alternating the phase of the $180°$ pulse which averages the phase anomolies [41]. When the delay between the pulses is in the order of, or longer than, T_2, the inhomogeneity of the magnet will relax the magnetisation, and no phase anomalies will be seen.

When using the inversion recovery sequence, the result can be affected by the time delay allowed for the system to equilibrate between sequences. Table 10.4 illustrates this point [41]. The T_1 of the ethylene glycol sample is $2.0 \pm 5\%$, significant errors can, however, be detected if a time less than $3\,T_1$ is allowed for the system to relax fully. Incomplete recovery is a major limitation for the IR technique, because in order to achieve an accurate result, a good estimate of the longest T_1 in the sample is necessary before starting; if, on the other hand, an arbitrarily long delay is used to circumvent this problem, the sequence becomes very long and hence inefficient.

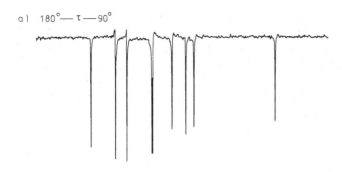

a) $180°\!-\!\tau\!-\!90°$

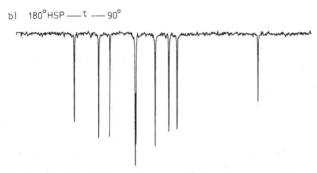

b) $180° \text{HSP}\!-\!\tau\!-\!90°$

Fig. 10.13. ^{13}C spectrum of nicotinamide in a $180°\!-\!\tau\!-\!90°$ experiment ($\tau =$ 0.5 s) (a) showing phase errors resulting from an inhomogeneous $180°$ pulse; (b) the spectrum given when a homogeneity spoiling pulse is applied after the $180°$ pulse.

TABLE 10.4 [41]

The effect of delay time on the measurement of T_1 by the conversion recovery technique[a]

Time between sequence i.e. $T_a + T_d = PD$	PD/T_1	Theoretical recovery (%)	T_1 measured (s)
15	7.5	99.9	1.90
10	5	99.3	2.05
8	4	98.2	2.00
6	3	95	1.95
5	2.5	91.8	1.65
4	2	86.5	1.60
3	1.5	77.7	1.32
2	1	63.2	1.30

[a] For 85% ethylene glycol at $38°$, measured at 67.9 MHz.

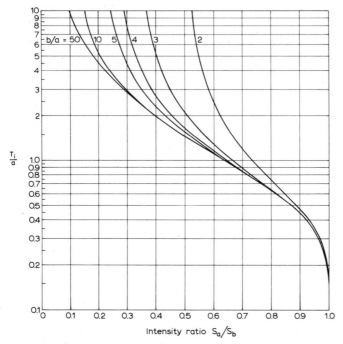

Fig. 10.14. Theoretical curves relating the ratio of intensities observed in progressive saturation experiments at two different pulse intervals (a and b) to the spin—lattice relaxation time (normalized with respect to the pulse interval a s). Each branch represents a different ratio of pulse intervals b/a. Only in the central region of this diagram are the pulse intervals correctly set up for an accurate determination of T_1. Pulse imperfections have been neglected. From R. Freeman, H.D.W. Hill and R. Kaptein, J. Magn. Res., 7 (1972) 327.

When calculating relaxation times from the spectral data it is normal to use peak heights as a measure of $M_t M_0$ etc. whereas peak areas are theoretically correct. The relationship between peak areas and height is affected by two major factors, resolution and, in the case of Fourier spectrometers, digitisation. Both factors affect the height but not the area. With regard to resolution, care must be taken that the resolution is *constant*, though not necessarily maximum, throughout the experiment. Digitisation affects peak heights when only a few data points represent a line; the peak 'intensity' is then sensitive to the 'position' of the data point on the line. The digitisation factor can be overcome either by using more data points per line (i.e. using smaller spectral widths or large data tables) or by broadening the lines with the use of a weighting function so that each line contributes to more data points [44]. The latter approach may be used to ease the resolution stability problem; if a large smoothing function is used, changes in line width due to resolution changes become insignificant and peak height becomes more reliable.

Having obtained the spectrum, the next aspect of the experiment is the determination of a T_1 value from the data. The classical method is a semi-log of $\ln(M_0 - M_t)$ against τ, T_1 being obtained from either the slope of the line or (by definition) the value of τ where $(M_0 - M_t)$ has fallen to e^{-1} (0.368) of its value at $\tau = 0$. Potentially in the simple IR sequence it is possible to use the property that $M_t = 0$ when $\tau = \ln 2\, T_1$ to evaluate T_1. The use of the 'zero signal' time should never be used for anything except crude estimation, as any systematic error which results in $M_{y'} = 0$ at the start, e.g. in pulse angle, will lead to large errors. Using a graphical method removes errors from this source since they appear in the ordinate but not the slope. The accuracy of any T_1 value obtained depends significantly on the value of M_0, as every plotted point involves this value. Great care must therefore be taken with the measurement of M_0, hence the development of the Freeman and Hill [36] modification to the IR sequence mentioned above. At the very least, the value of M_0 should be determined at the beginning and end of an experiment in order to monitor any change in conditions which have occurred during the experiment. The dependence of the measurement on the value of M_0 can be overcome by using a ratio method; if, for example, when using the progressive saturation technique two spectra are obtained using different values of τ (a and b), the ratio of the intensities is given by [42]

$$\frac{M_a}{M_b} = \frac{1 - \exp(-a/T_1)}{1 - \exp(-b/T_1)} \tag{10.45}$$

The graphical solution of eqn. 10.45 is given in Fig. 10.14 for different values of the ratio a/b. The T_1 values calculated from this graph are subject to errors in the amplitude ratio and are only reasonably accurate in the central part of the figure. The ratio method can be used as the basis for an automatic scheme to measure T_1 using any of the pulse sequences. The pulse interval is changed under computer control for successive experiments until intensity ratios suitable for an accurate determination of T_1 are found for all lines in the spectrum [42].

Because of the semilog nature of the plot used, the error distribution in M_t is skewed. This factor alone can lead to the determination of a T_1 which is shorter than the true value. The data points from long τ value tend to fall below the best straight line drawn through the points from short values of τ. For this reason values with τ greater than $1.5\,T_1$ are not generally useful except, of course, for the essential determination of M_0 [41]. If the spectrometer's computer is used to calculate T_1 by means of a 'least squares fit' programme, these effects should be borne in mind. If possible, the computer should perform an exponential curve fitting which is free from the problems of skew errors.

(c) The efficiency

Finally we must discuss the relative efficiency of the possible T_1 sequences. Which gives a value of T_1 quickest? Each sequence has n values of τ, each of which is itself the result of time averaging N transients. The value of N necessary to achieve the same S/N ratio differs for each sequence. In order to obtain a T_1 of comparable accuracy from any sequence one requires the same S/N ratio, at least to a first approximation; N will therefore vary depending on the sequence used. The number of pulses required using the IR sequence is one quarter that of the other two. The factor of four arises as the inversion recovery sequence does not work from a base of zero magnetisation but from a base of $-M_0$; the signal from IR is therefore twice that available from the PS and SR sequence. In a time averaging experiment, therefore, only a quarter the number of pulses are required. The difference in signal to noise ratio is illustrated in Fig. 10.15 which shows spectra for benzene/dioxane obtained by the IR and SR techniques using one sequence.

The relative efficiencies of the sequences can be evaluated using an idealised experiment, such as one requiring ten values of M_τ equally spaced in height. The detail of the result depends on the assumptions made in defining the experiment, e.g. the choice of T_a. The general result, however, is that the PS technique is always the fastest sequence. Values of $M_{(\tau)}$ below about 30% of M_0 frequently cannot be obtained without line broadening, which will result if τ and consequently T_a are reduced below that required by the sampling theory. This represents a severe limitation of the PS technique. The relationship between IR and SR is more complex and depends on T_1.

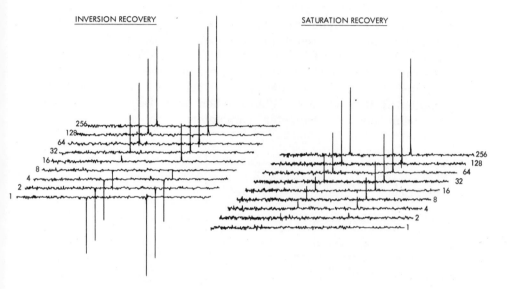

Fig. 10.15. $^{13}C\,T_1$ spectra for a 1:1 mixture of benzene and dioxane using the inversion recovery and progressive saturation sequences. Each spectrum is the result of a sequence.

For T_1 less than a second or so the *IR* sequence is most efficient; for longer T_1's *SR* becomes more efficient, despite needing four times as many pulses, since it does not require a delay of $4T_1$. When the T_1 being determined is in the order of 100 s *SR* can require only 70% the time of *IR*, and since such experiments are very long (days!) except on very concentrated samples, a 30% saving in time can be quite significant.

The efficiency of the Freeman Hill *IR* sequence is slightly lower than the simple *IR* experiment. Every alternate sequence only involves a 90° pulse, but its accuracy is probably higher as the required parameter $(M_0 - M_\tau)$ is determined during every sequence.

It is impossible to state that any one of the techniques discussed is the best way of measuring T_1. Probably the least satisfactory is the progressive saturation sequence. It is the fastest, but is very sensitive to experimental parameters and cannot measure short T_1's under high resolution conditions. For samples with long T_1's saturation recovery is most useful, but it has disadvantages arising from the homogeneity spoiling pulse affecting the spectrometer's field/frequency lock. Under most conditions the inversion recovery sequence is the most useful, provided the spectrometer has enough power to generate a sufficiently short 180° pulse. The pros and cons of the various T_1 sequences are summarised in Table 10.5.

It must be remembered that accurate, reliable T_1 values can only be obtained if great care is also taken with respect to sample preparation and spectrometer parameters such as temperature stability.

TABLE 10.5

Advantages and disadvantages of various T_1 measurement sequences

Sequence	Advantages	Disadvantages
Inversion recovery	Applicable to all ranges of T_1	Prior knowledge of T_1 (to set delay)
	τ independent of T_a	Requires 180° pulses (i.e. puts more demands on B_1 power)
	Insensitive to accuracy of pulses (if slope used to determine T_1)	Sensitive to off-resonance effects
	Gives quick, easy qualitative information on T_1	
Saturation recovery	Greater sensitivity than IR on long T_1 (especially where high S/N possible)	Requires extra hardware Can interfere with field/frequency lock, auto shim operation
	Requires only 90° pulses	
	Independent of T_a	
Progressive saturation	Most efficient sequence (on long runs)	Inapplicable to short $T_1(T_1 \leqq T_2)$
	Requires only 90° pulses and will work with any basic spectrometer	Sensitive to accuracy of 90° pulse Lowest value of τ limited by T_a

10.13 THE MEASUREMENT OF SPIN—SPIN RELAXATION TIMES

The measurement of spin—spin relaxation times is more difficult than the spin—lattice relaxation times discussed previously. The major cause of the increased difficulty, and, ironically, the major interest in T_2, is that spin—spin relaxation is sensitive to low and zero frequency magnetic fields. Any inhomogeneity in the basic magnetic field (and no magnet is perfectly homogeneous) results in an unwanted contribution to T_2. This contribution is referred to as R_2' and the normally observed value as R_2^* thus

$$R_2^* = R_2' + R_2 \qquad\qquad (10.46)$$

where R_2 is the 'natural' relaxation rate resulting from the mechanisms discussed in the previous sections. The value of R_2^* can be measured directly from any spectrum as the line width at half height of an n.m.r. line and is given by (πT_2^*). If R_2' is insignificant compared with R_2^*, as in the broad line of a quadrupolar nucleus, then T_2 can be read directly. The classical technique to overcome the unwanted effects of T_2' is the spin echo sequence originally proposed by Carr and Purcell [45]. An n.m.r. sample can be thought of as consisting of an assembly of extremely small regions called isochromats. Within an isochromat magnetic field inhomogeneities are negligible. Following a 90° pulse each isochromat will precess at its own specific Larmor frequency depending on its own local magnetic field. As time goes on the isochromats, which initially were all together, fan out; this is a T_2 process. This can be corrected for since the process is reversible by the use of a 180° pulse as is shown in Fig. 10.16. A 180° pulse has the effect of rotating all the isochromats to a mirror image position within the x'y' plane. Those isochromats which had precessed faster than the mean are now behind the mean, and vice-versa. If the time between the 90° and 180° pulses is τ s, then a further τ s later refocussing will occur and a spin echo results whose amplitude has decayed as T_2 alone.

After the first echo the isochromats lose phase coherence again, and can be refocussed by a further 180° pulse τ s later. Thus if a chain of pulses at times $(2n + 1)\tau$ are applied a train of echoes (called a Carr-Purcell train) at time $(2n + 2)\tau$ are formed. Amplitude of this echo train decays as T_2 provided that (a) τ is kept short enough that significant diffusion does not occur between pulses, as

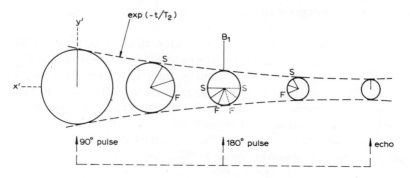

Fig. 10.16. The formation of a spin echo. All the spin isochromats are in phase following the 90° pulse. They then precess, those in a higher local magnetic field process faster (F) than the mean, those in a lower field slower (S). A 180° pulse reverses the order of the isochromats and then refocus and produce a spin echo.

this would inhibit echo formation, and (b) the r.f. pulses are homogeneous.

The problem of r.f. inhomogeneity, as in T_1 work, can be overcome by using phase shifted pulses. The sequence is then called the Carr-Purcell, Meiboom Gill (CPMG) [46] sequence which is

$$90^\circ_x - \tau - 180^\circ_y - 2\tau - 180^\circ_y - 2\tau - 180^\circ_y$$

The effect of the phase shift is shown in Fig. 10.17. The 90° pulse is applied along the x' axis. If the successive imperfect 180° pulses are also applied along the x' axis, as can be seen, the errors systematically add up. If, however, the imperfect 180° pulses are applied along the y' axis, their errors do not add up. The spin isochromats oscillate between two paths within the $x'y'$ plane, slightly displaced along the z axis.

If the duration between the pulses is decreased to zero, the experiment becomes the so called 'spin locking' technique. Here the spins are rotated with a 90°_x pulse onto the y' axis; then a field which is strong with respect to local magnetic inhomogeneities is applied along the y' axis, i.e. there is a 90° phase shift with respect to the pulse. In the presence of the locking field, T_1 type relaxation occurs, termed spin—lattice relaxation in the rotating frame or $T_{1\rho}$ [49]. On removal of the locking field normal relaxation (T_2^*) takes place. The value of $T_{1\rho}$ under high resolution conditions is equal to T_2.

Spin—spin relaxation times of individual lines can be obtained by

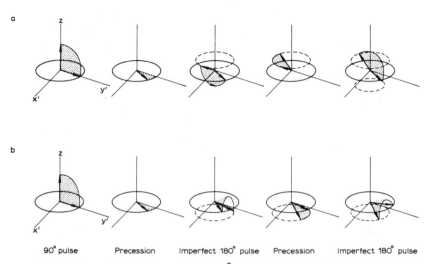

90° pulse   Precession   Imperfect 180° pulse   Precession   Imperfect 180° pulse

Fig. 10.17. The effect of imperfect 180° pulses on a Carr-Purcell experiment. (a) With the pulses along the x′ axis errors are progressive; (b) with the pulses along the y′ axis (CPMG sequence) errors result in a simple oscillation between two paths displaced about the x′y′ plane.

maintaining a CPMG train (or a locking field) for a variable time (t), after which the refocussing pulses are stopped and a normal f.i.d. is obtained, which is digitised, transformed etc. Peak heights in a series of spectra with an increasing duration of refocussing pulses decay as T_2. The experiment is shown in the top half of Fig. 10.19.

(a) J echo formation

The simple results described so far can only be obtained in the absence of homonuclear coupling. Consider the A spins of an AX system. After the 90° pulse, the two components of the A multiplet precess, each with its own Larmor frequency. The 180° pulse reverses the dephasing of each component but since the X part is also inverted it does *not* affect their relative precession frequency. Thus τ s later two echoes result from the A resonance which are phase shifted. This effect, echo modulation, is a major barrier to the use of spin echo techniques to the measurement of T_2 under high resolution conditions in the presence of homonuclear coupling.

Fig. 10.18a shows a set of ¹H spectra obtained from 1, 1, 2-trichloroethane using a CPMG sequence which was stopped after

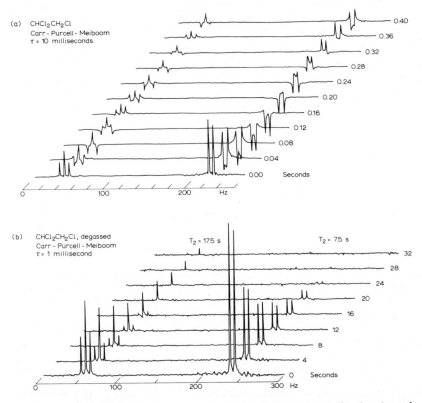

Fig. 10.18. ^{1}H spectra from 1, 1, 2,-trichloroethane [47] (a) showing phase modulation; (b) in the 'spin-locking limit' where phase modulation is removed.

successively longer periods of time. The CH_2 resonances and the outer lines of the CH triplet rotate their phases at $(180 \times J)^0$/s. The centre CH line, belonging to an anti-symmetric subspectrum, has a constant phase. In the general case, phase errors build up for A at a rate of $180J_{AX}I_{ZX}$ where I_{ZX} is the nett Z magnetisation quantum number of the X spins [48].

(i) In first order spectra with *only one* coupling constant, echo effects can be removed by choosing values of the delay time which are multiples of J, at which time the phase shifts are $360°$ [48].

(ii) One can use a very short interval between the $180°$ pulses, i.e. moving towards the spin locking limit (see Fig. 10.18b).

(iii) Magnitude spectra can be plotted which, by their very nature,

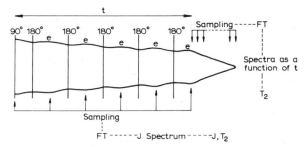

Fig. 10.19. Sampling techniques for T_2 and J spectra [50].

are phase-less. Their use suffers from the limitations imposed by this method of display with respect to overlapping lines.

(iv) One can use 'J spectra'.

(b) J spectra [50]

The most elegant solution to the echo modulation problem is to make use of the modulation itself. In this experiment, the lower part of Fig. 10.19, digitisation takes place only at the top of the spin echoes. Fourier transformation of this data yields J spectra consisting of lines which have their natural line width and are separated only by coupling constants. Chemical shifts, which are zero

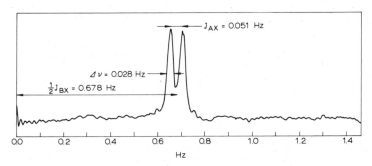

Fig. 10.20. Partial J spectrum of 3-bromothiophene-2-aldehyde. Carr-Purcell spin-echo responses of the aldehyde proton were detected selectively at a repetition rate of 2.5 Hz. 150 samples were taken at the tip of each echo and Fourier transformed. Resonances appear at $\frac{1}{2}(J_{bx} - J_{ax})$ and $\frac{1}{2}(J_{bx} + J_{ax})$ with a width corresponding to a T_2 of 11.2 ± 0.4 s. From H.D.W. Hill and R. Freeman, J. Chem. Phys., 54 (1971) 301.

frequency effects, do not appear as they are refocussed, along with inhomogeneity effects. The use of J spectra is the most sensitive method for measuring small spin coupling constants, such as the $0.051 \pm 0.002\,\text{Hz}$ coupling between the aldehyde and 4 protons in 3-bromothiophene-2-aldehyde demonstrated in Fig. 10.20.

10.14 ROTATING FRAME RESONANCE EFFECTS

Carbon spectra do not show problems from the T_2 point of view, due to homonuclear spin coupling. Echo problems do not arise due to heteronuclear spin coupling, as only one component of the spin multiplet is inverted by the $180°$ pulses. What might be initially unexpected effects do occur when heteronuclear decoupling is used due to rotating frame resonance affects. When the proton and carbon nuclei precess about an exciting field at the same frequency, as viewed in their individual rotating frames, i.e.

$$\bar{\gamma}_c B_1 = \gamma_H B_2 \qquad (10.47)$$

very efficient energy transfer is possible between the two spin systems and anomalously short T_2 values are observed. This was first described by Hartmann and Hart and it is the basis of the solid state spin transfer techniques [52].

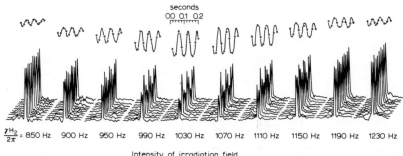

Fig. 10.21. Fourier transform carbon-13 signals from methyl iodide recorded after different periods (0.02—0.20 s) of forced precession in a strong radio-frequency field H_1. The signal intensities (circles) have been fitted to cosine waves, illustrating the modulation at twice the spinner frequency. This set of ten measurements has been repeated for several different levels of coherent proton irradiation (H_2), spanning the rotating frame resonance condition where the mean signal level is a minimum and the modulation component a maximum. From R. Freeman, H.D.W. Hill and J. Dadok, J. Chem. Phys., 58 (1973) 3107.

If T_2 for carbon is being measured and the decoupler power satisfies eqn. (10.47) anomalous values will result. If a noise decoupler is used which has a spread of B_2 values as a function of time, eqn. (10.47) must be satisfied for some period of the time. T_2 cannot be measured as the spectra of methyl iodide in Fig. 10.21 illustrate. Rotating frame resonance effects can be avoided even when using noise decoupling by gating the decoupler off only during the locking period and on during the f.i.d. Spectra which yield T_2 or $(T_{1\rho})$ can by obtained this way while maintaining the simplicity associated with noise decoupling. Also, with a suitable choice of delay times, the full n.o.e. can be achieved.

Use can be made of rotating resonance effects to enhance the sensitivity of dilute spins in solid samples (see section 5.6b) [53, 54]. The normal example is ^{13}C n.m.r. particularly in polymers [55]. ^{1}H—^{13}C dipolar broadening is removed by high power proton decoupling ($\gamma B_2 \sim 20$ kHz) and/or "magic angle" sample rotation, and since ^{13}C—^{13}C dipolar broadening is small due to isotropic dilution narrow lines can be obtained even in solid samples. The ^{13}C sensitivity is enhanced by energy transfer from the protons to the ^{13}C using the following sequence.

(i) The protons in the sample are strongly polarised by the spectrometer field B for a period T_d.

(ii) The protons are then spin locked; they are subjected to a 90° pulse after which the phase of B_2 is shifted by 90°, i.e. the decoupler is switched from the x' to the y' axis and left on. The proton vector now rotates about B_2 at the frequency $\gamma_c B_2$ Hz.

(iii) A rotating frame resonance condition is next established, i.e. eqn. (10.47) is satisfied by providing a B_1 field of the appropriate power for a period of τ. It is during this time that the energy is transferred from the abundant proton spins to the dilute carbon spins.

(iv) Finally the B_1 field is switched off and the decay of the carbon magnetisation digitised etc.

The sequence is repeated, the decay of the ^{13}C magnetisation time averaged and the spectrum produced by Fourier transformation. The gain in sensitivity achievable depends on the number of nuclei (N) of both types and their magnetogyric ratios. It is approximately given by $(N_{1_H}/N_{13_C})(\gamma_{1_H}/\gamma_{13_C})^2$ which is in the order of 30. A spectrum of solid adamantane (50 mg) gives an excellent spectrum, similar to that obtained in the liquid phase, in less than a second using this method [54]. The decoupler is on during data acquisition therefore $^1H-^{13}C$ dipolar coupling can be removed. However, unless 'magic angle' sample rotation is used, spectra obtained in this way show the effects of any chemical shift anisotropy present and therefore have complex line shapes.

REFERENCES

1 D. Debye, Polar Molecules, Dover Publications, New York, 1948.
2 N. Bloembergen, E.M. Purcell and R.V. Pound, Phys. Rev., 73, (1948) 679.
3 A. Abragam, The Principles of Nuclear Magnetism, Oxford University Press, London, 1961, ch. II and VIII.
4 T.D. Alger, S.C. Collins and D.M. Grant, J. Chem. Phys., 54, (1971) 2820.
5 I.D. Campbell and R. Freeman, J. Magn. Res., 11, (1973) 143.
6 I.D. Campbell and R. Freeman, J. Chem. Phys., 58, (1973) 2666.
7 T.D. Alger, R. Freeman and D.M. Grant, J. Chem. Phys., 57, (1973) 2168.
8 D. Doddrell, V. Glushke and A. Allerhand, J. Chem. Phys., 56, (1972) 3683.
9 W.T. Huntress, Jr., J. Chem. Phys., 48, (1968) 3524.
10 J.R. Lyerla, Jr. and D.M. Grant, in C.A. McDowell (Ed.), International Rev. Science, Phys. Chem. Series, Vol. 4, Medical and Technical Publishing Co., 1972, Ch. 5.
11 A.A. Maryott, T.C. Farrar and M.S. Malmberg, J. Chem. Phys., 54, (1971) 64.
12 T.C. Farrar and E.D. Becker, Pulse and Fourier Transform NMR, Academic Press, New York, 1971, Ch. 4.
13 G.C. Levy, D.M. White and F.A.L. Anet, J. Magn. Res., 6, (1972) 453.
14 R. Freeman and H.D.W. Hill, Molecular Spectroscopy 1971, Institute of Petroleum, London, 1971.
15 R. Freeman and H.D.W. Hill, J. Chem. Phys., 55, (1971) 1985.
16 R.R. Shouf and D.L. van der Hart, J. Amer. Chem. Soc., 93, (1971) 2053.
17 R.A. Ogg and J.D. Ray, J. Chem. Phys., 26, (1957) 1339.
18 W.B. Moniz and H.S. Gutowski, J. Chem. Phys., 38, (1963) 1155.
19 A.G. Massey, E.W. Randall and D. Shaw, Spectrochim. Acta 20, (1964) 379.
20 G.C. Levy, J.D. Cargioli and F.A.L. Anet, J. Amer. Chem. Soc., 95, (1973) 1527.
21 A. Carrington and A.D. McLachlan, Introduction to Magnetic Resonance, Harper and Row, London, 1967.
22 A. Abragam, The Principles of Nuclear Magnetism, Oxford University Press, London, 1961, Ch. IX.
23 R.A. Dwek, Nuclear Magnetic Resonance in Biochemistry, Oxford University Press, London, 1973.
24 J.R. Lyerla, Jr. and G.C. Levy in G.C. Levy (Ed.), Topics in ^{13}C NMR, Vol. 1, Wiley-Interscience, New York, 1974.
25 G.C. Levy and G.L. Nelson, Carbon-13 Nuclear Magnetic Resonance for Organic Chemists, Wiley-Interscience, New York, 1972.
26 F.W. Wehrli.
27 A. Allerhand, D. Doddrell and R. Komoroski, J. Chem. Phys., 55, (1971) 189.

28 D. Doddrell and A Allerhand, J. Amer. Chem. Soc., 93, (1971) 1556.

29 Y.K. Levine, N.J.M. Birdsall, A.G. Lee and J.C. Metcalfe, Biochemistry, 11 (1972) 1416.

30 A. Abragam, The Principles of Nuclear Magnetism, Oxford University Press, London, 1961, Ch. 11 and VIII.

31 G.C. Levy, D.M. White and F.A.L. Anet, J. Magn. Res. 6, (1972) 453.

32 J. Schaeffer in G.C. Levy (Ed.), Topics in ^{13}NMR, Vol. 1, Wiley-Interscience, New York, 1974.

33 J. Schaeffer, Macromolecules, 6, (1973) 882.

34 H.D.W. Hill, unpublished results.

35 R.L. Vold, J.S. Waugh, M.P. Klein and D.E. Phelps, J. Chem. Phys., 48, (1968) 3831.

36 R. Freeman and H.D.W. Hill, J. Chem. Phys., 54, (1971) 3367.

37 J.L. Markley, W.J. Horsley and M.P. Klein, J. Chem. Phys., 55, (1971) 3604.

38 G.G. McDonald and J.S. Leigh, Jr., J. Magn. Res., 9, (1973) 358.

39 R. Freeman, H.D.W. Hill and R. Kaptein, J. Magn. Res., 7, (1972) 327.

40 See section 3-5 of ref. 12.

41 G.C. Levey and I.R. Peat, J. Magn. Res.,

42 R. Freeman, H.D.W. Hill and R. Kaptein, J. Magn. Res., 7, (1972) 82.

43 D.E. Jones, J. Magn. Res., 6, (1972) 191.

44 I.M. Armitage, H. Huber, D.W. Live, W. Pearson and J.D. Roberts, J. Magn. Res., 15, (1974) 142.

45 H.Y. Carr and E.M. Purcell, Phys. Rev., 94, (1954) 630.

46 S. Meiboom and D. Gill, Rev. Sci. Instrum, 69, (1958) 688.

47 R. Freeman, unpublished results.

48 A.C. McLaughlin, G.G. McDonald and J.S. Leigh, J. Magn. Res., 11, (1973) 107.

49 I. Solomon, Compt. Rend., 248, (1959) 92.

50 H.D.W. Hill and R. Freeman, J. Chem. Phys., 54, (1971) 301.

51 R. Freeman, H.D.W. Hill and J. Dadok, J. Chem. Phys., 58, (1973) 3107.

52 S.R. Hartmann and E.L. Hahn, Phys. Rev., 128, (1962) 2042.

53 A. Pines, M.G. Bibby and J.S. Waugh, J. Chem. Phys., 56, (1972) 1776.

54 A. Pines, M.G. Bibby and J.S. Waugh, J. Chem. Phys., 57, (1973) 569.

55 J. Schaefer, E.O. Stejskal and R. Buchdahl, Macromolecules, 8, (1975) 291.

JEAN BAPTISTE JOSEPH FOURIER

Fourier was a fascinating person, and one of many talents. The work for which he is best known represents only a small fraction of his scientific work, and this in itself represents only a fraction of his achievements.

He was born, the son of a tailor, on the 21st March, 1768, in Auxerre in France. He wanted to be an artillery officer, but came from too low a social class to achieve his ambition. He was sent to school at St. Benoît-sur-Loire in the hope that he could persue his career later. The French Revolution, however, interfered with his plans, and he returned, in 1789, to be a teacher at his old school in Auxerre.

During the Revolution, Fourier was prominent in local affairs, and his courageous defence of some of the victims of terror led to his arrest in 1794. Appeals to Robespierre were unsuccessful, and he was due to be executed on the 28th July, 1794. Fortunately, Robespierre fell from power on 27th and Fourier was released. He then went as a student to Ecole Normale which opened and closed within the same year. The next year, when Ecole Polytechnique was started, he was appointed as Assistant Lecturer to support the teachings of Lagrange and Monge. There, ironically, he fell a victim of the reaction against the previous regime and was actually arrested as a supporter of Robespierre. His colleagues at the Ecole, however, successfully petitioned for his release. In 1789 Monge selected Fourier to join Napoleon's Egyptian campaign. There he became secretary of the newly-formed Institut d'Egypte. In Egypt, he gained a considerable reputation as a diplomat, and he may or may not — it does not appear to be clear — have been appointed to the governorship of Lower Egypt.

In 1801, he returned to France, wishing to work again at Ecole Polytechnique, but Napoleon had spotted his administrative genius, and appointed him Prefect of the department of Isère centred at

Grenoble, and extending to what was then the Italian border. Here, he made a very efficient administrator and engineer, the main achievement in the latter sphere being the drainage of a huge area of marshland near Bourgoin, turning it into useful farmland. Fourier also planned, and partly completed, the construction of the road from Grenoble to Turin. In 1808, Napoleon conferred a barony on him. The next year, he completed a work, started while in Egypt, on Egyptian history. Some of its historical details caused controversy at the time, but it was published with the strong support of Napoleon.

Fourier was still in Grenoble in 1814 when Napoleon fell from power, and on his way to Elba, Napoleon had to pass directly through this town. In order to avoid an embarrasing meeting with his former chief, Fourier negotiated a detour in the route of the cortège in order to bypass Grenoble. However, no such detour was possible during Napoleon's return in 1815 as he marched back to Paris. Fourier compromised, fulfilling his duties as a Prefect by ordering the preparation of the defences which he knew to be useless, and then leaving the town by the Lyons gate at the time Napoleon entered by another. He did however return, and the two friends met at Bourgoin. Fourier need have had no fears about his relationships with Napoleon who made him a Count and appointed him to be the Prefect of the neighbouring department of the Rhône, centred at Lyons. However, Fourier resigned from his post very shortly afterwards as a protest over the severity of the new regime.

Fourier then came to Paris to try to take up research as a full-time job. This was a very low point in Fourier's life. He had no job, only a small pension, and a very low political reputation. He was, however, able to get a job as Director of the Bureau Statistics, a post without many duties, but with a sufficient salary to enable him to persue his work. In 1816, Fourier proposed for membership of the re-constructed Académie des Sciences, but his nomination was refused by Louis XVIII, who could not accept somebody who had worked for Napoleon. Diplomatic negotiations, however, managed to smooth over this situation, and in 1817 he was elected unopposed. In 1822 he was elected permanent Secretary of the Académie de Sciences, with the support of the more physically orientated members like Laplace and despite the strong opposition from Poisson. In 1827, after more protests Fourier was elected to the Académie Française. About this period, he was also elected a foreign member of the Royal Society.

Fourier's spell in Egypt had a profound effect on his health. While there, he contracted a disease, possibly myxoedema, which probably was the cause of his death on the 16th May, 1830. He was buried in the 18th Division of the Cemetery of Père Lachaise, in Paris. Throughout his latter years, he lived a life of recluse in his quarters in Paris, taking to rather eccentric form of dress, probably to conceal the physical symptoms of his myxoedema.

Fourier had many varied mathematical interests. His famous concepts on the use of a series of trigonometrical functions to represent more general functions arose while working on the conduction of heat. The ideas began while he was in Egypt, but the main work was done during his stay in Grenoble. In 1807, he submitted a paper on the subject; three of the examiners, Laplace, Monge and Lacroix were in favour of accepting the work, but the fourth, Lagrange, stopped its publication. Lagrange's objection probably arose from his failure to apply successfully similar techniques to the vibrating string problem he studied in 1750. In 1810, a competition on the diffusion of heat was arranged, and Fourier resubmitted his 1807 paper, having substituted integrals, i.e. the transform, in place of the series approach used in the earlier work. The change was made in order to improve the generality of the treatment. He won the competition, but the jury (possibly on the insistance of Lagrange) published some criticisms on the work on the grounds of 'rigour and generality'. The origins of these criticisms were more probably jealousy and politics rather than science. The mathematical ideas were further expanded in his famous book, 'Theorie Analytique de la Chaleur' (Analytical Theory of Heat) in 1822.

Fourier also achieved significant advances in mathematical analysis and theory of the equations. While solving the equations for the conduction of heat in a cylinder, he derived what are now referred to as Bessel functions, several years before Friedrich Bessel. At the time of his death he was in the final stages of writing the mathematical textbook on the theory of equations. Fourier was more concerned than many of his contempories with the physical significance of the mathematical formulae derived. He believed that nearly all equations should have physical meanings, not necessarily all the intermediate steps, but all the final mathematical results should have physical meaning, a view not held by the establishment, particularly Poisson, at this period. Fourier's concern for the physical meaning of functions enabled him to see the potential in his formal techniques for checking the coherence of the groups of physical constants appearing in the exponentials of his Fourier integrals. From this came a full theory of units and dimensions which was the most significant advance in this area for many centuries. Fourier was also a superb master of analytical technique and notation; the symbol $\int_0^\infty$ is his invention, for example.

Jean Baptiste Fourier was indeed a remarkable person. The many modern uses of Fourier techniques in communications theory, electronics and even in n.m.r., make an interesting comment on the value of pure academic research. The techniques, being used with the aid of digital computers to solve these problems of practical relevance in the 20th Century, arose purely out of an academic challenge to solve equations which were necessary to describe the diffusion of heat.

APPENDIX 2

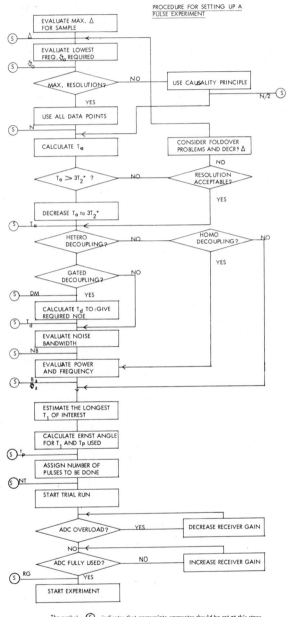

PROCEDURE FOR SETTING UP A
PULSE EXPERIMENT

The symbol ⓢ indicates that appropriate parameter should be set at this stage

RG = Receiver Gain   DM = Decoupler Mode

SUBJECT INDEX